DISTRIBUTED HYDROLOGICAL MODELLING

Water Science and Technology Library

VOLUME 22

Editor-in-Chief
V. P. Singh, *Louisiana State University,
Baton Rouge, U.S.A.*

Editorial Advisory Board

M. Anderson, *Bristol, U.K.*
L. Bengtsson, *Lund, Sweden*
A. G. Bobba, *Burlington, Ontario, Canada*
S. Chandra, *New Delhi, India*
M. Fiorentino, *Potenza, Italy*
W. H. Hager, *Zürich, Switzerland*
N. Harmancioglu, *Izmir, Turkey*
A. R. Rao, *West Lafayette, Indiana, U.S.A.*
M. M. Sherif, *Giza, Egypt*
Shan Xu Wang, *Wuhan, Hubei, P.R. China*
D. Stephenson, *Johannesburg, South Africa*

The titles published in this series are listed at the end of this volume.

DISTRIBUTED HYDROLOGICAL MODELLING

edited by

MICHAEL B. ABBOTT

*International Institute for Infrastructural,
Hydraulic and Environmental Engineering,
Delft, The Netherlands*

and

JENS CHRISTIAN REFSGAARD

*Danish Hydraulic Institute,
Hørsholm, Denmark*

KLUWER ACADEMIC PUBLISHERS
DORDRECHT / BOSTON / LONDON

Library of Congress Cataloging-in-Publication Data

```
Distributed hydrological modelling / edited by Michael B. Abbott and
  Jens Christian Refsgaard.
      p.    cm. -- (Water science and technology library ; v. 22)
    Includes index.
    ISBN 0-7923-4042-6 (hb : acid-free paper)
    1. Hydrology--Mathematical models.   I. Abbott, Michael B.
  II. Refsgaard, Jens Christian.   III. Series.
  GB656.2.M33D57  1996
  551.48'01'5118--dc20                                       96-14534
```

ISBN 0-7923-4042-6

Published by Kluwer Academic Publishers,
P.O. Box 17, 3300 AA Dordrecht, The Netherlands.

Kluwer Academic Publishers incorporates
the publishing programmes of
D. Reidel, Martinus Nijhoff, Dr W. Junk and MTP Press.

Sold and distributed in the U.S.A. and Canada
by Kluwer Academic Publishers,
101 Philip Drive, Norwell, MA 02061, U.S.A.

In all other countries, sold and distributed
by Kluwer Academic Publishers Group,
P.O. Box 322, 3300 AH Dordrecht, The Netherlands.

Printed on acid-free paper

All Rights Reserved
© 1996 Kluwer Academic Publishers
No part of the material protected by this copyright notice may be reproduced or
utilized in any form or by any means, electronic or mechanical,
including photocopying, recording or by any information storage and
retrieval system, without written permission from the copyright owner.

Printed in the Netherlands

Table of Contents

Foreword	vii
Chapter 1. The Role of Distributed Hydrological Modelling in Water Resources Management / J.C. REFSGAARD and M.B. ABBOTT	1
Chapter 2. Terminology, Modelling Protocol and Classification of Hydrolog-ical Model Codes / J.C. REFSGAARD	17
Chapter 3. Construction, Calibration and Validation of Hydrological Models / J.C. REFSGAARD and B. STORM	41
Chapter 4. Distributed Physically-based Modelling of the Entire Land Phase of the Hydrological Cycle / B. STORM and A. REFSGAARD	55
Chapter 5. Multi-species Reactive Transport Modelling / PETER ENGESGAARD	71
Chapter 6. Soil Erosion Modelling / J.K. LØRUP and M. STYCZEN	93
Chapter 7. Agrochemical Modelling / M. THORSEN, J. FEYEN and M. STYCZEN	121
Chapter 8. Weather Radar Precipitation Data and Their Use in Hydrological Modelling / C.G. COLLIER	143
Chapter 9. Application of Remote Sensing for Hydrological Modelling / F.P. DE TROCH, P.A. TROCH, Z. SU and D.S. LIN	165
Chapter 10. Geological Modelling / M. HANSEN and P. GRAVESEN	193
Chapter 11. Use of GIS and Database with Distributed Modelling / F. DECKERS and C.B.M. TE STROET	215
Chapter 12. An Engineering Case Study – Modelling the Influences of Gabcikovo Hydropower Plant on the Hydrology and Ecology in the Slovakian Part of the River Branch System of Zitny Ostrov / H.R. SØRENSEN, J. KLUCOVSKA, J. TOPOLSKA, T. CLAUSEN and J.C. REFSGAARD	233
Chapter 13a. A Discussion of Distributed Hydrological Modelling / K. J. BEVEN	255
Chapter 13b. Comment on 'A Discussion of Distributed Hydrological Modelling' by K. Beven / J.C. REFSGAARD, B. STORM and M.B. ABBOTT	279
Chapter 13c. Response to Comments on 'A Discussion of Distributed Hydrological Modelling' by J.C. Refsgaard et al. / K. J. BEVEN	289
Chapter 14. Hydrological Modelling in a Hydroinformatics Context / A.W. MINNS and V. BABOVIC	297
Index	313

DISTRIBUTED HYDROLOGICAL MODELLING

Foreword

It is the task of the engineer, as of any other professional person, *to do everything that is reasonably possible* to analyse the difficulties with which his or her client is confronted, and on this basis to design solutions and implement these in practice. The distributed hydrological model is, correspondingly, the means for doing everything that is reasonably possible - of mobilising as much data and testing it with as much knowledge as is economically feasible - for the purpose of analysing problems and of designing and implementing remedial measures in the case of difficulties arising within the hydrological cycle. Thus the aim of distributed hydrologic modelling is to make the fullest use of cartographic data, of geological data, of satellite data, of stream discharge measurements, of borehole data, of observations of crops and other vegetation, of historical records of floods and droughts, and indeed of everything else that has ever been recorded or remembered, and then to apply to this everything that is known about meteorology, plant physiology, soil physics, hydrogeology, sediment transport and everything else that is relevant within this context. Of course, no matter how much data we have and no matter how much we know, it will never be enough to treat some problems and some situations, but still we can aim in this way to do the best that we possibly can.

Now why should we need a model for this purpose; why, in particular, should we need a distributed model? The answer lies in our proviso of 'what we possibly can'. For we experience our outer world as being itself distributed in space and proceeding in time and we collect our data accordingly, while a large part of our knowledge is about how the quantities which enter our data change in space and time, both of themselves and in relation to one another. Thus, in order to comprehend this outer world we need a representation that is distributed, if only conceptually, in space, and which proceeds in time, with processes described correspondingly. Correspondingly again, our analyses, designs and remedial measures must also be described in a spatially-distributed and time-dependent way, and the need for economy of means and of thought then necessitate that this representation must be one that is expressed in terms of signs, and then, in our present time, of signs that are acceptable to a digital machine. Since these sign representations must then provide us with other, higher-order signs, which point the way to our analyses, our designs and our remedial measures, these must constitute models. Thus the spatially-distributed and time-dependent model becomes the *conditio sine qua non* for investigations in this area.

Given this position of distributed hydrological models, the question naturally arises of why they are so rarely applied to anything like their full potential to the multifarious problems of the hydrologic environment. The number and depth of applications appears to be quite out of proportion to the manifest needs for the results that these systems alone can provide. Of course, any number of reasons are adduced by organisations and individuals for this situation - 'insufficient data', 'insufficient knowledge of processes', 'unnecessary complication', and any number of others - but these have long since been exposed as being, for the most part, only surrogates for quite other kinds of difficulties. Some two decades of experience in the application of such systems has shown that the 'scientific' and 'technical' reasons that are presented are in fact surrogates for difficulties which have an institutional-political, administrative and, in a word, social origin. Modelling, as an integral part of hydroinformatics generally, is a *socio*technical activity, and the difficulties that confront distributed hydrological modelling arise primarily from the social (institutional and administrative) side. They are primarily functions of institutional and administrative structures, organisations and modes of collaboration.

An analysis of the many institutional setbacks experienced in the application of distributed models has shown a clear pattern in this direction and has allowed us to understand more clearly the origins of these reverses. Expressed in a nutshell: the distributed model puts demands upon the availability of data and of knowledge in hydrology that singularly few organisations *alone* are currently capable of supplying. Alongside this, and compounding its negative influences, there are few institutional arrangements in place for correcting this situation through a cooperation between organisations for the supply of hydrological data and knowledge. Thus, when seen by an individual or a group of persons situated within these organisations, it really does appear as though 'there is not enough data' and 'there is insufficient knowledge of processes'. Similarly, the application of a deterministic model must appear as an 'unnecessary inconvenience' when it appears to necessitate changes, both within the organisation itself and in its collaborative arrangements. But, of course, in a world of increasing commercial pressures on most organisations, few of these like to admit that it is they that do not have enough data or it is their organisations that cannot mobilise enough knowledge, so that they raise the spectre of insufficient data or insufficient knowledge or whatever else of this kind as a scientific or technical problem in its own right. The fact is then conveniently ignored that *if anything is to be done at all* - a reservoir to be extended, soil to restored after a chemical spill, a waste dump to be removed, a new water extraction policy to be introduced, etc, etc, - *the distributed model will nearly always narrow down the range of uncertainty in the outcome of the intervention*, and that, whatever the data and knowledge that is available, this represents the best that we can do for a given level of investment in data and knowledge. If we add to this the common observation that such studies point towards the most cost-effective means to reduce the range of uncertainty for a given extra investment in data and knowledge, the utility of such models is further emphasised.

Given this situation, the solution to the problem of encouraging the wider application of distributed hydrologic models must be sought in the experience of sociotechnology. This shows clearly that most problems that appear as 'social' can be overcome by the proper production and social employment of technical means. Thus the question concerning the future of this class of models devolves upon the question of elaborating an appropriate strategy of technological development, proceeding both on the side of the technology itself and on the side of its application in society. This, in its turn, directs attention to the encapsulation of distributed hydrologic models into such productions of hydroinformatics as real-time control, diagnostic and other model-based systems and decision-support environments. On one side (the 'input' side) this leads to means to encapsulate other data and knowledge than the hydrological and to combine and merge this in the (adaptive) objects of the systems and environments, and on the other side (the 'output' side) it leads to the further integration of the associated models with geographical information systems and other such standard tools of control and management practice. As this technology has now been rather thoroughly researched and applied in surrounding areas, such as for the real-time control of urban drainage systems and for the management of coastal-aquatic environments, the task of carrying it over to hydrologic applications is proving less onerous than would otherwise be the case. As the first such systems come to maturity, it is expected that distributed hydrologic modelling will emerge from the institutional-social impasse that has so long constrained it, so that it will finally come to attain to its full stature.

M.B. Abbott

J.C. Refsgaard

CHAPTER 1
THE ROLE OF DISTRIBUTED HYDROLOGICAL MODELLING IN WATER RESOURCES MANAGEMENT

J.C. REFSGAARD[1] AND M.B. ABBOTT[2]
[1] *Danish Hydraulic Institute, Hørsholm, Denmark*
[2] *International Institute for Infrastructural, Hydraulic and Environmental Engineering, Delft, The Netherlands*

1. Problems in Water Resources Management

"Scarcity and misuse of fresh water pose a serious and growing threat to sustainable development and protection of the environment. Human health and welfare, food security, industrial development and the ecosystems on which they depend, are all at risk, unless water and land resources are managed more effectively in the present decade and beyond than they have been in the past". (ICWE, 1992)

The present status and the future challenges facing hydrologists and water resources managers are summarized in this way in the introductory paragraph of the Dublin Statement on Water and Sustainable Development (ICWE, 1992). The Dublin Statement was adopted by government-designated experts from 114 countries and representatives of 80 international, intergovernmental and non-governmental organizations at the International Conference on Water and the Environment (a preparatory conference for the UNCED conference held in Rio de Janeiro in June 1992).

Since the ancient civilizations of Persia, Egypt and Babylon some 4000 years ago, water resources and water supply technology have played foundational roles in the development and organisation of many societies.

However, rapid population growth and the industrial development during the past few decades have caused an increasing pressure on land and water resources in almost all regions of the world. Due to increasing demands for water for domestic, agricultural, industrial, recreational and other uses and due to an increasing pollution of surface and groundwater, water resources have become scarce natural resources.

The availability of good-quality water is critical for human survival, economic development and the environment. Yet, water resources are not being managed in an efficient and sustainable manner. At the ICWE and UNCED conferences focus was put on past experiences of water resources management and new principles outlining improved future approaches were agreed upon. The World Bank in a follow-up policy paper (World Bank, 1993) emphasizes three problems which in particular need to be addressed:

* Fragmented public investment programming and sector management, that have failed to take account of the interdependencies among agencies, jurisdictions, and sectors
* Excessive reliance on overextended government agencies that have neglected the

need for economic pricing, financial accountability, and user participation and have not provided services effectively to the poor
* Public investments and regulations that have neglected water quality, health and environmental concerns.

Central elements in the World Bank's new policy are adoption of a comprehensive policy framework and the treatment of water as an economic good, combined with decentralized management and delivery structures, greater reliance on pricing, and fuller participation of stakeholders.

Such new approach to water resources management requires, first of all, combined efforts of professionals from a large range of disciplines such as economists, administrators, engineers, hydrologists and ecologists, as well as a cross-sectoral integration in the planning and management process. As the traditions for cooperation and integration among these various disciplines and sectors have generally not been very strong, this challenge is very large and crucial.

Additionally, the increased water resources problems and the new management approach require improved water resources management tools based on sound scientific principles and efficient technologies. Key characteristics of such improved technologies are that they to a larger extent than the existing tools must facilitate a holistic view of water resources as well as cooperation among different disciplines and sectors involved in water resources management. This involves, amongst others, an integrated description of the entire land phase of the hydrological cycle, an integrated description of water quantity, quality and ecology, and integration of hydrological, ecological, economical and administrative information in information systems specifically designed for decisions makers at different levels.

The role of distributed hydrological models should be seen in this context. As will be described later in this chapter and in other chapters of this book distributed hydrological models are essential elements comprising some of these required capabilities. Hence, distributed hydrological models are important and necessary, but far from sufficient, tools in improving the future water resources management.

In Section 2 of this chapter a brief review is made of important present problems and trends related to water resources. Section 3 provides a review of the state-of-the-art in hydrological modelling aiming to assist in the analysis and management of these problems.

Section 4 contains a discussion on which factors limit the practical use of distributed hydrological modelling in water resources management.

2. Key Issues and Trends in Water Resources

2.1. EFFECTS OF EXPLOITATION OF WATER RESOURCES

In 1940 the total global water use was about 1,000 km^3 per year. It had doubled by 1960 and doubled again by 1990 (Clarke, 1991). In most countries of the world there is not enough readily available water of sufficient quality for another such doubling. Developments in some countries, such as China and India, suggest that this may be experienced within the first decade of the next century, unless some major improvements are made in water use efficiency.

One common result of a continuous exploitation of surface and groundwater sources is a periodic or permanent lowering of the groundwater table and/or water level in surface water bodies, thus limiting both the quantity and quality of water available for other users.

Heavy and sustained extraction of water and the resulting changes in its quantity, quality and accessibility can have irreversible effects on the flora and fauna in the affected area. For example, the Niger River dried out for the first time in history in 1986, due to the combined effects of drought and the expansion of areas under irrigation leading to increased water losses due to evaporation. Other major, regional examples are the Colorado River (Carrier, 1991) and the Aral Sea, which both have seriously deteriorated due to extensive water extraction in the upstream catchment area.

In addition to the lowering of the water table, other adverse effects of excessive groundwater abstraction include increased concentrations of pollutants in the aquifer due to reduced flow rates or changed geochemical conditions, increased risk of salt water intrusion and land subsidence. The latter phenomenon may be illustrated by Bangkok, where groundwater abstractions over decades for urban water supply has resulted in land subsidence, in some places by more than 10 cm per year, thus contributing to serious flooding (BMA, 1986).

2.2. IRRIGATION

The purpose of irrigation is to enable cultivation in regions, and during periods, otherwise unsuited for farming and to stabilize crop production in regions with large fluctuations in rainfall. Common sources of irrigation water are streams and rivers, downstream releases from reservoirs or pumping from groundwater aquifers. In addition, conjunctive use of surface and groundwater is particularly attractive in regions where dry season irrigation is impossible from surface water sources alone.

By the mid-1980s the irrigated agricultural land in the world amounted to approximately 220 million hectares. Only 15% of the world's crop land is irrigated but it contributes 30-40% of all agricultural production (FAO, 1990).

Irrigated agriculture accounts for about 70% of water withdrawals in the world, but the current overall performance of many irrigation systems is very poor. Inadequate operation and maintenance and inefficient management contribute to many environmental problems. In many irrigation schemes, 60% of the water diverted or pumped for irrigation is 'lost' on its way from the source to the plant. Hence, the potential for conserving water by increasing irrigation efficiency is tremendous.

The major environmental problems result from the excessive application of irrigation water to land with poor or non-existing drainage facilities. Under such conditions the groundwater table rises and the land finally becomes waterlogged with increasingly saline water and reduced crop yields. Once-thriving civilizations in such areas as Mesopotamia and ancient Sri Lanka were destroyed as a result of waterlogging and salinization, and the phenomenon is widespread today along the Indus, Nile, Tigris, Euphrates and in other semi-arid and arid regions of the world. According to FAO

(1990), around 15% of the world's irrigated land is severely affected by salinity, and an additional 30%, approximately, are affected to some noticeable degree.

2.3. LAND DEGRADATION AND SOIL EROSION

Land degradation and soil erosion is another major problem worldwide. Soil erosion leads to the loss of valuable topsoil and causes silting, sedimentation, turbidity problems and pollution in downstream areas.

In Europe, for instance, erosion rates on agricultural land in the hilly areas of the Mediterranean and on sandy, loamy and chalky soils in northern Europe can reach 10-100 t/ha/year (Morgan et al, 1992). These rates should be compared with a value of 1 t/ha/year which is generally considered the maximum allowable for the control of pollution and the preservation of soil resources (Evans, 1981).

In some developing countries, land degradation is accelerated by increasing human and livestock populations, resulting in overgrazing, bushfires, exploitation of croplands and deforestation due to demand for firewood. In semi-arid and arid regions, such degradation is called desertification. According to FAO (1990), desertification affects nearly 75% of all productive rainfed lands and 60% of the rural population (280 million people) living is these areas.

Negligence and lack of knowledge of the importance of upland catchments in soil and water conservation do not only affect people living in these areas, but also result in considerable damage and losses for lowland populations due to the flooding and sedimentation of reservoirs. As a result, an estimated 160 million hectares of upland catchments in tropical developing countries have been seriously degraded, affecting approximately 20% of the world's population (Danida, 1988).

2.4. SURFACE AND GROUND WATER POLLUTION

Until a few decades ago water quality was relatively unimportant, except in arid lands where salinization occurred. Population growth, urbanisation and industrialization have now resulted in such increased water demands and such levels of contamination of water from the disposal of wastes that the use of water today is limited by its quality rather than the quantity available in many areas.

Major issues in surface water pollution are pathogenic agents, organic pollution, heavy metals, pesticides and industrial organics, acidification and eutrophication (WHO, 1991).

The extent and severity of the contamination of unsaturated zones and aquifers have been underestimated in the past due to the relative inaccessibility of the aquifers and the lack of reliable information on aquifer systems generally.

In many industrial countries groundwater pollution has become a key issue within the last two decades. For example in Denmark, where more than 99% of the water used is abstracted from groundwater, it was a general belief 15 years ago that the risk of groundwater pollution was small. Today, surveys have indicated the existence of more than 10,000 'hot spot' point sources for groundwater pollution in terms of landfills, chemical dumps, leaky oil tanks and many others and more than 400 million DKK (about 70 million US$) is spent annually for monitoring and remediation in order to save the country's groundwater resources for the future. Furthermore, nitrogen and pesticide pollutions due to agricultural activities have been recognized as very serious threats, and these will necessitate significant, and potentially very expensive, changes in agricultural practices, if these effects are to be reduced in the future.

2.5. FLOODS AND DROUGHTS

Economic losses from natural disasters increased three-fold between the 1960s and the 1980s (ICWE, 1992), while floods and droughts kill more people and cause more damage than do any other natural disasters (Rodda, 1995).

In spite of a continuous and ongoing construction of flood control structures such as reservoirs and dikes the flood damages in many countries have continued to increase. With an increasing pressure on land, also in flood plains, and an increasing awareness of the potential negative ecological impacts of large reservoirs and other regulation measures, these flood damages can be expected to increase significantly in the future. The recent (1993, 1994, 1995) very large floods in the continental rivers Mississippi and Rhine, both of which were considered to be 'well controlled', have generated considerable uncertainty about whether the hydrological regime has changed due to climate change or changes in land use, or both simultaneously.

2.6. AQUATIC ECOSYSTEMS

Water is a vital part of the environment and a home for many forms of life on which the well-being of humans also ultimately depends. Disruption of flows has reduced the productivity of many such ecosystems, devastated fisheries, agriculture and grazing, and marginalized the rural communities which rely on these. Various kinds of pollution exacerbate these problems, degrading water supplies, requiring more expensive water treatment, destroying aquatic fauna, and denying recreation opportunities (ICWE, 1992)

As an example, the destruction of wetlands has generally taken place at a dramatic rate during this century, and especially so because these were formerly viewed, quite wrongly, as wastelands. Thus, in the USA, 54% of the original wetlands had been lost by the mid-1970s (Tiner, 1984), and similar figures apply for several countries in Europe (Adams, 1986; Dugan, 1993).

2.7. POSSIBLE CLIMATE CHANGES

It is presently accepted that significant man-induced climate change may occur over the next decades, in particular related to the increase of CO_2 concentrations in the atmosphere.

Among the most important impacts of climate change will be its effects on the hydrological cycle and its associated water management systems. Hence, there is a danger that the resources of some areas will be reduced while others will suffer an increase in flood damage.

3. Model Applications in Water Resources Management - State-of-the-Art

Many reviews of hydrological models and their applicability to various water resources problems exist in the literature, e.g. Stanbury (1986), Bowles and O'Connel (1988), De Coursey (1988), Mangold and Tsang (1991) and Feddes et al (1988). Existing model reviews generally focus on the scientific and to some extent the technological aspects. However, only a few reviews focus on the status of practical applications of models.

In a review prepared for the Commission for the European Communities (SAST, 1992) the state-of-the-art of the existing modelling techniques was characterised both with respect to the scientific and the technological status. Furthermore SAST (1992) provided an overview of the status of practical use of models. An updated version of this status is given in Table 1. For each of the potential fields of model application the present status has been qualitatively assessed, ranging from "practically no operational applications" to "standard professional tool in many regions of the world". Furthermore, the major constraints for practical applications of models within the various fields have been identified.

It is realized that the scientific and technological status of the various modelling systems presently applied at different institutions varies considerably. The status given in Table 1 refer to the state-of-the-art versions of modelling systems. Some of the statements in Table 1 are justified and elaborated in the following subsections 3.1 - 3.9.

3.1. WATER RESOURCES ASSESSMENT

Water resources assessment is the determination of the quantity, quality and availability of water resources, on the basis of which an evaluation of the possibilities for their sustainable development, management and control can be made.

Sound water resources assessment requires both access to good hydrological data and the application of suitable modelling techniques. In cases where the focus is concentrated completely on surface water or completely on groundwater the relevant modelling tools are often rainfall-runoff models of the lumped conceptual type or traditional two-dimensional groundwater models, respectively. In cases where surface water and groundwater interaction is important, more comprehensive modelling tools are required, such as distributed physically-based integrated catchment models.

In general, for whatever kind of application, adequate model codes exist and are being used in many cases. There is however a need and a potential for a significant increase in model application for water resources assessment. The main constraint in this respect is most often an administrative tradition. However, technological innovations such as improved user friendliness and computer-assisted parameter estimation methods are also important.

TABLE 1. Status of application of hydrological modelling systems to various problem types

Field of Problem	STATUS	OF	APPLICATION		
	Adequacy of Scientific Basis	Scientifically Well Tested ?	Validation on Pilot Schemes ?	Practical Applications	Major Constraint for Practical Application
Water resources assessment					
* Groundwater	Good	Good	Adequate	Standard/Part	Administrative
* Surface water	Very good	Very good	Adequate	Standard/Part	Administrative
Irrigation	Good	Good	Partially	Very limited	Techno/Admin
Soil erosion	Fair	Fair	Very limited	Nil	Science
Surface water pollution	Good	Good	Adequate	Some cases	Administrative
Groundwater pollution					
* Point sources (landfills)	Good	Good	Partially	Standard/Part	Techno/Admin
* Non-point (agriculture)	Fair	Fair	Very limited	Very limited	
On-line forecasting					
* River flows/water levels	Very good	Very good	Adequate	Standard	Nil
* Surface water quality	Good	Good	Adequate	Standard/Part	Data/Admin
* Groundwater heads/w.table	Very good	Very good	Partially	Very limited	Data/Techno
* Groundwater quality	Fair	Fair	Nil	Nil	Science
Effects of land use change					
* Flows	Good	Fair	Fair	Very limited	Science
* Water quality	Fair	Fair	Fair	Nil	Science
Aquatic ecology	Fair	Fair	Very limited	Very limited	Science/Techno
Effects of climate change					
* Flows	Good	Good	Fair	Very limited	Science
* Water quality	Fair	Fair	Nil	Nil	Science

LEGEND:
Adequacy of scientific basis
- Poor: Large and crucial needs for improvements in scientific basis
- Fair: Considerable needs for improvements in scientific basis
- Good: Some needs for improvements in scientific basis
- Very good: No present significant need for improvements in scientific basis

Scientifically well tested ?
- Poor: Large needs for fundamental tests of scientific method
- Fair: Considerable needs for testing (some) of the scientific basis
- Good: Some needs for testing of the scientific basis
- Very good: No present significant need for testing of the scientific basis

Validation on pilot schemes ?
- Nil: No successful validation on well controlled pilot schemes so far - urgent need for validation on pilot schemes
- Very limited: A few (a couple of) validation cases - considerable needs for more validation projects on pilot schemes
- Partially: Some cases with successful validation on pilot schemes - some needs for further validations
- Adequate: Many good validation - no further present needs

Practical applications
- Nil: Practically no operational applications
- Very limited: A few well proven cases of operational practical applications
- Some cases: Some cases of well proven operational practical applications
- Standard/Part: Standard professional tool in some regions
- Standard: Standard professional tool in many regions of the world

Major constraint for further practical application
- Data: Data availability a major constraint
- Science: Inadequate scientific basis is a major constraint
- Technology: A technology push is required in order to make well proven methods more widely applicable
- Administrative: Administrative tradition or missing economical motivation is a major constraint

3.2. IRRIGATION

The irrigation sector is, with few exceptions, dominated by a low technological level as far as hydrology is concerned. Thus, whereas the potential for improved irrigation management by use of modern technology is tremendous with regard to positive economic and environmental impacts, few attempts have been made so far to exploit this.

The requirements for the modernization of irrigation management include the following key elements:

* Improved data acquisition techniques including modern sensors, on-line data transmission and spatial information from remote sensing data.
* Detailed hydrological/hydraulic modelling enabling full descriptions of soil moisture and groundwater conditions on a spatially distributed basis in the command area as well as dynamic modelling of water flows and storages in the distribution and drainage channel system. One of the first attempts in this regard was that made by Lohani et al. (1993).
* Optimization techniques for managing the operation of reservoirs and other hydraulic control structures.

The scientific basis appears adequate, while the two major constraints are the administrative and engineering tradition in combination with a lack of well proven, fully integrated, and user friendly technological solutions.

3.3. SOIL EROSION

The soil and water conservation sector is also characterized by a low technological level as far as hydrological modelling is concerned. The most common method of estimating soil erosion from a catchment is still the 'Universal Soil Loss Equation (USLE)' (Wischmeier and Smith, 1965) which is a very simple, empirical equation originally developed for hand calculations (see also Lørup and Styczen, Chapter 6).

A number of physically-based soil erosion models are being developed but more research is required on process descriptions before large scale applications will be feasible.

3.4. SURFACE WATER POLLUTION

The state-of-the-art in surface water quality modelling is comparatively good both with regard to the scientific and technological status and models are being applied extensively.

3.5. GROUNDWATER POLLUTION

The main problem in studies of groundwater pollution relates to obtaining a sufficiently detailed three-dimensional description of the geology and in this way obtaining data on the spatial variability of the hydraulic parameters which ultimately determine the transport and spreading of contaminants (Hansen and Gravesen, Chapter 10). In general, the modelling technology with regard to groundwater flow and transport is well advanced in comparison to the data acquisition problem (Storm and Refsgaard, Chapter 4).

With regard to groundwater quality processes, research is still required on process identification and associated parameter assessment for organic and inorganic geochemical processes and their interactions (Engesgaard, Chapter 5). With regard to non-point agrochemical pollution, some basic process descriptions in the root zone still require research, especially with regard to the importance of agricultural management techniques, such as tillage (Thorsen et al., Chapter 7).

3.6. ON-LINE FORECASTING

Hydrological models in combination with modern on-line data acquisition systems are being used as standard tools for real-time flood forecasting purposes. At present, lumped conceptual rainfall-runoff models in combination with hydrodynamic river routing models represent the state-of-the-art for this type of application, and it is not likely that more sophisticated distributed physically-based models in general can provide significantly better levels of accuracy and overall reliability.

The scientific and, to a large extend, the technological basis also exist for applying on-line forecasting systems in the fields of surface water quality and groundwater flow and head estimation. In these areas the administrative tradition represents a constraint upon practical applications.

3.7. EFFECTS OF LAND USE CHANGE

Prediction of effects of land use change on water quantity and water quality represent an area of increasing importance. For instance, the possible effects of urbanisation or deforestation on floods and droughts and the possible effects of changed cropping pattern and other agricultural practises on soil erosion and ground water quality are key issues in water resources management in many areas (Lørup and Styczen, Chapter 6; Thorsen et al., Chapter 7).

Whereas the existing modelling tools are very useful in addressing these issues, the major constraint for widespread model applications in these subjects is a lack of basic knowledge on process descriptions and parameter values.

3.8. AQUATIC ECOLOGY

Modelling of wetlands and aquatic ecology has traditionally been dominated by very simple hydrological modelling tools, because the ecological processes themselves are extremely complex and data demanding and because the most advanced hydrological models, until recently, have not been able to provide sufficiently detailed descriptions for ecological modellers. Hence, comprehensive modelling of aquatic ecology has until now been carried out in relatively few cases.

However, a prerequisite for establishing a predictive capability for management purposes within aquatic ecology is to make use of the most advanced of the distributed physically-based modelling systems. An example of a comprehensive floodplain model is given by Sørensen et al. (Chapter 12).

Thus, the main constraints presently are of scientific and technological nature. These constraints may be expected to be reduced as inverse methods, such as those effected by Kalman filtering, neural networks, and genetic algorithms are brought into regular use, while more advanced knowledge-based ecological models will no doubt also contribute further to advancing this area of application (e.g. Abbott et al., 1994).

3.9. EFFECTS OF CLIMATE CHANGE

Prediction of the hydrological effects of climate change is maybe one of the most difficult issues with which hydrology is confronted. With todays technology the impact of specified changes in precipitation, temperature and evaporation on river runoff, soil moisture regime and groundwater recharge can be calculated with reasonable accuracy, and any number of simple models can be constructed that will reproduce the one or the other aspect of climate change.

However, a climate change will, gradually, also result in successions of vegetation types, changes in agricultural practises etc, which may be expected to have significant effects on water resources. Thus, a considerable amount of further research is required on these issues.

A major weakness in the present generation of climate models is their very simple description of land surface processes, especially soil moisture and its spatial variations, which to a large extent control the land surface - atmosphere interactions. Distributed hydrological models appear to be suitable for this purpose, but have not been used so far maybe due to lack of interdisciplinary interactions.

3.10. HISTORICAL RECONSTRUCTIONS OF THE IMPACTS OF HUMAN ACTIVITY

A considerable interest accrues also to the reconstruction of the hydrology of river basins and other areas as these have changed due to changing land use practises and the construction of structures over considerable periods of time. Indeed, so extended are these periods that one can perhaps better speak of 'hydrological archaeology'. This is necessary to determine the causes and possible remedies for often catastrophic flood events for the purposes of risk assessment, insurance and investment planning, legislation and litigation.

4. A discussion on Constraints for the Practical Use of Distributed Hydrological Modelling in Water Resources Management

4.1. THE NEED FOR DISTRIBUTED HYDROLOGICAL MODELLING

It is evident from Section 3 that there is a growing need for advanced distributed physically-based models. The traditional hydrological models of lumped conceptual type are well suited to deal with the main part of current water resources assessment and flood and drought forecasting, but more advanced tools are required for the remaining problems. It is noticed that for many of the problem areas the need for the distributed models reflects a demand for predictive capability on the effects of man-induced impacts. Thus, there is a growing need to use distributed models as a management tool.

In the current discussion on the role and capabilities of distributed models versus lumped models (Beven, Chapter 13A; Refsgaard et al., Chapter 13B) much focus is put on rainfall-runoff modelling, whereas it should be recognized that the main challenge and potential applications for distributed models lie much beyond this field of application - in more difficult areas.

4.2. CONSTRAINTS FOR APPLICATIONS OF DISTRIBUTED HYDROLOGICAL MODELS

Computer-based hydrological modelling has been carried out for more than three decades. However, as elaborated by Klemes (1988) in a discussion of the (lack of) scientific tradition in hydrology, the traditional deterministic hydrological models of lumped conceptual type, such as the Sacramento model, are technological tools which can not rightfully be claimed to be scientifically sound. Klemes (1988) generally questioned the reliability of predictions made by such models. Similarly, Abbott (1972) in a survey of hydrological models found little of predictive value.

The needs and the concept for a physically-based distributed catchment model were initially outlined by Freeze and Harlan (1969). With Freeze's pioneering work as a particular inspiration, three European organizations in 1976 started the development of the Système Hydrologique Européen (SHE) (Abbott et al., 1986a,b). Since then, a large number of other distributed models have been developed.

Nevertheless, in spite of two decades of modelling development, distributed hydrological models are today being used in practise only at a fraction of their potential. For example an initial marketing survey conducted for the Commission of the European Communities by the organisations behind the SHE in 1978 indicated a market in the order of 21 billion ECU for works that would benefit from the application of such modelling technology. With this background a distribution of the two present SHE versions, MIKE SHE and SHETRAN, to not more than some 60 organisations worldwide by the end of 1994, is far below the initial expectations of its developers. This naturally calls for an explanation. With our background in developments and applications of the SHE over the past two decades we shall present our perception of the main reasons, or the constraints, that have given rise to this slower-than-expected development.

4.2.1. Data availability
A prerequisite for making full use of the distributed physically-based models is the existence and easy accessibility of a large amount of data, including detailed spatial information on natural parameters such as geology, soil and vegetation and man-made impacts such as water abstractions, agricultural practices and discharge of pollutants.

In many cases all such relevant data do not exist and even the existing data is most often not easily accessible due to the lack of suitably computerized data bases. A further complication in this regard is the administrative problem created by the fact that these models, in addition to the traditional hydrometeorological data, require and can make use of many other data sources, such as those arising from agricultural, soil science and geological investigations.

Thus, distributed models have had to work with such data as happens to be available, which are rarely if ever collected with a view to their compatibility with distributed hydrological modelling activities. Hence, most distributed model codes are able to run with different levels of data availability, and indeed in many cases (topographic data, vegetation maps, etc.) are the only means whereby this data can be introduced directly into decision-making processes.

For many years, high expectations have been directed to remote sensing techniques for providing spatial data of use in distributed hydrological models. However, so far operational use of remote sensing data are, with the exception of satellite-inferred snow cover data and land use/vegetation mapping, not common practise. With the launching of new satellites in recent years improved possibilities arise and, as discussed by De Troch et al. (Chapter 9), there are now good reasons to expect the long awaited breakthrough for large scale operational application of remote sensing data jointly with distributed hydrological models.

Another important development gradually improving the availability of data is the application of GIS technology, which is particularly suitable for couplings with distributed hydrological models (Deckers and Te Stroet, Chapter 11).

4.2.2. Lack in scientific-hydrological understanding
With the introduction of a new modelling paradigm and concurrent research in process descriptions, new shortcomings in the scientific-hydrological understanding have emerged, especially with regard to flow, transport and water quality processes at small scales and their up-scaling to describe larger areas. Some of the key scientific problems are highlighted in several chapters of this book.

These scientific shortcomings have, on the one hand, constrained the practical applications of distributed hydrological models and, on the other hand, the existence and application of such models have put a new focus on some of these problems, thus contributing to advances in the scientific-hydrological understanding.

4.2.3. Traditions in hydrology and water resources engineering
The distributed physically-based model codes such as the SHE constituted a 'quantum jump' in complexity as compared with any other code so far known in hydrology. Moreover, it used technologies, such as had been developed in computational

hydraulics, with which few hydrologists were familiar. Although the numerical-algorithmic problems could be largely overcome by development of the codes into user-friendly fourth generation modelling systems with well proven algorithms, such as the MIKE SHE, the problem was then shifted back to one of comprehending the fully integrated complexity of the physical system that was being modelled together with the constraints that were inherent in the modelling procedure. Very few professional engineers and managers were, and still are, educated and trained with the necessary integrated view of hydrological processes in anything like their real physical complexity.

This difficulty is exacerbated by the very nature of hydrology itself, whereby most professionals posses only limited view on the totality of the physical processes. Soil physicists, plant physiologists, hydrogeologists and others usually have only a very partial view on the whole system, while there are few organisations that have available both the full range of such specialists and the more broader-ranging professionals that are needed in many situations to exploit the potential of distributed physically-based codes to such a degree that this exploitation is economically justified.

4.2.4. Technological constraint

In order to achieve a large dissemination of modelling technology to a considerable part of the professional community (and not only to experienced hydrological modellers) experience from hydraulic engineering shows that so-called fourth generation modelling systems (user-friendly software products) are required. Furthermore, it is believed that fifth generation systems are required in order for the modelling technologies to achieve their full potential in terms of practical application. The fifth generation systems are hydroinformatics-based including some of the elements outlined in Section 4.3. More exact definitions of fourth and fifth generation systems are given by Refsgaard (Chapter 2). At present only a few fourth generation distributed physically-based hydrological modelling systems exist and fifth generation systems are still at an experimental stage. Experience in hydraulics however, where more than 2000 organisations already make use of fourth generation systems, suggests that this situation will change in the foreseeable future.

4.3. THE FUTURE ROLE OF HYDROINFORMATICS

In order to ensure a better use of the existing (and coming), most scientifically advanced models for practical application by a wide group of professionals and managers it will be necessary to integrate the hydrological model codes with new hydroinformatics technologies including, amongst others, the following elements (see also Babovic and Minns (Chapter 14):

Standards for "open" modelling systems. The more widespread application of modelling technology necessitates the development of standards defining common user interfaces and model application interfaces, so that several modelling systems can be easily coupled for specific applications. For instance, a hydrological flow model could in this way be coupled with a soil erosion model and a river sediment transport model, developed at different institutions, without changing anything in the programme codes, i.e these should be mutually full compatible. The standards should also ensure full compatibility and easy use of other commonly used software, such as the various classes of data bases, and then especially Geographical Information Systems (GISs).

Logical modelling techniques. The application of logical modelling techniques, that is, the coupling of numerical and logical programming, as already well advanced in hydraulics, must be expected to have its own impact on the knowledge engineering aspects of hydrology.

Knowledge base systems. The application of proprietorial knowledge base systems for various fields within water resources management can be expected to spread correspondingly. Hence, knowledge elicitation technologies, making use of computer aided knowledge engineering/elicitation tools, will also need to be applied in this field also.

Systematic calibration. The application of methods for calibration of hydrological models on an objective basis making use of both the available data and the hydrologist's experience may be expected to advance further. This will unavoidably involve many elements of inverse modelling and introducing and combining many aspects of expert-systems.

Optimization methods integrated with advanced models. The application of optimization methods which are fully integrated both with advanced models and with the above indicated hydroinformatics tools can be expected to be developed in hydrology as they are now currently being applied in hydraulics.

Decision methods. The application of decision methods and the integration of such methods with hydrological modelling systems for use in water resources control and management will have also to be advanced. At this stage the hydrological model becomes integrated in the new kinds of architectures and paradigms (of object orientation and agent orientation) that are now becoming established in other fields.

The ultimate output of such research activities is a fifth generation modelling environment, which may be described as a *virtual hydrological environment*. It will make possible a combined access to several powerful features such as:
- standard interfaces to the most common data bases and GIS's, giving the user access to necessary specific data,
- a choice of alternative and compatible hydrological modelling systems,
- graphical interface,
- utilities for model calibration, optimization, decision making and other methods, and
- enable professionals from other fields (agronomists, ecologists, meteorologists, etc) to access integrated hydrological data and knowledge resources in an efficient and responsible way.

The declarations adopted by the ICWE-Dublin and the UNCED-Rio conferences call for the abandoning of traditional sectoral approaches and the adoption of more integrated water resources management strategies. As stated by Matthews (1994) in a description of the World Bank's new policy, the need for "use of hydroinformatics for the implementation of the Dublin and Rio declarations becomes evident".

5. References

Abbott, M.B. (1972) The use of digital computers in hydrology. In *Teaching Aids in Hydrology*, Technical papers in Hydrology 11, UNESCO, LC No. 72-87901.

Abbott, M.B., Bathurst, J.C., Cunge, J.A., O'Connel, P.E. and Rasmussen, J. (1986a) An Introduction to the European Hydrological System - Systeme Hydrologique Europeen, "SHE", 1: History and Philosophy of a Physically-based, Distributed Modelling System. *Journal of Hydrology*, 87, 45-59.

Abbott, M.B., Bathurst, J.C., Cunge, J.A., O'Connel, P.E. and Rasmussen, J. (1986b) An Introduction to the European Hydrological System - Systeme Hydrologique Europeen, "SHE", 2: Structure of a Physically-Based, Distributed Modelling System. *Journal of Hydrology*, 87, 61-77.

Abbott, M.B., Babovic, V., Amdisen, L.K., Baretta, J. and Dørge, J. (1994) Modelling ecosystems with intelligent agents. In *Verwey, A., Minns, A., Babovic V. and Maksimovic C. (Eds) Proceedings, Hydroinformatics '94*. Balkema, Rotterdams, 179 - 186.

Adams, W.M. (1986) *Nature's Place: Conservation Sites and Countryside Change*. Allen & Unwin, London.

BMA (1986) *Bangkok Flood Management Model - Feasibility Study*. Report prepared by Acres International Ltd and Asian Institute of Technology for Bangkok Metropolitan Administration.

Bowles, D.S. and O'Connel, P.E. (Eds) (1988) *Recent Advances in the Modelling of Hydrologic Systems*. Proceedings from the NATO ASI, Sintra, Portugal. Kluwer Academic Publishers.

Carrier, J. (1991) The Colorado. A river drained dry. *National Geographic*, June 1991, 4 - 35.

Clarke, R. (1991) *Water: The International Crises*, Earthscan Publications Ltd, London.

Danida (1988) *Environmental issues in water resources management*. Danish Ministry of Foreign Affairs, Copenhagen.

De Coursey, D.G. (Ed) (1988): *Proceedings of the International Symposium on Water Quality Modelling of Agricultural Non-Point Sources*. Utah, June 19-23. U.S. Department of Agriculture. Vol 1, 422 pages. Vol 2, 459 pages.

Dugan, P. (1993) *Wetlands in Danger*. Mitchell Beazley, London.

Evans, R. (1981) Potential soil and crop losses by soil erosion. *Proceedings, SAWMA Conference on Soil and Crop Loss: Development in erosion control*. National Agricultural Centre, Stoneleigh.

FAO (1991) *An international action programme on water and sustainable agricultural development. A strategy for the implementation of the Mar del Plata Action Plan for the 1990s*. UN Food and Agricultural Organisation

Feddes, R.A., Kabat, P., van Bakel, P.J.T., Bronswijk, J.J.B. and Halbertsma, J. (1988) Modelling Soil Water Dynamics in the Unsaturated Zone - State of the Art. *Journal of Hydrology*, 100, 69 - 112.

Freeze, R.A. and Harlan, R.L. (1969) Blueprint for a physically-based digitally-simulated hydrological response model. *Journal of Hydrology*, 9, 237-258.

ICWE (1992) *The Dublin Statement and report of the conference*. International Conference on Water and the Environment: Development issues for the 21st century. 26-31 January 1992, Dublin, Ireland.

Klemes, V. (1988) A hydrological perspective. *Journal of Hydrology*, 100, 3-28.

Lohani, V.K., J.C. Refsgaard, T. Clausen, M. Erlich and B. Storm (1993) Application of the SHE for irrigation command area studies in India. *Journal of Irrigation and Drainage Engineering*, 119 (1), 34-49.

Matthews, G.J. (1994) Hydroinformatics and the World Bank. In *Verwey, A., Minns, A., Babovic V. and Maksimovic C. (Eds) Proceedings, Hydroinformatics '94*. Balkema, Rotterdams.

Morgan, R.P.C, J.N. Quinton and R.J. Rickson (1992) EUROSEM Documentation Manual, Version 1. Silsoe College, UK.

Rodda, J.C. (1995) Whither world water? *Water Resources Bulletin*, 31 (1), 1-7.

SAST (1992) *Research and technological development for the supply and use of freshwater resources. Report on monitoring and modelling*. Expert report prepared by I Krüger Consult AS and Danish Hydraulic Institute for the Monitor - SAST Project No 6. Commission of the European Communities.

Stanbury, J. (Ed) (1986) *Proceedings of the International Conference on Water Quality of the Inland Natural Environment*. Bournemouth, England, June 10-13. BHRA.

Tiner, R.W. (1984) *Wetlands of the United States: Current status and future trends*. U.S. Fish and Wildlife Service, Washington D.C.

WHO (1991) *Water quality. Progress in implementing the Mar del Plata Action Plan and a strategy for the 1990s*. Report sponsored by UN/GAPD/DIESA, UN/DTCD, UNDP, UNEP and WHO and prepared by GEMS MARC and WHO. World Health Organisation.

World Bank (1993) *Water resources management. A World Bank policy paper*. Washington D.C.

Wischmeier, W.H. and Smith, D.D. (1965) Predicting rainfall erosion losses from cropland East of the Rockey Mountains. Agricultural Handbook No 282. Agricultural Research Service, USDA, Purdue Agricultural Experimental Station.

CHAPTER 2
TERMINOLOGY, MODELLING PROTOCOL AND CLASSIFICATION OF HYDROLOGICAL MODEL CODES

J.C. REFSGAARD
Danish Hydraulic Institute

1. Introduction

All hydrological models are simplified representations of the real world. Models can be either physical (e.g. laboratory scale models), electrical analogue or mathematical. The physical and analogue models have been very important in the past. However, the mathematical group of models is by far the most easily and universally applicable, the most widespread and the one with the most rapid development with regard to scientific basis and application. The present book is devoted entirely to mathematical models.

The present book deals with simulation models in contrary to optimization models. In Section 2 a consistent, general terminology terminology and methodology applicable within the whole range of hydrological modelling is presented. Important elements related to the methodology are described within the modelling context. Alongside this, a terminology that is more suited to hydroinformatics applications, where such process models form only part of more general information and knowledge-based systems, is also introduced.

To a user, a hydrological model is composed of two main parts, namely a hydrological core and a technological shell. The hydrological core is based on a certain hydrological scientific basis providing the definitions of variables, the process descriptions and other aspects. The technological shell is the programming, user interface, pre- and postprocessing facilities etc. These two different and equally important aspects are addressed in Section 3 and Section 5, respectively. Section 4 is devoted to the problems caused by spatial variability of hydrological parameters and the fundamental different ways that different model types takes this aspect into account.

2. Terminology and Methodology

2.1. DEFINITIONS OF BASIC TERMS

No unique and generally accepted terminology is presently used in the hydrological community. Therefore, definitions of some of the most common terms used in hydrological modelling and in hydroinformatics and employed in the following text are given here.

The *natural system* that is considered here is the hydrological cycle or parts of it as we currently conceive it. This conception is naturally a function of our own social environment, historical developments, linguistic traditions and other variables that influence all disciplines, and not just hydrology. We shall suppose however that this process of conceptualisation is uniform over current hydrological practice. Thus, to use language of computer science that is also adopted by hydroinformatics, we shall assume that there is 'one universe of discourse', and that this is uniform.

A *hydrological model* is a simplified representation of the natural system. From a hydroinformatics point of view, a model is a collection of signs that serves as a sign, so that the hydrological model is the set of signs (symbols and other tokens) that serve as a representation of the natural system or some aspects of it.

A *mathematical model* is a set of mathematical expressions and logical statements combined in order to simulate the natural system. This is then, hydroinformatically, a model in which the signs are symbols organised within definite formal systems called mathematical languages.

Simulation is the time-varying description of the natural system computed by the hydrological model. A simulation may be seen as the model's imitation of the behaviour of the natural system. In hydroinformatics we speak of 'virtual worlds' that provide signs that mirror (albeit possibly in a simplified or distorted way) the signs that we ourselves experience as 'the world of nature'. Our 'time consciensness' can then also be mirrored, whether by a succession of images or through other, static-graphical measures.

A *routine*, a *component*, or a *submodel* is part of a more comprehensive model, e.g. the snowmelt simulation part of a model for the complete land phase of the hydrological cycle. Hydroinformatics speaks, with computer science generally, of *an object*, as a representation of anything to which our thoughts can be directed. When an object encapsulates knowledge that acts upon information so that it both acts upon and is itself acted upon by other objects, it is called an *agent*.

A *parameter* is a constant in the mathematical expressions or logical statements of the mathematical model. It remains constant in virtual time. A *variable* is a quantity which varies in space and time. It can be a series of inputs to and outputs from the model, but also a description of conditions in some component of the model. In hydroinformatics it is an indicative sign (in the sense of Husserl), the indication of which is towards a state of affairs that varies in time.

A *deterministic model* is a model where two equal sets of input (i.e. collections of signs) always yield the same output sign if run through the model under identical conditions. A deterministic model has no inner operations with a stochastic behaviour.

A *stochastic model* has at least one component of random character which is not explicit in the model input, but only implicit or 'hidden'. Therefore, identical inputs will generally result in different outputs if run through the model under, externally seen, identical conditions. This notion can be extended to models in which the input has a direct stochastic character.

We conceive a *catchment* as a region of physical space over which flows occur that are collectively of concern to us. A catchment is thus an object towards which the universe of discourse of hydrology becomes directed.

A *lumped model* is a model where the catchment is regarded as one unit. The variables and parameters are thus representing average values for the entire catchment. Thus the virtual world is reduced to just one object.

A *distributed model* take account of spatial variations in all variables and parameters.

By a *physical process* we understand a representation in our own minds of an event occurring in our outer world of sense experience.

A *black box* or an *empirical model* is a model developed without any consideration of the physical processes that we otherwise associate with the catchment. The model is merely based on analyses of concurrent input and output time series.

A *conceptual model* is one that is constructed on the basis of the physical processes that we 'read' into our observations of the catchment. In a conceptual model, physically sound structures and equations are used together with semi-empirical ones. However, the physical significance is not usually so clear that the parameters can be assessed from direct measurements. Instead, it is necessary to estimate the parameters from calibrations, applying concurrent input and output time series. A conceptual model, which is usually a lumped-type model, is often called a *grey box model*.

A *physically-based model* describes the natural system using the basic mathematical representations of the flows of mass, momentum and various forms of energy. For catchment models, a physically-based model in practice also has to be fully distributed. This type of model, also called a *white box model*, thus consists at its most basic 'human-friendly' level of a set of linked partial differential, integral-differential and integral equations together with parameters which, in principle, have direct physical significances and can be evaluated by independent measurements.

In connection with the technological classification, a distinction is made between a model and a modelling system. A *model* is defined as a particular hydrological model established for a particular catchment. A *modelling system*, on the other hand, is defined as a generalized software package, which, without program changes, can be used to establish a model with the same basic types of equations (but allowing different parameter values) for different catchments. The term *model code* is often used synonymously with the term modelling system. Thus, most of the models referred to in the present book are in fact generated using modelling systems. However, as this technological distinction between model and modelling system is not recognized rigorously throughout the hydrological scientific community, it is not maintained strictly in this book either.

2.2. GENERAL TERMINOLOGY FOR MODEL CREDIBILITY

Hydrological models are being developed and applied in increasing number and variety. At the same time contradictions are emerging regarding the various claims of model applicability on the one hand and the lack of validation of these claims on the other hand. Hence, the credibility of the advanced models can often be questioned, and often with good reason.

Many different definitions of model validation are presently used. A consistent terminology with a set of definitions for terms such as conceptual model, computerized model, verification, validation, domain of applicability and range of accuracy is given by Schlesinger et al. (1979). Slightly different definitions are used by other authors e.g.

Konikow (1978), Tsang (1991), Flavelle (1992) and Anderson and Woessner (1992).

Oreskes et al. (1994), using a philosophical framework, states that verification and validation of numerical models of natural systems is theoretically impossible, because natural systems are never closed and because model results are always non-unique. Instead, in this view models can only be confirmed.

The following terminology is based on the general terminology proposed by Schlesinger et at. (1979), but is relativised by references to current usage in hydroinformatics (e.g. Abbott, 1991, 1993, 1994). The elements in the terminology and their interrelationship are illustrated in Fig. 1.

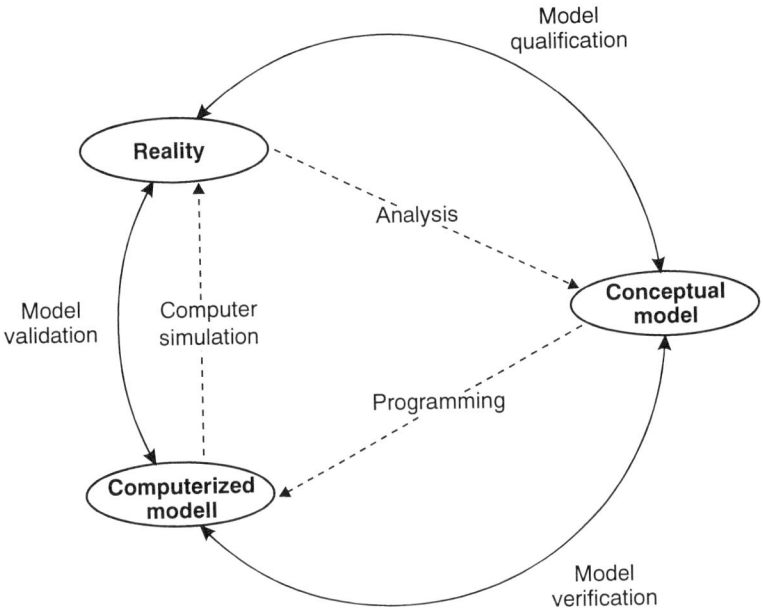

Figure 1. Elements of modelling terminology and their interrelationships. After Schlesinger et al. (1979).

Schlesinger et al. (1979) defines the different terms in a manner that appears the most suited to current hydrological thinking, while hydroinformatics is obliged to define them more generally, as follows:

Reality: The natural system, understood here as the hydrological cycle or parts of it. For hydroinformatics (as in philosophy and theology) reality and truth must be defined together. Thus (Abbott, 1994): "*Reality* is the name that we give to the interface between our outer and our inner worlds and a *truth* is an intimation of the oneness of these two worlds". Here our outer world is the world given to us by our senses and our inner world is our world of knowledge, including perceptual knowledge.

Conceptual model: Verbal descriptions, equations, governing relationships, or 'natural laws' that purport to describe reality. This is then the sign representation of that place where our outer and inner worlds intersect.

Domain of intended application (of a conceptual model): Prescribed conditions for which the conceptual model is intended to match reality. This is formulated differently in hydroinformatics, where reality in this sense is considered dynamic, and hence this domain is viewed as that to which the intentions of the user of the model is directed.

Level of agreement (of the conceptual model): Expected agreement between the conceptual model and reality, consistent with the domain of intended application and the purpose for which the model was built. This is often expressed in terms of performance criteria. This then becomes the observed agreement that can be expressed between the virtual world that the mathematical model is capable of reproducing when translated into code and the world of nature as we perceive it.

Model qualification: An estimation of the adequacy of the conceptual model to provide an acceptable level of agreement for the domain of intended application.

Computerised model: Anyone of an (in principle) infinite number of operational computer programs which implements a given conceptual model.

Model verification: Substantiation that a computerized model is in some sense a true representation of a conceptual model within certain specified limits or ranges of application and corresponding accuracy.

Domain of applicability (of a computerised model): Prescribed conditions for which the computerised model has been tested, i.e. compared with reality to the extent that is practically possible and judged suitable for use (through the process of model validation, described below).

Range of accuracy (of computerised model): Demonstrated agreement between the computerised model and reality within a stipulated domain of applicability. Since the introduction of Computational Hydraulics (e.g. Abbott, 1979) this has also been called the 'performance envelope' of the computerised model.

Model validation: Substantiation that a computerised model within its domain of applicability possesses a satisfactory range of accuracy consistent with the intended application of the model.

Although hydroinformatics can maintain some more or less tenuous relations to current practice in hydrology up to this point, as illustrated by the above definitions, at this point on it proceeds along a different track. As it makes recourse to earlier and longer established definitions, it treats *validation* in the manner set down by the Scholastics of the XIIth and XIIIth centuries and as carried over in the modern existentialistic movement by Kierkegaard. Modifying this point of view to suit the present context, there is a relation between ourselves and our outer world and another relation between our model and its outer world. The first is the relation established by our own sense experiences of the world of nature and the second is the relation expressed through the virtual world that the model presents to us. The process of validation then corresponds to a working out of the relation between these two relations so as to bring them as closely as possible into harmony (see Abbott, 1991), as schematised in Fig. 2

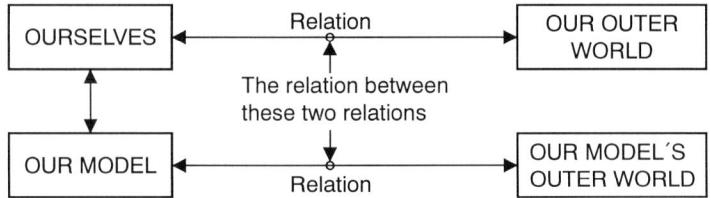

Figure 2. Schematic illustration of the term validation in traditional hydroinformatics sense as working out the relation between two relations.

It is generally understood, whether from theology, anthropology, mathematical logic or whatever other discipline that investigates this process, that this working out process is unending (e.g. Abbott, 1994). Thus models generally can only be more or less validated, but never 'absolutely validated'. This is to say that 'reality' and 'truth' can never coincide in any human construction.

Certification documentation: Documentation intended to communicate knowledge and information concerning a model's credibility and applicability, containing, as a minimum, the following basic elements:
(1) Statements of the purposes for which the model has been built.
(2) Verbal and analytical descriptions of the conceptual model and the corresponding computerised model.
(3) Specification of the domain of applicability and range of accuracy related to the purpose for which the model is intended.
(4) Description of tests used for model verification and model validation and a discussion of their adequacy.

Model certification: Acceptance by the model user of the certification documentation as adequate evidence that the computerised model can be effectively used for a specific application.

Computer simulation: Exercise of a tested and certified computerized model to 'gain insight into' reality or rather, in the general terms of computer semiotics (e.g. Anderson, 1990), to modify the user's perception of reality.

In the above definitions the term conceptual model should not be confused with the word conceptual used in the traditional classification of hydrological models ("lumped, conceptual" rainfall-runoff models), cf Subsection 2.1 and Section 3.

In practice the computerised model is usually not programmed separately for every case, but most often prepared on the basis of a generalised software package. It is therefore important to distinguish between the terms modelling system and model, as defined in Subsection 2.1 and Section 5.

In this context it should be observed that a modelling system or a code can itself be verified. A code verification involves comparison of the numerical solution generated by the code with one or more analytical solutions or with other numerical solutions. Verification ensures that the computer program solves the equations that constitute the mathematical model with an accuracy that is deemed adequate for the proposed application.

Similarly, a model is said to be validated if its accuracy and predictive capability throughout the process of validation has proven to lie within acceptable limits or errors. It is important to notice that the term 'model validation' refers to a site specific validation of a model. This must not be confused with a more general validation of a generalised modelling system, which in principle is not possible.

2.3. MODELLING PROTOCOL

There are numerous pitfalls into which the modeller can fall when using hydrological models. In this context it is essential that the user both has a thorough knowledge of the hydrological processes being modelled and has a solid experience in modelling. A third crucial factor is a good and rigorous procedure for applying models. Such a procedure, which comprises a sequence of steps in a hydrological model application, is often referred to as a modelling protocol.

In principle, such a protocol should be flexible, adaptive, open to new insights and, even when it becomes formalised as a specific list of actions, quite 'opportunistic' in its application. The protocol described below is a translation of the general terminology and methodology defined in Subsection 2.2 into the field of hydrological modelling. It is furthermore inspired by the modelling protocol suggested by Anderson and Woessner (1992), but modified concerning certain steps.

The protocol is illustrated in Fig. 3 and described step by step in the following. For each step the *reference to Fig. 3 are shown with italic text, [while the references to the general, current hydrological terminology in Fig. 1 and Subsection 2.2 are shown in brackets with italic text]*.

(1) The first step in a modelling protocol is to *define the purpose* of the model application. Examples of purposes are rainfall-runoff simulation in a gauged catchment, prediction of changes in runoff pattern due to changes in land use, and the prediction of migration of contaminants. Other examples include interpretation of flow and transport pattern as a framework for assembling and organizing field data and formulating ideas about system dynamics. An important element in this step is to give a first assessment of the desired accuracies of the model outputs. Once the purpose has been sufficiently clearly defined, it may be obvious which type of modelling system is required in order to solve the specific problem. Of course it may happen that the results of the study indicate other possibilities and the possible need to use other tools, so that it may be necessary to return to this step during any one project. *[According to the terminology outlined in Subsection 2.2 this step corresponds to defining the domain of intended application.]*

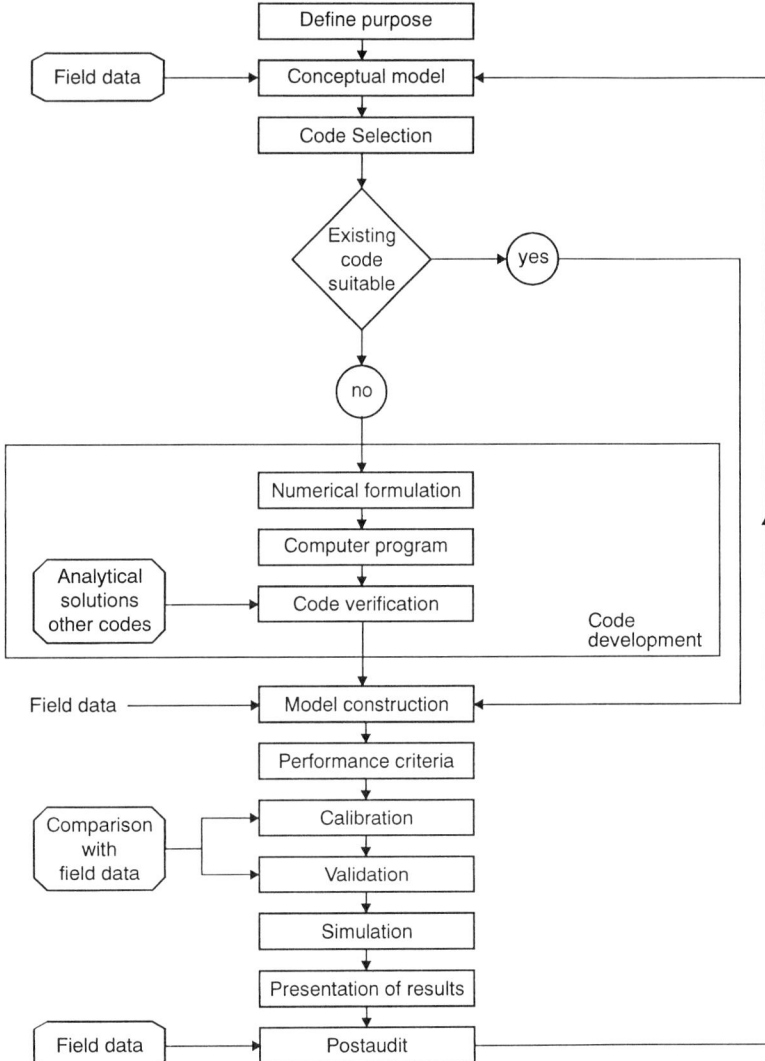

Figure 3. The different steps in hydrological model application - a modelling protocol. Modified after Anderson and Woessner (1992).

(2) Based on the purpose of the specific problem and an analysis of the available field data, the user must establish a *conceptual model*. In case of groundwater modelling this may, amongst others, involve investigating the geological conditions and decide the range of complexation to be included in the geological model. In the case of combined groundwater and surface water hydrology studies, it may, for instance, involve assessments of whether processes such as macropore

flow and hysteresis in the unsaturated zone are important and need to be explicitly modelled. In other words, a conceptual model comprises the user's perception of the key hydrological processes in the catchment and the corresponding simplifications and numerical accuracy limits which are assumed acceptable in the mathematical model in order to achieve the purpose of the modelling. *[According to the terminology outlined in Fig. 1, this step corresponds to preparing a conceptual model on the basis of an analysis of the current perception of reality.]*

(3) After having defined the conceptual model, a suitable computer program has to be selected. In principle, the computer program can be prepared specifically for the particular purpose. In practice, a *code* is often *selected* among existing generalized modelling systems. In this case it is important to ensure that the selected code has been successfully verified for the particular type of application in question. *[The terminology in Subsection 2.2 does not explicitly consider selection of an existing code, but rather programming which is development of a new code.]*

(4) In case no existing code is considered suitable for the given conceptual model a a *code development* has to take place. The computer code is a computer program that contains an algorithm capable of solving the mathematical model numerically. The term computer code is here used synonymously with the term generalised modelling system. In order to substantiate that the code solves the equations in the conceptual model within acceptable limits of accuracy a *code verification* is required. In practise, code verification involves comparison of the numerical solution generated by the model with one or more analytical solutions or with other numerical solutions. *[According to the terminology outlined in Fig. 1, this step comprises programming and model verification.]*

(5) After having selected the code and collected the necessary field data, a *model construction* has to be made. This involves designing the model with regard to the spatial discretisation of the catchment, setting boundary and initial conditions and making a preliminary selection of parameter values from the field data. In the case of distributed modelling, the model construction involves parametrisation. This process is described in more details in Chapter 3 of this book. *[This step, together with step 7 below, corresponds to the process of establishing a computerised model referred to in Fig. 1.]*

(6) The next step is to define *performance criteria* that should be achieved during the subsequent calibration and validation steps. When establishing performance criteria, due consideration should be given to the accuracy desired for the specific problem (as assessed under step 1) and to the realistic limit of accuracy determined by the field situation and the available data (as assessed in connection with step 5). If the performance criteria are specified unrealistically high, it will either be necessary to modify the criteria or to collect more and possibly quite other field data. *[According to the hydrological terminology outlined in Section 2.2, this step corresponds to specifying the level of agreement.]*

(7) *Model calibration* in general involves manipulation of a specific model to reproduce the response of the catchment under study within the range of accuracy specified in the performance criteria. In practice this is most often done by trial-

and-error adjustment of parameters, but automatic parameter estimation methods may also sometimes be used. It is important to assess the uncertainty in the estimation of model parameters, for example from sensitivity analyses. These calibration techniques are discussed further in Chapter 3 of this book. *[This step, together with step 5 above, corresponds to the process of establishing a computerised model referred to in Fig. 1.]*

(8) *Model validation* is the process of demonstrating that a given site specific model is capable of making sufficiently accurate predictions. This implies the application of the calibrated model without the changing of the parameter values that were set during the calibration when simulating the response for another period than the calibration period. The model is said to be validated if its accuracy and predictive capability in the validation period have been proven to lie within acceptable limits or to provide acceptable errors (as specified in the performance criteria). Validation schemes for different purposes are outlined in Chapter 3 of this book. *[It is this step that is also referred to as model validation in Fig. 1.]*

(9) *Model simulation* for prediction purposes is often the explicit aim of the model application. In view of the uncertainties in parameter values and, possibly, in future catchment conditions, it is advisable to carry out a predictive sensitivity analysis to test the effects of these uncertainties on the predicted results. *[This is referred to as computer simulation in Fig. 1.]*

(10) *Results* are usually *presented* in reports. However, with the newest information technology that is now emerging it is also possible to display modelling results as (colour) animations. Furthermore, in certain cases, the final model is transferred to the end user for subsequent day-to-day operational use. *[This step is not explicitly included in the general terminology in Section 2.2.]*

(11) An extra possibility of validation of a site specific model is a socalled *postaudit*. A postaudit is carried out several years after the modelling study is completed and the model's predictions can be evaluated. Examples of postaudits from groundwater modelling are given in Konikow (1986), Alley and Emery (1986) and Konikow and Person (1985). *[This step is not included in the general terminology in Section 2.2.]*

3. Classification according to Hydrological Process Description

3.1. CLASSIFICATION

Many attempts have been made to classify hydrological models, see e.g. Fleming (1975) and Woolhiser (1973). The present classification, illustrated in Fig. 4, is applicable to hydrological simulation models. The classification is in principle applicable both to catchment models and to single component models such as groundwater models. It is emphasized that the classification is schematic, and that many model codes do not fit exactly into one of the given classes. The construction of such a classification corresponds to the introduction of a taxonomy into the hydrological universe of discourse.

The two classical types of hydrological models are the deterministic and the stochastic. During the 1960s and 1970s these two fundamentally different approaches were developed within two more or less separate "hydrological schools". However, over the last decade, increasingly more interplay has occurred and a joint stochastic-deterministic methodology today provides a very useful framework for addressing some of the fundamental problems in hydrology such as taking spatial variability (scale problems) into account and assessing uncertainties in modelling.

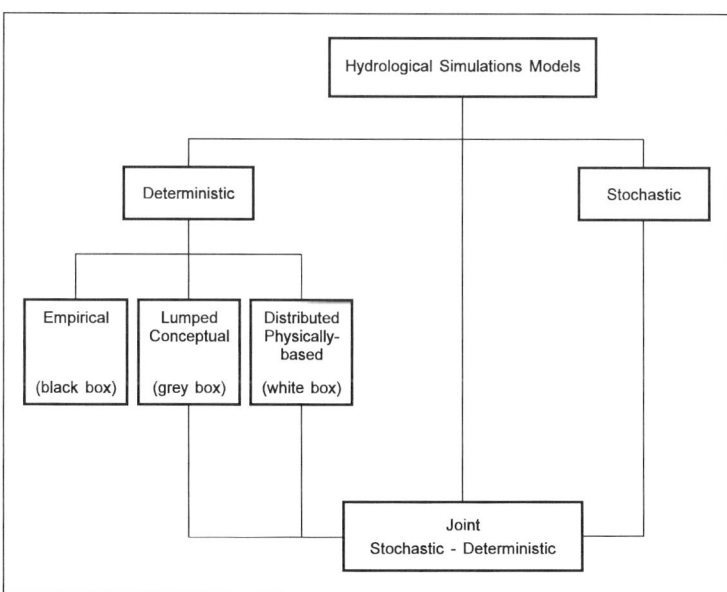

Figure 4. Classification of hydrological models according to process description.

3.2. DETERMINISTIC MODELS

Deterministic models can be classified according to whether the model gives a lumped or a distributed description of the considered area, and whether the description of the

hydrological processes is empirical, conceptual, or more physically-based. As most conceptual models are also lumped, and as most physically-based models are also distributed, the three main groups of deterministic models shown in Fig. 4 emerge:
* Empirical models (black box).
* Lumped conceptual models (grey box).
* Distributed physically-based models (white box).

3.2.1. Empirical (black box) models
Black box models are empirical, involving mathematical equations that have been assessed, not from the physical processes in the catchment, but from analyses of concurrent input and output time series. Black box models may be divided into three main groups according to their origin, namely empirical hydrological methods, statistically based methods and hydroinformatics based methods.

Empirically hydrological methods. Probably, the best known among the black box models in hydrology is the unit hydrograph model and the models applying the unit hydrograph principles, Sherman (1932), Nash (1959). Today, empirical hydrological methods are often used in some components of more comprehensive models, e.g. the unit hydrograph is often used for streamflow routing and a linear reservoir is often used to represent the groundwater system in conceptual rainfall-runoff models.

Statistically based methods. Traditional statistical methods in hydrology have been developed extensively with support from basic statistical theory. These methods are often mathematically more advanced than the above empirical hydrological methods.

An important type of statistically based methods is comprised of the regression and correlation models. Linear regression and correlation techniques are standard statistical methods used to determine functional relationships between different data sets. The relationships are characterized by correlation coefficients and standard deviations and the parameter estimation is carried out using rigorous statistical methods involving tests for significance and 'validity' of the chosen model. Regression and correlation models are often used as so-called 'transfer function models' converting input time series to output time series. Examples of such models include:
* Autoregressive Integrated Moving Average (ARIMA) models (Box and Jenkins, 1970), which amongst others have been used extensively in surface water hydrology for establishing relationships between rainfall and discharge.
* The Constrained Linear Systems (CLS) model (Todini and Wallis (1977), which is basically a composition of different linear regression relationships valid for intervals between certain thresholds, so that it altogether functions as a non-linear model.
* Gauge to gauge correlation methods (e.g. WMO, 1994), which previously have constituted the most used method for flood forecasting in large rivers, e.g. in India and Bangladesh.
* Antecedant Precipitation Index (API) model (e.g. WMO, 1994) correlating rainfall volume and duration, the past days' rainfall, and the season of the year to runoff.

Hydroinformatics based methods. A new group of 'transfer function models' based on methods introduced more generally in hydroinformatics is now emerging. One class of techniques is based on neural networks, while another class is based on evolutionary algorithms. Both methods have been successfully tested on a trial basis for rainfall-runoff modelling, but they have not yet been sufficiently tested to be applied in practice. However, the potential for such methods appears to be significant, especially as a replacement of the statistically based methods, and for this reason they are introduced in Chapter 14 of this book.

3.2.2 Lumped conceptual models

Models of the lumped conceptual type are mainly found within the field of rainfall-runoff modelling. Lumped conceptual models operate with different but mutually-interrelated storages representing physical elements in a catchment. The mode of operation may be characterized as a bookkeeping system that is continuously accounting for the moisture contents in the storages.

Due to the lumped description, where all parameters and variables represent average values over the entire catchment, the description of the hydrological processes cannot be based directly on the equations that are supposed to be valid for the individual soil columns. Hence, the equations are semi-empirical, but still with a physical basis. Therefore, the model parameters cannot usually be assessed from field data alone, but have to be obtained through the help of calibration.

The Stanford modelling system (Crawford and Linsley, 1966) is the classical representative of this model type. The structure of the Stanford system is shown schematically in Fig. 5.

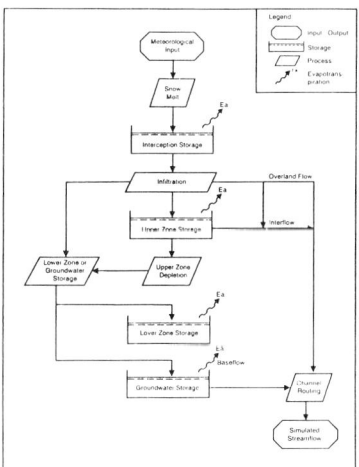

Figure 5. Structure of the Stanford model

Numerous other rainfall-runoff modelling systems of the lumped conceptual type exist. A brief description of 19 different systems is given by Fleming (1975). More comprehensive and recent descriptions of a large number of modelling systems are provided in Singh (1995).

The lumped, conceptual models are especially well suited for the simulation of the rainfall-runoff process when hydrological time series exist that are sufficiently long for a model calibration. Thus, typical fields of application include the extension of short streamflow records based on long rainfall records and real-time rainfall-runoff simulations for flow forecasting.

Although no significant improvements have been made during the last decade with regard to the fundamental structure and functioning of these models, many of the codes have undergone a comprehensive technological development (see Section 4) enabling them to be disseminated widely for practical application by a large group of users.

3.2.3 Distributed physically-based models

The principal mode of operation of a distributed physically-based model is illustrated in Fig. 6. Contrary to the lumped conceptual models, a distributed physically-based model does not consider the water flows in an area to take place between a few storage units. Instead, the flows of water and energy are directly calculated from the governing continuum (partial differential) equations, such as for instance the Saint Venant equations for overland and channel flow, Richards' equation for unsaturated zone flow and Boussinesq's equation for groundwater flow.

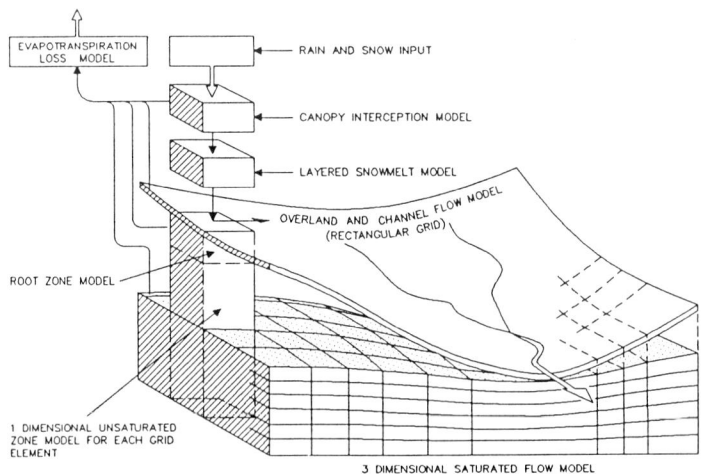

Figure 6. Schematic diagram of a catchment and the MIKE SHE quasi three-dimensional distributed physically-based model (DHI, 1993).

Distributed physically-based models have been used for a couple of decades on a routine basis for the simulation of single hydrological processes, so that almost all groundwater models and most unsaturated zone flow models belong to this type. A first attempt to outline the potentials and some of the elements in a distributed physically-based model on a catchment scale was made by Freeze and Harlan (1969). Today, several general purpose catchment model codes of this type exist such as SHE (Abbott et al., 1986), MIKE SHE (Refsgaard and Storm, 1995), IHDM (Beven et al., 1987) and THALES (Grayson et al., 1992).

Distributed physically-based models give a detailed and potentially more correct description of the hydrological processes in the catchment than do the other model types. Moreover, they are able to exploit the quasi-totality of all information and all knowledge that is available concerning the catchment that is being modelled. The distributed physically based models can in principle be applied to almost any kind of hydrological problem. However, in practice, they will be used complementary to the other model types for cases where the other models are not suitable. Some examples of typical applications are:

* Prediction of the effects of catchment changes due to human interference in the hydrological cycle, such as changes in land use (including urbanization), groundwater development and irrigation. Since parameters of the model tend to be more physically based, the change in parameter values corresponding to the catchment changes can some times be estimated directly.
* Prediction of runoff from ungauged catchments and from catchments with relatively short records. As opposed to the lumped conceptual models, which require long historical time series of rainfall, runoff and evaporation data for the assessment of parameters, the parameters of the distributed physically-based models may be assessed directly from intensive, short-term field investigations.
* Water quality and soil erosion modelling for which a more detailed and physically correct simulation of water flows is important.

However, even with the most advanced of the distributed, physically based models, significant deficiencies still persist at the level of the process descriptions. Hence, the term 'white box', suggesting as it does that everything within the model is transparent and correct, is not fully justified, and 'light grey' might be more appropriate.

Due to the significant progress being made within the scientific community in the direction of an enhanced physical understanding, this model type will surely continue to be improved over the next many years.

3.3 STOCHASTIC TIME SERIES MODELS

A stochastic time series model treats the sequence of events that it comprehends as time-dependent. Traditionally, a stochastic model is derived from a time series analysis of the historical record. The stochastic model can then be used for the generation of long hypothetical sequences of events with the same statistical properties as the historical record. The technique of generating several synthetic series with identical statistical properties is denoted the *Monte Carlo technique*. These generated sequences of data can then be used in the analyses of design variables and their uncertainties, such as these may arise, for example, when estimating reservoir storage requirements. An overview of stochastic time series modelling systems is given in Salas (1992).

With regard to process description, the classical stochastic simulation models are comparable to the empirical, black box models described in Subsection 3.2.1 above. Hence the stochastic time series models are in reality composed of a simple deterministic core (the black box model) contained within a comprehensive stochastic methodology.

3.4 JOINT STOCHASTIC-DETERMINISTIC MODELS

On one hand, a substantial part of the hydrological processes, including the spatial and temporal variations of hydrological parameters and variables, can today be described using deterministic simulation models. On the other hand, the available information on parameter values and input variables will always be incomplete. This lack of knowledge is an important source of uncertainty in hydrological simulation.

Acknowledging this duality, several model types based on a joint stochastic-deterministic approach have been developed. These models are composed of two, in principle equally important parts, namely a deterministic core within a stochastic frame. In contrast to the stochastic simulation models described in Section 3.3 above, the deterministic core is here composed of more comprehensive models, whether of the lumped conceptual or the distributed physically-based type.

In the following, some examples illustrating different types of joint deterministic-stochastic models are given.

3.4.1 State space formulations - Kalman filtering
The state space theory and the Kalman filtering technique (e.g. Gelb, 1974) are powerful mathematical tools originally developed within the field of statistical control theory for linear systems, but later extended to comprehend non-linear systems also. They are now being applied increasingly also in hydrology.

The key model variables are recognized as stochastic variables that are parameterized in terms of their mean and standard deviations. The input variables (e.g. rainfall data) are similarly described by a mean value (the recorded value) and a standard deviation (the uncertainty). In this way it is possible to calculate how uncertainty on, for example, input data propagates through the model and causes uncertainties on model state variables and output results.

Combinations of Kalman filtering and lumped, conceptual rainfall-runoff models have been realised, amongst others for the Sacramento modelling system (Kitanidis and Bras, 1978; Georgakakos et al., 1988) and for the NAM system (Refsgaard et al., 1983; Storm et al., 1988).

3.4.2 Spatial variabilities of parameter values and stochastic PDE's
Subsurface hydraulic parameters exhibit very large spatial variabilities even on very small scales. Therefore, it is not realistic to give a correct complete deterministic description in all details. In most cases, however, knowledge of the detailed flow pattern is of little interest; it is rather the impact of the spatial heterogeneity on the

overall flow pattern which is important.

One way of treating this problem is to represent (some of) the hydraulic parameters as realizations of stochastic variables with known statistical properties (probability distribution type, mean, standard deviation, spatial autocorrelation). As a result of following this approach, the governing equations become what are often described as stochastic partial differential equations (PDEs), which are of course usually much more difficult to solve than are traditional deterministic PDEs. Two possible approaches to the stochastic PDE's are:

* The Monte Carlo technique, whereby the deterministic model is run several times using different (equally probable) realizations of the parameter field. Finally, statistical analyses are made of the results from all the deterministic simulation runs. The advantage of this approach is that the deterministic model can be preserved, whereas the main disadvantage is the very large CPU requirement. Classical examples of this approach are provided by Smith and Freeze (1979a,b) and Freeze (1980) for groundwater flow and rainfall-runoff processes, respectively. Zhang et al. (1993) demonstrated a methodology of using the Monte Carlo technique to determine the effect of uncertainty in model parameters and rainfall on the uncertainty of model responses and the further impact of these uncertainties on sample size requirements for simulating solute transport through soils.
* The stochastic PDE is simplified and solved analytically. This is often carried out by elegant mathematical solution techniques and has proven very useful for obtaining general insights into fundamental research problems. However, this methodology puts so significant limitations on the range of data input, the size of the problem, the boundary conditions and other aspects that it appears to have had limited practical application. This methodology has been used especially within the fields of unsaturated zone and groundwater flow and transport for research purposes, see, for example, Gelhar (1986) and Dagan (1986). Jensen and Mantoglou (1992) applied a stochastic theory to modelling of a small experimental field. This theory was subsequently incorporated into MIKE SHE and applied to catchment size problems by Sonnenborg et al. (1994).

The joint stochastic-deterministic methods may be seen as extensions of the deterministic simulation models described in Section 3.2. The output result of a deterministic model simulation is in principle one time series for each of the model variables. The joint stochastic-deterministic models, on the other hand, produce uncertainty bands (such as are described by mean values and standard deviations) for each of the predicted time series. Thus, the joint stochastic-deterministic models are able to transfer the inherent uncertainty on the input variables or effects of non-described variability of spatial parameter values to probabilistic descriptions of the output variables.

A more comprehensive approach also considering the uncertainty in the model structure and process equations is the GLUE technique introduced by Beven and Binley (1992).

4. Modelling of Spatial Variability of Hydrological Parameters

Hydrological parameter values exhibit very large spatial variations in nature. Thus, detailed field measurements indicate that, for instance, the hydraulic conductivity may vary by several orders of magnitude within a small field which is traditionally characterised as 'homogeneous' and belonging to the same soil type (Nielsen et al., 1973). One of the key differences between lumped and distributed models are the fundamentally different ways in which they take the spatial variability into account.

Consider for illustrational purposes how the Stanford and the MIKE SHE, as typical representatives for the lumped conceptual and the distributed physically-based model types, describe the infiltration process in the unsaturated zone over a catchment.

In the Stanford an assumption of spatial variability of the infiltration capacity is in a simplified way built into the infiltration equation of the model code. Thus, the spatial variability of the key soil parameter is implicitly taken into account. As a result the mechanism for generation of overland flow functions as a contributing area approach.

In MIKE SHE the infiltration calculations are based on Richards' equation which is assumed to be valid theoretically for single-domain porous media flow. Being a distributed model the catchment is divided into a number of grids, and the infiltration equation is thus used on a number of unsaturated zone columns, each of which is characterised by soil hydraulic parameters. Thus, in a distributed model the spatial variability is taken into account explicitly through the variability of the model parameter values. In principle, the model parameters are different in the different soil columns. However, in practice the parametrization procedure usually prescribes identical parameter values for all grid points with the same soil type, and furthermore there will always be significant variability within a grid which cannot be accounted for by a deterministic approach. Thus, unless the variability among soil groups is significantly larger than the variability within soil groups (and this is often not the case) MIKE SHE does not explicitly take all the spatial variability into account. The result is that the overland flow generation in the catchment theoretically becomes too much of an 'on/off' process. In practice, this effect is often overshadowed by other dominating processes or it can be compensated through calibration.

Hence, even in fully physically-based models the spatial variability of hydrological parameters will most often not be at all fully described. In order to take the variability more fully into account, a joint stochastic-deterministic model as outlined in Subsection 3.4.2 may be adopted.

Another approach is adopted in TOPMODEL (Beven, 1986), where the spatial variability of soil properties is built into the process equations. TOPMODEL does not fit into one of the schematic model classes defined in Section 3, but may be characterized as a semi-distributed semi-physically-based modelling system. The advantages of TOPMODEL as compared to traditional lumped conceptual systems are the explicit accounting of the spatial variability and the direct use of spatial data such as topography and channel system together with semi-distributed calculations of hydrological variables.

5. Classification according to Technological Level

A practice of modelling can be categorized according to Abbott et al. (1991) according to its technological level in different generations of modelling systems:

First Generation - Computerised Formulae. The first generation dates back to the early 1950s. The introduced computer codes mainly aimed at making the methods of numerical calculations that had been developed for human computation easier and quicker in through the application of the very simplest numerical methods. At one time it was popular saying that this was to use the computer as a 'super slide rule' but nowadays, of course, few persons know anymore what even a slide rule was.

Second Generation - One-Off Numerical Models. The second generation of modelling appeared in the 1960s. It was involved in construction, applying and developing models for the most part by university or research institutes, for solving one specific problem. The practice of the second generation of modelling was generally restricted to the group of persons who had developed the code for one particular geographical domain, but the models so developed were not generally applicable to use by other persons or to similar problems under other external conditions. Second generation models generally required comprehensive user experience within the fields of computers (hardware and software) and the related numerical techniques.

Third Generation - Generalised Numerical Modelling Systems. Third generation modelling was directed to constructing generalised modelling systems with which many different problems could be solved by the use of one and the same computer code. Third generation modelling systems emerged in the 1970s in hydrodynamics. In hydrology the first distributed physically-based systems were made in the beginning of the 1980s, while the much simpler lumped conceptual systems appeared in the 1960s. These systems were mainly applied by computer-experienced specialists. Some basic knowledge of numerical methods was also required by the user. Traditionally, third generation systems have been based on main frame computers, but most of them can run on PC's today. However, this approach is now largely superceded by fourth generation modelling.

Fourth Generation - The Industrial User-friendly Software Product. These are user-friendly software products that can be applied by engineering and scientific professionals. Fourth generation systems differ from third generation systems in the following ways:

* They are usually PC (or workstation) based.
* They are fully menu based, providing interactive execution with on-line help menus.
* They provide comprehensive error messages and automatic checks for obviously erroneous input data.
* They provide more powerful graphics facilities both on monitor, printer and plotter devices.
* They are generally much better documented and more well-proven due to a larger installation base.
* They are more easily transferable and installable.
* They have a much wider circulation (being distributed in one or two orders of magnitude more copies).

These systems therefore make little or no demands on user experience in computer systems or numerical techniques, but they still require user experience in modelling. The first fourth generation systems appeared in the mid 1980s.

Fifth Generation - 'Intelligent' Modelling System. These begin with modelling systems designed for technically-skilled but non-expert users. They include numerical stability monitoring and diagnosis tools, interfaces to standard CAD and data base systems, and knowledge base system frames applicable in decision making. They merge into hydroinformatics tools generally, such as diagnostic and real-time control systems and management support systems (e.g. Abbott, 1991; IAHR, 1994; Verwey et al, 1994).

At present only a few fourth generation systems are operational and fifth generation systems are still mainly at an experimental stage outside of real-time control applications for urban drainage systems (e.g. Gustafsson et al., 1993). By way of an example the development history of DHI's river modelling system, MIKE 11, is illustrated in Fig 7. In order to achieve a large dissemination of modelling technology to a considerable part of the professional community (and not only to modelling experts) experience shows that fourth generation systems are required. Furthermore, it is believed that fifth generation systems extending to hydroinformatics systems are required in order for the modelling technologies to achieve their full potential in terms of practical application.

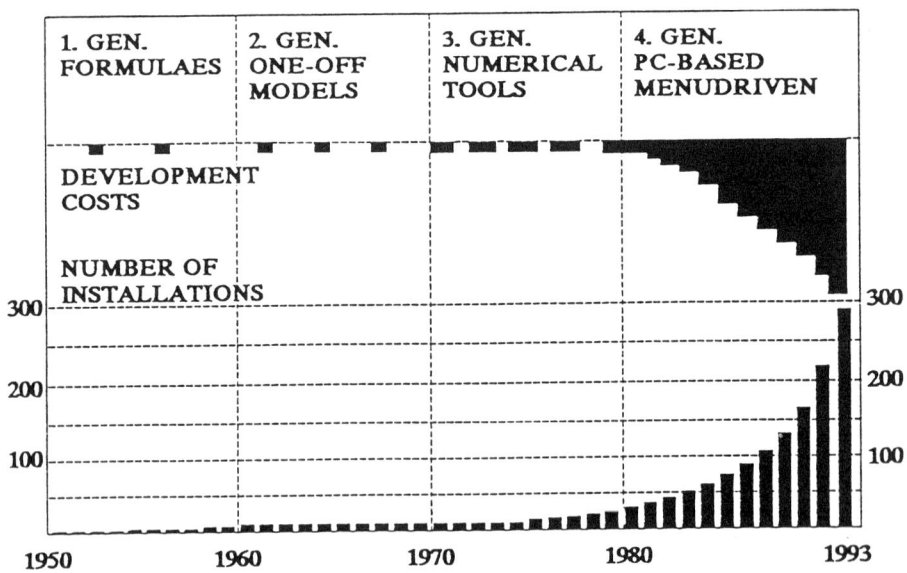

Figure 7. The development history of DHI's river modelling system MIKE11.

6. Acknowledgement

Some of the ideas presented on the terminology is the result of input from and discussion with M.B. Abbott, who is also thanked for a carefull review of this chapter.

7. References

Abbott, M.B. (1979) *Elements of the theory of free surface flows*. Pitman, London. Second edition.
Abbott, M.B. (1991) *Hydroinformatics: information technology and the aquatic environment*. Avebury, Aldershot.
Abbott, M.B. (1993) The encapsulation of knowledge in hydraulics, hydrology and water resources. *Advances in Water Resources*, 16, 21-39.
Abbott, M.B. (1994) Hydroinformatics and Copernicam revolution in hydraulics. *Extra issue on hydroinformatics, Journal of Hydraulic Research*, 32, 3-14.
Abbott, M.B., Bathurst, J.C., Cunge, J.A., O'Connel, P.E. and Rasmussen, J. (1986a) An Introduction to the European Hydrological System - Systeme Hydrologique Europeen, "SHE", 1: History and Philosophy of a Physically-based, Distributed Modelling System. *Journal of Hydrology*, 87, 45-59.
Abbott, M.B., Bathurst, J.C., Cunge, J.A., O'Connel, P.E. and Rasmussen, J. (1986b) An Introduction to the European Hydrological System - Systeme Hydrologique Europeen, "SHE", 2: Structure of a Physically-Based, Distributed Modelling System. *Journal of Hydrology*, 87, 61-77.
Abbott, M.B., Havnø, K and Lindberg, S. (1991) The fourth generation of numerical modelling in hydraulics. *Journal of Hydraulic Research*, 29(5), 581-600.
Abbott, M.B. and Minns, A. (1995) *Computational hydraulics*. Avebury, Aldershet.
Alley, W.M. and Emmery, P.A. (1986) Groundwater model of the Blue River Basin, Nebraska - twenty years later. *Journal of Hydrology*, 85, 225-250.
Anderson, M.P. and Woesner, W.W. (1992) The role of postaudit in model validation. *Advances in Water Resources*, 15, 167-173.
Anderson, P.B. (1990) *A theory of computer semantics*. Cambridge Univ., Cambridge, UK.
Beven, K. (1986) Runoff production and flood frequency in catchments of order n: An alternative approach. In V.K. Gupta et al. (Eds) *Scale problems in hydrology*, 107-131. D. Reidel Publishing Company.
Beven, K. and Binley, A. M. (1992) The future role of distributed models: model calibration and predictive uncertainty. *Hydrological Processes*, 6, 279-298.
Beven, K., Calver, A., and Morris, E.M. (1987) The Institute of Hydrology distributed model. Institute of Hydrology Report 98, Wallingford, UK.
Box, G.E.P. and Jenkins, G.M. (1970) *Time series analysis, forecasting and control*. Holden-Bay Inc., San Francisco.
Crawford, N.H. and Linsley, R.K. (1966) Digital simulation in hydrology, Stanford watershed model IV. Department of Civil Engineering, Stanford University, Technical Report 39.
Dagan, G. (1986) Statistical theory of groundwater flow and transport: pore to laboratory, laboratory to formation, and formation to regional scale. *Water Resources Research*, 22(9), 120-134.
DHI (1993) MIKE SHE WM - A short description. Danish Hydraulic Institute.
Flavelle, P. (1992) A quantitative measure of model validation and its potential use for regulatory purposes. *Advances in Water Resources*, 15, 5-13.
Fleming, G. (1975) *Computer simulation techniques in hydrology*. Elsevier, New York.
Freeze, R.A. (1980) A stochastic-conceptual analysis of rainfall-runoff processes on a hillslope. *Water Resources Research* 16(2), 391-408.
Freeze, R.A. and Harlan, R.L. (1969) Blueprint for a physically-based digitally-simulated hydrological response model. *Journal of Hydrology*, 9, 237-258.
Gelb, A. (Ed.) (1974) *Applied optimal estimation*. MIT Press, Cambridge, Mass.

Gelhar, L.W. (1986) Stochastic subsurface hydrology. From theory to applications. *Water Resources Research*, 22(9), 135-145.

Georgakakos, K.P., Rajaram, H. and Li, S.G. (1988) On improved operational hydrologic forecasting of streamflows. IIHR Report No 325. Department of Civil and Environmental Engineering and Iowa Institute of Hydraulic Research, The University of Iowa.

Grayson, R.B., Moore, I.D., and McHahon, T.A. (1992) Physically based hydrological modelling. 1. A terrain-based model for investigative purposes. *Water Resources Research*, 28(10), 2639-2658.

Gustafsson, L.-G., Lumley, D.G., Persson, B. and Lindeborg, C. (1993) Development of a catchment simulator as an on-line tool for operating a wastewater treatment plant. ICUSD '93, 6th International Conference on Urban Storm Drainage, Niagara Falls, Ontario, Canada, September 12-17, 1993.

IAHR (1994) Hydroinformatics, extra issue. *Journal of Hydraulics Research*.

Jensen, K.H. and Mantoglou, A. (1992) Application of stochastic unsaturated flow theory, numerical simulations and comparison to field observations. *Water Resources Research*, 28(1), 269-284.

Kitanidis, P.K. and Bras, R.L. (1978) Real time forecasting of river flows. Technical Report 235. Ralph M. Parson's Laboratory for Water Resources and Hydrodynamics, MIT, Cambridge, Massachusetts.

Konikow, L.F. (1978) Calibration of groundwater models. In *Verification of Mathematical and Physical Models in Hydraulic Engineering*, American Society of Civil Engineers, New York, 1978, 87-93.

Konikow, L.F. (1986) Predictive accuracy of groundwater model - lessons from postaudit. *Ground Water*, 24, 173-184.

Konikow, L.F. and Person, M. (1985) Assessment of long-term salinity changes in an irrigated stream-aquifer system. *Water Resources Research*, 21, 225-250.

Nash, J.E. (1955) Systematic determination of unit hydrograph parameters using method of moments. *Journal of Geophysical Research*, 64, 111-115.

Nielsen, D.R., Biggar, J.W. and Ehr, K.T. (1973) Spatial variability of field measured soil-water properties. *Hilgardia*, 42(7), 215-260.

Oreskes, N., Shrader-Frechette, K. and Belitz, K. (1994) Verification, validation and confirmation of numerical models in the earth sciences. *Science*, 264, 641-646.

Refsgaard, J.C., Rosbjerg, D., and Markussen, L.M. (1983) Application of the Kalman filter to real-time operation and to uncertainty analyses in hydrological modelling. *Proceedings of the Hamburg Symposium, Scientific Procedures Applied to the Planning, Design and Management of Water Resources Systems, August 1983*. IAHS Publication 147, 273-282.

Refsgaard, J.C. and Storm, B. (1995) MIKE SHE. In V.J. Singh (Ed) *Computer models in watershed hydrology*. Water Resources Publications.

Salas, J.D. (1992) Analyses and modelling of hydrologic time series. In D.R. Maidment (Editor in Chief): *Handbook of Hydrology*. McGraw-Hill.

Schlesinger, S., Crosbie, R.E., Gagné, R.E., Innis, G.S., Lalwani, C.S., Loch, J., Sylvester, J., Wright, R.D., Kheir, N. and Bartos, D. (1979) Terminology for model credibility. SCS Technical Committee on Model Credibility. *Simulation*, 32(3), 103-104.

Sherman, L.K. (1932) Stream flow from rainfall by the unitgraph method. *English News Record*, 108, 501-505.

Singh, V.J. (Ed) (1995) *Computer models of watershed hydrology*. Water Resources Publications.

Smith, L. and Freeze, R.A. (1979a) Stochastic analysis of steady state flow in a bounded domain 1. One-dimensional simulations. *Water Resources Research*, 15(3), 521-528.

Smith, L. and Freeze, R.A. (1979b) Stochastic analysis of steady state flow in a bounded domain 2. Two-dimensional simulations. *Water Resources Research*, 15(6), 1543-1559.

Sonnenborg, T.O., Butts, M.B. and Jensen, K.H. (1994) Application of stochastic unsaturated flow theory. *Proceedings of the Nordic Hydrological Conference, Thorshavn*.

Storm, B., Jensen, K.H., and Refsgaard, J.C. (1988) Estimation of catchment rainfall uncertainty and its influence on runoff prediction. *Nordic Hydrology*, 19, 77-88.

Todini, E., and Wallis, J.R. (1977) Using CLS for daily or longer period rainfall-runoff modelling. In: *Mathematical Models for Surface Water Hydrology*. Wiley, New York.

Tsang, C.-F., (1991) The modelling process and model validation. *Ground Water*, 29, 825-831.

Verwey, A., Minns, A., Barbovic, V. and Maksimovic, C. (Eds) (1994) *Proceedings, Hydroinformatics 94*. Balkena, Rotterdam.

WMO (1994) Guide to hydrological practices. WMO-No. 168. fifth edition. World Meteorological Organization, Geneva.

Woolhiser, D.A. (1973) Hydrologic and watershed modelling - State of the art. *Transactions of the ASAE*, 16, 533-559.

Zhang, H, Haan, C.T and Nofziger, D.L. (1993) An approach to estimating uncertainties in modelling transport of solutes through soils. *Journal of Contaminant Hydrology*, 12, 35-50.

CHAPTER 3
CONSTRUCTION, CALIBRATION AND VALIDATION OF HYDROLOGICAL MODELS

J.C. REFSGAARD AND B. STORM
Danish Hydraulic Institute

1. Introduction

In Chapter 2 of this book a terminology was defined and the different steps in a modelling application was put into the framework of a modelling protocol. In the present chapter some of the elements in such protocol are addressed in more details with special reference to distributed models.

The different types of error sources in hydrological simulation are outlined in Section 2. Section 3 gives an introduction to goodness of fit and accuracy criteria. Section 4 deals with the procedure of model construction including parametrisation. Section 5 provides an overview of different procedures adopted for model calibration, while the model validation issue is dealt with in Section 6. Finally, the issue of credibility or degree of validity of a generic modelling system is addressed in Section 7.

2. Reasons of Uncertainty in Hydrological Modelling

2.1 DETERMINISTIC MODELLING CONCEPT

The concept of deterministic mathematical modelling can be illustrated as in Fig. 1, where the physical system, in this case a catchment, is shown to the left. The mathematical model, which is the sign-representation of the physical system is shown to the right. Both the quantification of the physical system through monitoring and the representation through mathematical modelling are subject to errors.

The temporal and spatial variability of inflows, outflows and internal conditions are quantified from measurements at discrete locations. Unfortunately these measurements alone can neither provide us with any complete picture of the conditions inside the catchment area nor an understanding of the mass, energy and other exchanges between the catchment and its surroundings. Firstly, because it is not feasible to measure all the spatial and temporal variations of flows and state variables, and secondly because the measurements include errors. Consequently, the model will be using an approximate dataset which to some degree contain sampling errors. On the other hand, by definition, the physical system experiences the real data and responds to this real input. However, the response is measured with some uncertainty.

When the hydrological model is used to simulate the behaviour of the physical system, it produces output data, which is affected by the approximate input data. In order to test the accuracy of the developed model simulated output is then compared with the recorded data which also is subject to uncertainties. A desired model accuracy is achieved by changing the parameter values used in the model until a satisfactory agreement between simulated and the recorded variables is obtained. This parameter adjustment process is the process of calibration.

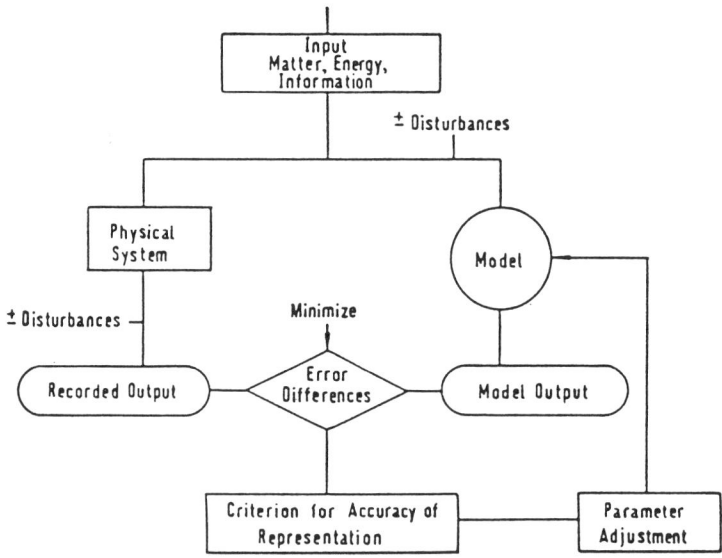

Figure 1. The concept of deterministic mathematical modelling. After Fleming (1975).

2.2 SOURCES OF UNCERTAINTY

Differences between recorded data and simulated model output arise basically from four sources of uncertainty:
(1) Random or systematic errors in the input data, i.e. precipitation, temperature and evapotranspiration etc. used to represent the input conditions in time and space over the catchment.
(2) Random or systematic errors in the recorded data, i.e. the river water levels, groundwater heads, discharge data or other data used for comparison with the simulated output.
(3) Errors due to non-optimal parameter values.
(4) Errors due to an incomplete or biased model structure.

Although a disagreement between simulated and recorded data is the combined effect of all four error sources, only error source (3) can be minimized during the calibration process. The measurement errors, error sources (1) and (2), are the "background noise" which determine the maximum achievable agreement, which the modeller can hope to obtain. Changes in parameter values or model structure cannot and should not improve the results beyond that. The objective of a calibration process is therefore to reduce the error source (3) until it becomes insignificant compared to the data error sources (1) and (2).

During a calibration process it is of utmost importance to ensure that a clear distinction is drawn between the different error sources, so no attempt is made to compensate for errors from one source by adjustments within another source, such as compensating for a data error by parameter adjustments. Otherwise the calibration will approach curve fitting, which may result in a reasonable fit within the calibration period but may incidently provide unreliable predictions.

It is also important that the modeller recognises the limitations of a chosen modelling approach and is not trying to impose a certain conceptual representation of the physical system because it fits to the mathematical formulation in the model code applied.

3. Goodness of fit and accuracy criteria

During the calibration procedure different accuracy criteria can be used to compare the simulated and measured data. This allows us to define an objective measure of the goodness of fit associated with each set of model parameters and estimate the parameter values which provide the best overall agreement between model output and measured data. However, the selection of an appropriate criterion is complicated by the variation in the sources of error discussed in the previous section. It further depends on the objective of the model simulation (e.g. to simulate flood peaks or low flows) and on the type of information available to check model output variables, such as phreatic surface levels, soil moisture contents, stream discharges or stream water levels.

No single criterion is entirely suitable for all variables, and even for a single variable it is not always easy to establish a satisfactory criterion. Hence a large number of different criteria has been developed. Green and Stephenson (1986) discuss the problem in detail and list 21 approaches which can be used for single-event simulations depending on the simulation objectives, range of simulation conditions and other factors. Criteria for the detection of systematic errors relevant to long-term simulations are discussed by Aitken (1973).

It is possible to calibrate a model by optimizing just one criteria. However, a calibration based on a 'blind' optimization of a single numerical criterion can produce physically unrealistic parameter values which, if applied to a different time period, may give poor simulation results. Green and Stephenson (1986) conclude that no single criterion is sufficient to assess adequately the overall measure of fit between a computed and an observed hydrograph, particularly in view of the many objectives behind hydrological modelling. At the same time, it should be remembered that the criteria measure only the correctness of the estimates of the hydrological variables generated by the model and not the hydrological soundness of the model relative to the processes being simulated. It is therefore recommended that, in a calibration, numerical criteria be used for guidance only.

Finally, the benefit of using graphical comparison of e.g. simulated and observed hydrographs should be emphasized. Although the analysis becomes more subjective, graphical comparisons provide a good overall indication of the models capabilities, they are more easily assimilated and may impart more practical information than statistical functions alone. Fig. 2 illustrates the combined use of numerical and graphical comparisons of runoff data. In the figure the flow duration curve error index, EI, is a numerical measure of the difference between the flow duration curves of the simulated and observed daily flows (perfect agreement for EI = 1). Similarly, R2 refers to the Nash-Sutcliffe coefficient (Nash and Sutcliffe, 1970), computed on the basis of the sequence of observed and simulated monthly flows.

4. Model construction

Model construction is the process of preparing the data in a correct format and entering them into a set of input data files required by the model code so that the first model run can be made as the first step in the subsequent model calibration.

For lumped conceptual model codes or similar type of model codes which only requires few parameter values for a set of empirical equations, the model parameters can generally not directly be estimated from catchment characteristics, but must be estimated through calibration. The data requirements are therefore very small and the data preparation is usually confined to construction of data files containing time series of climatic input data and a corresponding time series of control data e.g. river discharges. These can usually be prepared very quickly.

For distributed physically-based models, the data preparation phase is more complicated and involves often a comprehensive programme of work. A fully distributed physically-based model contains only parameters which can be assessed from field data, so that, in principle, a calibration would not be necessary if sufficient data were available. In practice, distributed physically-based model codes are most often applied at scales on which the parameter values cannot be directly assessed, and will therefore require calibration. In the subsequent calibration phase, allowed parameter variations may be restricted to relatively narrow intervals compared to parameter values related to the empirical functions in empirical or lumped conceptual models.

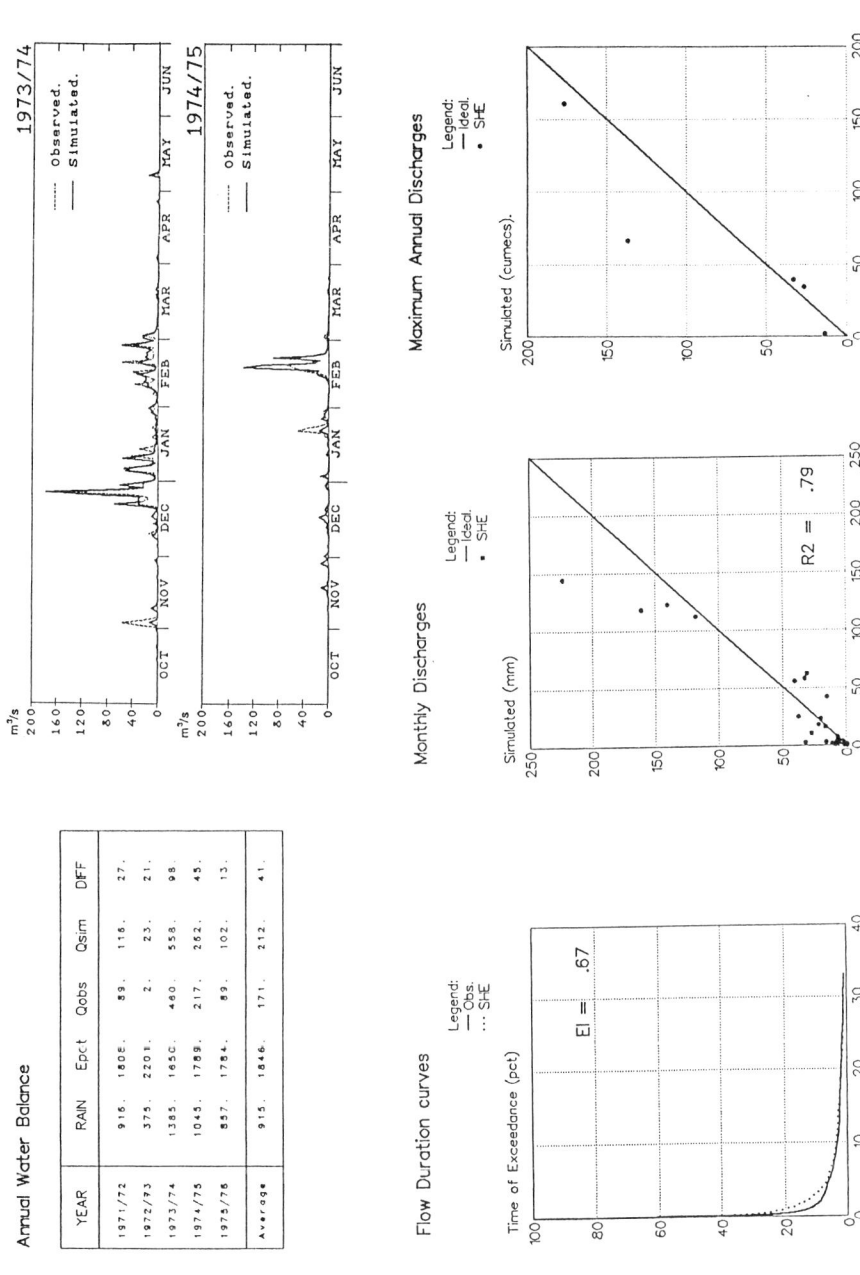

Figure 2. Example of combined use of numerical and graphical performance criteria. Example from proxy-basin validation test of MIKE SHE to the Lundi catchment in Zimbabwe (DHI, 1993).

Distributed models describe the spatial variations in catchment characteristics and input data in a network of grids points. In order to provide a sufficient detail the spatial resolution often requires several thousands of grid points, each of which is characterized by one or more parameters. Although the parameter values in principle (as in nature) vary from grid point to grid point, it is neither feasible nor desirable to allow the parameter values to vary so freely. Instead, a given parameter should only reflect the significant and systematic variation described in the available field data, as exemplified by the practice of using representative parameter values for individual soil types, vegetation types or geological layers. This process of defining the spatial pattern of parameter values, which is denoted *parametrisation*, effectively reduces the number of free parameter coefficients which needs to be adjusted in the subsequent calibration procedure.

An example of a parametrisation procedure used for constructing SHE models for catchments in India is presented in Refsgaard et al. (1992) and Jain et al. (1992). As an example the 820 km^2 Kolar catchment is parameterised into three soil classes and 10 land use/soil depth classes. For the soil type classes calibration was allowed for the hydraulic conductivity in the unsaturated zone (for each soil type class the conductivity could vary among three different land uses => nine parameter values). For the land use/soil depth classes the calibration parameters comprised soil depths (10 parameters in total) and the Strictler overland flow coefficients for four land use types (four parameters in total). Further three parameters were subject to calibration (hydraulic conductivity in the saturated zone, an (empirical) by-pass coefficient and a surface retention parameter; all kept constant throughout the catchment). Although the 26 calibration parameters could not be assessed from field data alone, but had to be modified through calibration, the physical realism of the parameter values resulting from the subsequent calibration procedure could be evaluated from available field data.

An example of parametrisation procedure for a combined groundwater/surface water model for an 800 km^2 area near Aarhus, Denmark is given by Refsgaard et al. (1994).

A rigorous parametrisation procedure is crucial in order to avoid methodological problems at the subsequent phases of model calibration and validation. The following points are important to consider in the parametrisation procedure:

* The parameter classes (soil types, vegetation types, climatological zones, geological layers, etc.) should be selected so that it becomes easy, in an objective way, to associate parameter values. Thus the parameter values in the different classes should to the highest possible degree be assessable from available field data.
* It should explicitly be evaluated which parameters can be assessed from field data alone and which need some kind of calibration. For the parameters subject to calibration physically acceptable intervals for the parameter values should be estimated.
* The number of real calibration parameters should be kept low, both for practical and methodological points of view. This can be done, for instance, by fixing a spatial pattern of a parameter but allowing its absolute value to be modified through calibration.

5. Calibration methods

In a calibration procedure, values are estimated for those parameters which cannot be assessed directly from field data. In principle, three different calibration methods can be applied:
(1) Trial-and-error, manual parameter adjustment.
(2) Automatic, numerical parameter optimization.
(3) A combination of (1) and (2).

In Subsection 5.1 the three methods as well as their advantages and disadvantages are described. The specific status on calibration techniques within the areas of lumped conceptual and distributed physically-based modelling are summarized in Subsections 5.2 and 5.3.

5.1. ADVANTAGES AND DISADVANTAGES OF USING TRIAL-AND-ERROR AND AUTOMATIC METHODS.

5.1.1 Trial-and-error manual calibration.
The trial-and-error method implies a manual parameter assessment through a number of simulation runs. This method is by far the most widely used and is the most recommended method, especially for the more complicated models. A good graphical representation of the simulation results is a prerequisite for the trial-and-error method.

5.1.2 Automatic parameter optimization.
Automatic parameter optimization involves the use of a numerical algorithm which finds the extremum of a given numerical objective function. The purpose of automatic parameter optimization is to search through as many combinations and permutations of parameter levels as possible to achieve the set which is the optimum or 'best' in terms of satisfying the criterion of accuracy.

The *advantages of automatic parameter optimization* over the trial-and-error method are:
* Automatic optimization is fast, because almost all work is carried out by the computer.
* Automatic optimization is less subjective than the trial and error method, which to a large degree depends on visual hydrograph inspection and the personal judgement of the hydrologist.

The *disadvantages of automatic parameter optimization* include:
* The criterion to be optimized has to be a single numerical criterion based on a single variable. As discussed in Section 3, though, the selection of an appropriate criterion under these constraints is a complicated and, in its turn, often quite subjective task.
* If the model contains more than a very few parameters, the optimization will probably result in the location of a local optimum instead of the global one.
* Most theories behind the search algorithms assume that the model parameters are mutually independent. This assumption is usually not justified.
* An automatic routine cannot distinguish between the different error sources mentioned in Section 2. Accordingly, an automatic optimisation algorithm may try

to compensate, for example, for data errors by parameter adjustments, with the results that the parameter values often become physically unrealistic and give poor simulation results when applied to a period different from the calibration period.

To these purely technical considerations may be added a more general one (Todini, 1988). To calibrate the parameters of a model by the optimization of an objective function means adopting some statistical technique (least squares, linear or non-linear regression, maximum likelihood, etc) based on an analysis of residuals, while completely neglecting the physical characteristics of the model. In other words, a procedure of automatic calibration, rather than capitalising on prior knowledge intrinsic to the structure of the model, avoids it, and thus emphasises the uncertainty inherent in every statistical analysis.

5.1.3 Combination of trial-and-error and automatic parameter optimization.
A combination could involve, for example, an initial adjustment of parameter values by trial-and-error to delineate rough orders of magnitude, followed by a fine adjustment using automatic optimization within the delineated range of physically realistic values. The reverse procedure is also possible: first carrying out sensitivity tests by automatic optimization to identify the important parameters and then calibrating them by trial-and-error. The combined method can be very useful but does not yet appear to have been widely adopted in practice.

5.2. CALIBRATION OF LUMPED CONCEPTUAL MODELS

The most widely used method for the calibration of conceptual rainfall-runoff models is still the manual trial-and-error method. These models are simple and fast to operate. However their calibration requires an experienced user. An experienced hydrologist can usually achieve a calibration using visual hydrograph inspection within 5-15 trial runs.

Much research work during the past two decades has concentrated on establishing appropriate automatic parameter estimation methods based on traditional numerical search algorithms. One of the oldest methods, which was very popular in the 1970s, is Rosenbrock's method (Rosenbrock, 1960). A comparison of different search optimization algorithms and a discussion of some problems relating to automatic calibration are reported in Gupta and Sorooshian (1985) and Sorooshian et al. (1993).

A different method, which has emerged within the last few years, is based on expert system technology. An expert system can be defined as a computer program in which knowledge is introduced within a first-order predicate logical frame that allows it to operate at an expert level. The idea in this approach is to let an experienced user teach the computer how he or she, in a trial-and-error process, decides which parameter adjustments to carry out before the next run. One example of a rule-based expert system is given by Azevedo et al (1993).

Automatic parameter optimization was extensively attempted in practice during the 1970s, but is not used very much today owing to the problems outlined in Subsection 5.1.

5.3. CALIBRATION OF DISTRIBUTED PHYSICALLY-BASED MODELS

Comprehensive research has been carried during the past decade on establishing and testing methods for automatic parameter optimization methods in groundwater modelling (described as inverse modelling), see e.g. Keidser and Rosbjerg (1991). Computer programmes comprising inverse methods applicable for two-dimensional flow problems are available but are not much used in practice. For solute transport problems and for three-dimensional models the inverse methods are still at the experimental stage.

For distributed hydrological catchment modelling systems such as MIKE SHE, which simulates many catchment processes and the dynamic interactions between them, suitable optimization methods are not yet available, and indeed are not even at a research stage. There are many problems related to the application of automatic parameter optimization methods in such complex models. One important problem relates to the necessity of having a multicriteria objective function, containing, for example, references to discharge, soil moisture, groundwater levels and possibly also concentration values at several points within the catchment.

However, the massive investments in inverse methods in other fields, such as meteorology and oceanography may be expected to provide tools that will be applicable to hydrology also. Certainly, inverse methods currently appear as the most promising candidates for promoting the practice of distributed hydrological models over the next decade.

6. Model validation

6.1. INTRODUCTION

If a model contains a large number of parameters it is often possible to find a combination of parameter values which provides an acceptable match between measured data and simulated output data for a short simulation period. This may be true even if the structure of the model is inappropriate or the conceptual understanding of the hydrological system is incorrect. A good match does not necessarily guarantee that a correct set of parameter values has been found, because the calibration may have been achieved purely by numerical curve fitting without considering whether the parameter values obtained are physically reasonable. This may also be illustrated by the fact that if identical models were calibrated by different people apparently equally good calibrations would be achieved based on different combinations of parameter values.

In order to assess whether a calibrated model can be considered valid for subsequent use it must be tested (validated) against data different from those used for the calibration (e.g. Stephenson and Freeze, 1974). Klemes (1986) states that a simulation model should be tested to show how well it can perform the kind of task for which it is specifically intended.

According to the methodology established in Chapter 2 of this book model validation implies substantiating that a site specific model can produce simulation results within the range of accuracy specified in the performance criteria for the particular study. For this purpose, a general scheme for validation tests is outlined in Subsection 6.2, while a discussion of different validation requirements for lumped and distributed models is given in Subsection 6.3.

6.2. SCHEME FOR CONSTRUCTING SYSTEMATIC VALIDATION TESTS

Klemes (1986) has proposed a hierarchical scheme for systematising validation tests of hydrological simulation models. The scheme is said to be hierarchial because the modelling tasks are ordered according to their increasing complexity and the demands of the testing increase in the same direction. The scheme was originally developed for rainfall-runoff modelling, but the methodology can be applied more generally.

Klemes (1986) distinguished between simulations conducted for the same station (catchment) as was used for calibration and simulations conducted for ungauged catchments. He also distinguished between cases where catchment conditions (climate, land use, ground water abstraction etc.) are stationary and cases where they are not.

This division leads to the definition of four basic categories of typical modelling tests:

A: *Split-sample test:* Calibration of model based on, say, 3-5 years of data, and validation on another period of similar length.

B: *Differential split-sample test:* Calibration of a model based on data before catchment change occurred, the adjustment of model-parameters to characterise the change and a validation based on the subsequent period.

C: *Proxy-basin test:* No direct calibrations are allowed but advantage is taken of information available from other gauged catchments. Hence, validation will comprise the identification of a gauged catchment deemed to be of a similar nature to the validation catchment, the initial calibration, the transfer of the model, including the adjustment of parameters to reflect actual conditions within the validation catchment, and validation.

D: *Proxy-basin differential split-sample test*: Again no direct calibration is allowed but information from other catchments may be used. Hence, validation will comprise an initial calibration on other relevant catchment, the transfer of the model to the validation catchment, the selection of two parameter sets to represent the periods before and after the change, and subsequent validations on both periods.

6.2.1. Split-sample test

The split-sample test is the classical test, being applicable to cases where there is sufficient long time series of control data for both calibration and validation, and where the catchment conditions remain unchanged, i.e. are stationary.

The available data record is divided into two parts. A calibration is carried out in turn for each part and then validated against the other part. Both these calibration and validation exercises should give acceptable results.

The main problem associated with the split-sample test is that not all the available data are used for the calibration. Therefore the data record should be of such length that, when split into parts, these parts can support an adequate calibration. On the other hand, if both split-sample tests produce acceptable results, a final calibration of the model can make use of the full record.

6.2.2. Proxy-basin test

This test should be applied when there is not sufficient data for a calibration of the catchment in question. If, for example, streamflow has to be predicted in an ungauged catchment Z, two gauged catchments X and Y within the region should be selected. The model should be calibrated on catchment X and validated on catchment Y and vice versa. Only if the two validation results are acceptable and similar can the model command a basic level of credibility with regard to its ability to simulate the streamflow in catchment Z adequately.

6.2.3. Differential split-sample test

This test should be applied whenever a model is to be used to simulate flows, groundwater levels and other variables in a given gauged catchment under conditions different from those corresponding to the available data. The test may have several variants depending on the specific nature of the modelling study.

If for example a simulation of the effects of a change in climate is intended, the test should have the following form. Two periods with different values of the climate parameters of interest should be identified in the historical record, such as one with a high average precipitation, and the other with a low average precipitation. If the model is intended to simulate streamflow for a wet climate scenario, then it should be calibrated on a dry segment of the historical record and validated on a wet segment.

Similar test variants can be defined for the prediction of changes in land use, effects of groundwater abstraction and other such changes. In general, the model should demonstrate an ability to perform through the required transition regime.

6.2.4. Proxy-basin differential split-sample test

This is the strongest test to pass for a hydrological model, because it deals with cases, where there is no data available for calibration, and where the model is intended for prediction of non-stationary conditions.

6.3. DIFFERENT VALIDATION REQUIREMENTS FOR LUMPED AND DISTRIBUTED MODELS

The validation procedure is basically the same for lumped and distributed model codes, but because of the differences in model structures, modes of operation and objectives of application, the validation requirements are much more comprehensive for distributed models. Traditional validation based on comparing simulated with observed outflows at the catchment outlet still remains the only option in many practical cases. However, as emphasized by Rosso (1994) this method is poorly consistent with spatial distributed modelling. The differences are summarized in Table 1, which clearly illustrates the need for multicriteria, multi-scale validation criteria.

TABLE 1. An illustration of the need for the incorporation of multicriteria and multi-scale aspects in methodologies for the validation of distributed models.

	LUMPED CONCEPTUAL	DISTRIBUTED PHYSICALLY-BASED
Output	At one point: * Runoff	At many points: * Runoff * Surface water level * Ground water head * Soil moisture
	=> *single variable*	=> *multi variable*
Success criteria (excl problem of selecting which statistical criteria to use)	Measured <=> simulated * Runoff, one site	Measured <=> simulated * Runoff, multi sites * Water levels, multi sites * Groundwater heads, multi sites * Soil moisture, multi sites
	=> *single criteria*	=> *multi criteria*
Typical model application	Rainfall-runoff * stationary conditions * calibration data exist	Rainfall-runoff, unsaturated zone, groundwater, basis for subsequent water quality modelling Impacts of man's activity * non-stationary conditions some times * calibration data do not always exist
Validation test	Usually "Split-sample test" is sufficient	More advanced tests required: * Differential split sample test * Proxy basin test
	=> *well defined practise exist*	=> *need for rigorous methodology*
Modelling scale	Model: catchment scale Field data: catchment scale	Model: depends on discretization Field data: many different scales
	=> *single scale*	=> *multi scale* problems

The table shows that because the project objectives and the output requirements are more demanding for distributed models, the success criteria has to be more rigorous. Validation against one single discharge station (e.g. at the catchment outlet) is sufficient if the purpose is to generate streamflow values at that location, but not sufficient to draw any conclusions about the internal representation of the flow conditions within the catchment. Based on the single discharge station the total water balance may be correct, but for example a systematic underestimation of the actual evapotranspiration could lead to unrealistic trends in the groundwater heads over the entire or parts of the catchment.

Basically, a success criterion needs to be fulfilled for each output variable which we intend to make predictions for. Multi-site calibration/validation is needed if spatially distributed predictions are required, and multi-variable checks are required if predictions of the behaviour of individual sub-systems within the catchments are needed.

Styczen and Storm (1995) showed that calibration of both streamflow and the trend in groundwater heads was required to ensure that the simulated groundwater recharge rates for estimation of nitrate leakage was correct.

In Storm and Punthakey (1995) MIKE SHE was used to simulated the groundwater table variations in an irrigation district. Data from a very detailed groundwater observation network was available for the calibration, but no drainage flow data was available. Although the model was used to estimate groundwater accessions from the land surface, they could not guarantee that the actual flow passing through the groundwater system was simulated correctly.

7. The Credibility of a Generic Modelling System

According to the terminology defined in Chapter 2 of this book a modelling system can, in principle, never be validated. Instead of a full validation we can think of the degree of validity as the *credibility of a given modelling system*. The degree of validity of a modelling system is expressed in the first and most immediate place by the sum of all successful validations of all models that have been constructed and operated to date using the modelling system. As the number of such successful model increases, so the credibility of the system itself grows in strength.

Behind this, most superficial of views, lies the assumption that the modelling system is in fact being improved on the basis of the operating experience; that it is functioning within its market, tracking the needs of that market and thus learning from this market. From this point of view, the development of a modelling system is not one that leads directly to a finished, rounded and complete product, but it is rather a *process of adaption through evolution*. Thus, although the modelling system is indirectly a product, it is one that is constantly evolving, so that its evolution corresponds to a process. The general principles of such a development have been expressed by Floyd (1987).

8. References

Aitken, A.P. (1973) Assessing systematic errors in rainfall-runoff models. *Journal of Hydrology*, 20, 121-136.

Azevedo, L.G, Fontane, D.G. and Porto, R.L. (1993) Expert system for the calibration of SMAP. *Water International*, 18, 103-109.

DHI (1993) Validation of hydrological models, Phase II. Unpublished research report. Danish Hydraulic Institute, Hørsholm.

Fleming, G. (1975) *Computer Simulation Techniques in Hydrology*. Elsevier.

Floyd, C. (1987) Outline of a paradigm change in software engineering. In Bjerknes, G., Eha, P. and Kyng, M. (Eds.) *Computers and democracy*. Avebury, Aldershot, UK, and Brookfield, USA.

Green, I.R.A. and Stephenson, D. (1986) Criteria for comparison of single event models. *Hydrological Sciences Journal*, 31 (3), 395-411.

Gupta, V.K. and Sorooshian, S. (1985) The automatic calibration of conceptual catchment models using derivative-base optimization algorithms, *Water Resources Research*, 21, 473-486.

Jain, S.K., Storm, B., Bathurst, J.C., Refsgaard, J.C and Singh, R.D. (1992) Application of the SHE to catchments in India - Part 2: Field experiments and simulation studies on the Kolar Subcatchment of the Narmada River. *Journal of Hydrology*, 140, 25-47.

Keidser, A. and Rosbjerg, D. (1991) A comparison of four inverse approaches to groundwater flow and transport parameter identification. *Water Resources Research*, 27, 2219-2232.

Klemes, V. (1986) Operational testing of hydrological simulation models. *Hydrological Sciences Journal*, 31, 13-24.

Nash, I.E. and Sutcliffe, I.V. (1970) River flow forecasting through conceptual models, Part I. *Journal of Hydrology*, 10, 282-290.

Refsgaard, A., Refsgaard, J.C., Jørgensen, G.H., Thomsen, R. and Søndergaard, V. (1994) A hydrological modelling system for joint analyses of regional groundwater resources and local contaminant transport. Unpublished note. Danish Hydraulic Institute, 28pp.

Refsgaard, J.C., Seth, S.M., Bathurst, J.C., Erlich, M., Storm, B., Jørgensen, G.H., and Chandra S. (1992) Application of the SHE to catchment in India - Part 1: General results. *Journal of Hydrology*, 140, 1-23.

Rosenbrock, K.H. (1960): An automatic method for finding the greatest or least value of a function. *The Computer Journal*, 7 (3).

Rosso, R. (1994) An introduction to spatially distributed modelling of basin response. *In Rosso, R., Peano, A., Becchi, I. and Bemporad, G.A. (Eds): Advances in Distributed Hydrology*, Water Resources Publications, 3-30.

Sorooshian, S., Duan, Q. and Gupta, V.K. (1993) Calibration of rainfall-runoff models: Application of global optimization to the Sacramento soil moisture accounting model. *Water Resources Research*, 29, 1185-1194.

Stephenson, G.R. & Freeze, R.A. (1974) Mathematical simulation of subsurface flow contributions to snowmelt runoff, Reynolds Creek Watershed, Idaho. *Water Resources Research*, 10, 284-294.

Storm, B. and Punthakey J.F. (1995) Modelling of environmental change in the Wakool Irrigation District. *MODSIM 95, International Conference*, Newcastle, Australia.

Styczen, M. and B. Storm (1995): Modelling the effects of Management Practices on Nitrogen in Soils and Groundwater. *In: P.E. Bacon (Ed.) Nitrogen Fertilization in the Environment*, Marcel Dekker, Inc. New York.

Todini, E. (1988) Rainfall-runoff modelling: past, present and future. *Journal of Hydrology*, 100, 341-352.

CHAPTER 4

DISTRIBUTED PHYSICALLY-BASED MODELLING OF THE ENTIRE LAND PHASE OF THE HYDROLOGICAL CYCLE

B. Storm and A. Refsgaard
Danish Hydraulic Institute
Agern Allé 5
DK-2870 Hørsholm
Denmark

1. Introduction

Physically-based distributed hydrological model codes have been developed from a need to analyze and solve specific hydrological problems often required in multi-objective and multi-decision management investigations. These problems may differ in type and scale, but have usually one thing in common, namely that in order to obtain a useful outcome of the modelling exercise, variations in state-variables over space and time need to be considered and realistic representations of internal flow processes have to be computed.

Different types of models, categorized as physically-based and fully distributed, have been developed and successfully applied to describe individual processes of the hydrological cycle. Two important examples of such models are: soil water flow models, e.g. based on the one-dimensional Richard's equation, to simulate soil moisture conditions in a profile, and groundwater models for simulating groundwater flow and head in aquifer systems. When it comes to provide an integrated description over catchment areas, there seems to be strong and diverse opinions among theoretical oriented people about the validity and appropriateness of such physically-based and fully distributed model codes. A general argument put forward is that the spatial resolution used to represent flow processes in the models are only valid at small scale and the variations which can be accounted for in the models at catchment scale is far too coarse compared to the natural conditions. The results are therefore flawed to an extend that the reliability of the simulation results may be questionable.

Despite theoretical and philosophical discussions of what type of codes should be defined as physically-based model codes, and when and how they can be used (e.g. Refsgaard et al., 1995b), professionals involved with water resources problems recognize that the hydrological issues they are concerned with are introduced in a spatially distributed manner, and a conceptualization which mirrors the prototype conditions as closely as possible is required.

This is supported by the fact that groundwater model codes, which generally are accepted as being 'physically-based' models have been successfully applied in thousands of projects around the world in the past twenty five years. Even though many of these model applications have included crude approximate descriptions of the hydrogeological settings and the groundwater flow, for example three-dimensional flow regimes have been treated as two-dimensional, they have served the purpose to provide valuable information for decision making in connection with groundwater planning, management and protection.

One particular problem, which many of the users of these 'traditional' groundwater codes have experienced, is to define realistic boundary conditions, for example temporal and spatial pattern of groundwater recharge, flow exchange with rivers and channels systems, and appropriate dynamic conditions in areas with shallow water table. It is under such circumstances that integration of processes, covering surface water as well as subsurface water becomes important.

This Chapter gives a brief presentation of the concept and use of physically-based distributed models. We will use MIKE SHE as an example, but many of the conclusions drawn may be equally representative for other types of distributed model codes. A brief presentation of the hydrological processes in MIKE SHE will be given, and we will share some of our experiences in using MIKE SHE in connection with different types of water resources projects.

2. The MIKE SHE hydrological modelling system

Although it is nearly twenty years since the development of the Système Hydrologique Européen - SHE (Abbott et al., 1986a,b) was initiated, the model code MIKE SHE (a further development of SHE, (Refsgaard and Storm, 1995) is today one of the few catchment modelling codes available for project work, which may be categorized as a physically-based and fully distributed hydrological model code. A large number of other model codes have been developed, but are still mainly applied in research context, e.g. IHDM (Beven, 1985; Calver, 1988; Beven and Binley, 1992) and SWAGSIM (Prathapar et al., 1995). Despite discussions of the limitations and validity of e.g. MIKE SHE (Grayson et al., 1992; Smith et al., 1994), this modelling system has obtained a successful record of applications within the last few years.

MIKE SHE was developed as an alternative to the lumped conceptual rainfall-runoff models (e.g. Stanford model (Crawford and Linsley, 1966), NAM model (Nielsen and Hansen, 1973) etc.) to provide a rigorous approach based on accepted theories of the physically processes of surface and subsurface water and solutes. However, its current use in water resources and environmental projects has in many cases been generated from a wish to overcome some of the above mentioned difficulties that arise from using traditional groundwater models. In fact there is a great tendency among model developers to improve 'traditional' groundwater model codes to accommodate features that are included in the MIKE SHE.

The experience of the authors from using MIKE SHE in local and regional catchment studies under various data availability has been that the physically-based description provides an excellent framework for investigating a range of water resources problems on different temporal and spatial scales. Most of the process descriptions, which theoretically only are fully valid at small scale, provides often an acceptable conceptual framework on regional scale.

It is important to notice that although all natural systems exhibit pronounced spatial heterogeneity to an extent we never can hope to describe (neither in the model nor in the measured data), the underlying processes are often satisfactorily valid. However, it is the responsibility of modellers to ensure that approximations introduced in modelling applications, in terms of conceptualization (ie. description of the natural system), calibration accuracy, and predictions, are adequately reported and give advise on the limitations and the confidence we should put into the simulation results.

3. Description of Water Flow Processes

The basic description of the water flow in the surface- and subsurface water systems has changed very little from the original SHE code (Abbott et al., 1986a,b). The international collaboration effort in connection with its development made it necessary to design a modular programme structure comprising six individual process-oriented components, each describing one major hydrological process in the hydrological cycle. Individually, they can describe parts of the hydrological cycle, but combined, they provide a complete integrated description of the land phase part of the hydrological cycle, Fig. 1.

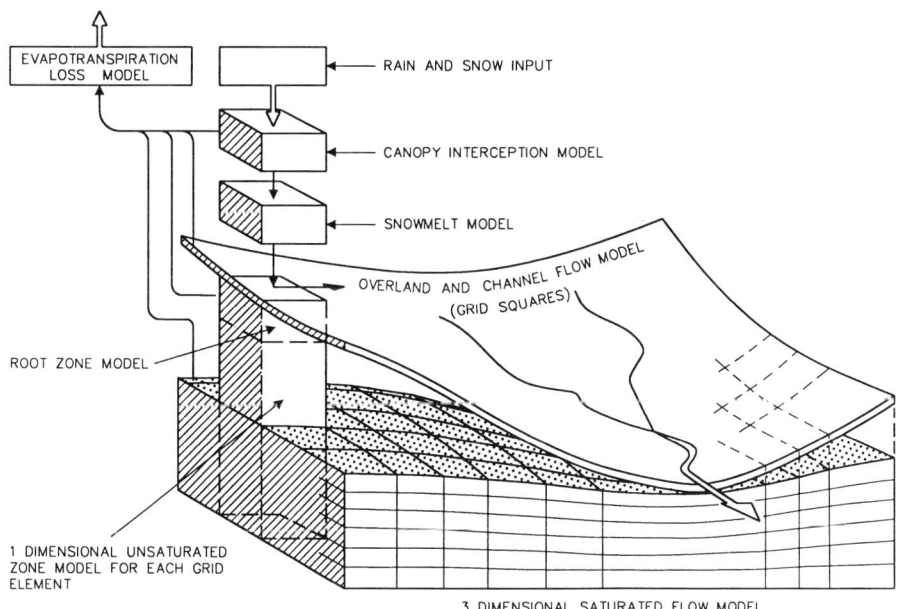

Figure 1   Schematic representation of the flow components of MIKE SHE

This modular structure has today become an important feature because many studies only require use of some of the process components. This may be relevant in studies where either data availability, time-frame or study objective does not justify a complete integrated description.

During the initial testing, and subsequent use in project work some of the process descriptions have been modified to enhance numerical efficiency, and accomodate new flow features, e.g. fractured flow (Brettman et al., 1993). However, an important part of the software development was devoted to produce an extensive user-interfaces to facilitate data processing and presentation of simulation results, but also to make the software more transferable, which some see as a potential risk for misuse (Grayson et al., 1992).

MIKE SHE was originally developed with a view to describe the entire land phase of the hydrological cycle in a given catchment with a level of detail sufficiently fine to be able to claim a physically-based concept. The equations used in the model are with few exceptions non-empirical and well-known to represent the physical processes at the appropriate scales in the different parts of the hydrological cycle. The parameters in these equations can be obtained from measurements as long as they are compatible with the representative volumes for which the equations are derived and used on the appropriate scale in the model application.

The flow processes represented in MIKE SHE and other physically-based distributed modelling systems include: snow melt, interception and evapotranspiration, overland and channel flow, vertical flow in the unsaturated zone, and groundwater flow. In MIKE SHE, individual processes operates at time steps consistent with their own temporal scales. For example, unsaturated flow cannot be expected to realistically represent the infiltration process and movement of sharp wetting fronts in the root zone if calculated at time steps of days or weeks which may be appropriate for groundwater flow. However, because of the computational demands of distributed models it is important to use as large time steps as possible, and this is often dictated by the actual hydrological conditions.

Below is given a brief description of the individual processes. For a more detailed review see Refsgaard and Storm (1995).

Variations in hydraulic heads, flows and water storages on the ground surface, in rivers and in the unsaturated and saturated zones are modelled in a network of grid squares. All spatial variations in input data and catchment characteristics can be represented by the spatial resolution given by the modeller.

Flow in the unsaturated zone is commonly described by the one-dimensional Richard's equation and solved in a finite difference scheme in the vertical as illustrated in Fig. 2, (Jensen, 1983). A similar approach is used in other unsaturated zone model codes using Richard's equation, e.g. SWATRE (Feddes, 1988). In MIKE SHE the solution technique has an advanced feature which estimates the exchange of water between the saturated zone and the unsaturated zone based on the retention, and adjusts the phreatic surface accordingly (Storm, 1991). The simulation of the unsaturated flow plays a central part in most model applications, e.g. recharge estimation, transpiration, surface-groundwater interaction and fate of pollutants.

The use of Richard's equation is probably the most controversial issue in connection with catchment application, and in many investigations this approach would not be necessary. SWAGSIM (Prathapar et al., 1995) uses an analytical solution to calculate the flow in the unsaturated zone, which provides a much faster model use. Today alternative versions, e.g. neglecting the capillary forces, are included in MIKE SHE, which under many conditions provide sufficient accurate estimates and considerable computational savings.

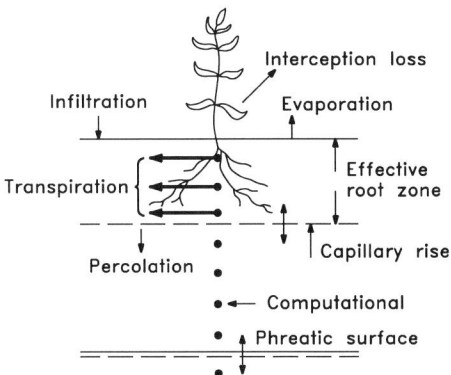

Figure 2   Illustration of the node representation for the unsaturated zone

Actual evapotranspiration is calculated from potential evaporation data. Two methods are included, which are either based on an empirical formula (Kristensen and Jensen, 1975) or on the Penman-Monteith Equation (Monteith, 1965). Both methods use the actual soil moisture/retention conditions in the root zone to calculate the actual evapotranspiration loss. The amount of water that can be drawn out of the root zone depend on crop and soil properties. The interception/evapotranspiration component is an integral part of the unsaturated zone component to determine the timing and magnitude of recharge and overland flow generation.

Overland flow is generated when the top-soil becomes saturated. The routing of the water is computed using a two-dimensional diffusive wave approximation of St. Venant's equation (Preissman and Zaoui, 1979). Net rainfall, evaporation and infiltration are introduced as source/sinks allowing the surface to dry out on more permeable soil areas. The solution assumes sheet flow conditions, which under regional applications is a very crude approximation recognised for example when MIKE SHE provide the framework for soil erosion modelling (Styczen and Nielsen, 1989; Hasholt and Styczen, 1993). Local depressions in the ground surface and physical barriers (e.g. levees, roads etc.) are conceptually modelled using a detention storage allowing water only to evaporate or infiltrate as long as the water level is below a specified threshold.

Routed excess water on the ground surface enters the river system as lateral inflow. Flow in channels and streams is simulated using a branched and looped one-dimensional diffusive wave approximation of the St Venant equation. The representation of the river in the models is approximated to run along the boundaries of individual grid squares.

Groundwater flow is calculated from a three-dimensional governing equation (Refsgaard and Storm, 1995). The equation is solved numerically by a finite difference method using a modified Gauss-Seidel implicit, iterative scheme (Thomas, 1973). The spatial resolution in the vertical can either follow a rigid network for fully three-dimensional flow or follow the geological layering for a quasi-three-dimensional (or in special cases two-dimensional) flow. Discharge to or recharge from the river system occurs from all computational cells located along the river links. The integrated description provide a more comprehensive interaction with the river, accounting for the dynamic variations in river water level. A feature which is often lacking in traditional groundwater models, e.g. MODFLOW (McDonald and Harbaugh, 1988), etc.

3. Description of solute transport processes

Since the initial development of SHE more attention has been given to water quality aspects, especially in the groundwater used for water supply purposes. Leakage of pollution from waste disposal sites, polluting industries, and agricultural areas etc. has increased the need for modelling tools which can assist in detection and cleaning up of contaminated areas. Transport models are important tools for prediction of future likely pollution patterns, designing optimal remediation schemes, or designing water quality monitoring network. Models to locate new waste disposal sites to prevent further deterioration of the water quality in water supply areas is another important aspect.

Water quality issues are not confined to the groundwater only, but often experienced in soils and receiving rivers as well. The interaction between surface water and subsurface water becomes therefore an important aspect also in many environmental problems. For example infiltration of poor water quality river water may lead to problems in connection with river bank filtration schemes for water supply, or use of fertilizer can deteriorate the water quality of the groundwater and the receiving rivers dramatically.

The advection-dispersion module developed as an add-on module for MIKE SHE can describe the transport processes for all the water flow components in MIKE SHE. This means that transport of solvents can be simulated for the entire hydrological cycle in a catchment as illustrated in Fig. 3. This is important in the above mentioned cases, but provide also a powerful option in connection with e.g. groundwater management. A regional model of the hydrological system can be developed and form basis for detailed local models including solute transport simulation, (Refsgaard et al., 1994).

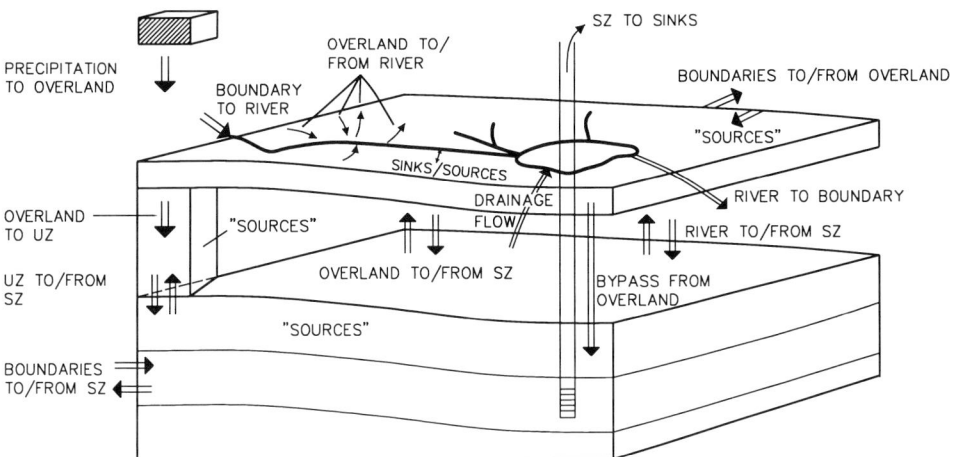

Figure 3 Schematic representation of the transport components of MIKE SHE

Basically or microscopically, solute transport in porous media is governed by three processes: advection, molecular diffusion and kinematic or mechanical dispersion. Advection describes the movement of solutes carried by the flowing water which is the flow velocities computed by the water flow module in MIKE SHE. As a consequence of this advection-dispersion simulations reflect to a large degree the uncertainties and errors in the flow description. Diffusion describes the spreading of solute molecules by virtue of their kinetic motion the diffusive flux normally being described as proportional to the concentration gradients. The kinematic dispersion is the spreading of the solutes caused by microscopic variations in the flow velocity field which usually is not represented by the spatial resolution used in the models.

The governing equations used in the different components of MIKE SHE are similar to the ones normally used in transport models. They are all based on the general advection-dispersion equation (ADE), see e.g. Bear and Verruijt (1987). The difference between the numerical equations is reflected in the dimensionality, the formulation of the dispersion coefficients and the numerical method applied.

Most attention has been given the groundwater transport component as it includes both a solution to the three-dimensional formulation of the ADE, a particle tracking solution, and an option for modelling transport in a dual porosity medium, (Brettmann et al., 1993). The dispersion term in the groundwater component includes options for isotropic conditions and anisotropic conditions with axial symmetry around the vertical axis, see Bear and Verruijt (1987). The ADE is solved using an explicit scheme called the QUICKEST scheme (Vested et al., 1992). The solute transport in the unsaturated zone is solved using a one-dimensional version.

Transport in water on the ground surface is computed using a two-dimensional formulation of the ADE. Solutes in rain and exchanges of water and solutes with the unsaturated zone and the groundwater (e.g. seepage) are introduced as sources/sinks. Evaporation of ponded water may lead to high concentrations and solutes are allowed to precipitate if the concentration exceeds a certain level and dissolve again when the dilution becomes high enough again. The ADE is also solved using the QUICKEST scheme.

In the channel transport component, the ADE is simplified to a one-dimensional formulation. Lateral inflow from the overland and base flow/leakage loss are treated as source/sinks. The equation is solved using a modified Lax scheme (Abbott and Cunge, 1982).

4. Types of application

A number of possible application areas where physically-based distributed models codes would be applicable and provide a better approach than traditional lumped or semi-distributed conceptual models have been suggested by e.g. Abbott et al. (1986a,b), Bathurst and O'Connell (1986, 1992). Many of those have been demonstrated in subsequent studies as described below. Others, for example prediction of impacts of land-use changes, runoff prediction from ungauged catchment, still need to be shown. In fact, the above references may have been too optimistic regarding the direct applicability of point measurements to estimate model parameter values and our ability to relate changes in catchment conditions to changes in the parameter values used in model codes.

Refsgaard et al. (1995b) showed that focusing on only runoff simulation from either gauged or ungauged catchments, a physically-based model (MIKE SHE) did not prove significantly better than a simple lumped conceptual model - NAM. For ungauged conditions, the physically-based model may narrow the error-band of the prediction, but considering the effort required to set up the model, the simple lumped conceptual model would normally be the most appropriate.

Bathurst and O'Connell (1992) found that for event modelling, physically-based approaches such as SHE provides a reliable basis for model prediction as soon as a short period (single event) is available for model calibration. However, our findings suggest that this may not be generally true. From comparisons of long time series of runoff, there seems no evidence that the rigorous approach provide significantly better prediction than a simple lumped conceptual model, unless there are features in the runoff simulation which is neglected by the lumped conceptual model. The latter was demonstrated on some stream systems in connection with simulation of summer runoff, where a substantial loss from the stream occurred due to potential evapotranspiration rates from low-lying areas close to the stream. The ability of MIKE SHE to account for the spatial variations in depths to groundwater table was important for calculating the capillary rise in these areas.

It is also our experience that prediction of variables which are not directly tested (calibrated against) may include considerable uncertainties. It should be stressed, that this does not limit the applicability of physically-based distributed models, and as illustrated below, there is a wide range of water resources investigations, for which this type of modelling is not only warranted but also the only alternative. This concerns in particular analysis of localised human activities such as groundwater abstraction schemes in environments where the interaction between surface water and subsurface water is a crucial issue.

A large number of applications confirm the wide range of water resources problems for which a model code like MIKE SHE is a suitable tool. They also reveal that most of the studies have focused on the groundwater system, and the benefits from choosing MIKE SHE was primarily the ability to integrate the groundwater flow with the surface water system. It is worth mentioning that a large number of other MIKE SHE applications, which are not directly known to the authors, have been carried out, mainly in engineering projects.

Early applications of MIKE SHE were mainly concerned with runoff predictions, see e.g. Jain et al. (1992) and Refsgaard et al. (1992). The model applications were part of a technology transfer project, and gave important operational experience in simulations on medium-scale catchments (up to 5000 km^2). In general, limited data were available, obtained mainly from literature, but as part of the project work, data became available through dedicated field investigations to enhance the model accuracy and reliability.

A number of studies have been concerned with groundwater development and its impact on groundwater heads and river environment (Refsgaard et al., 1994; Refsgaard and Sørensen, 1994; and Refsgaard, 1994). These models generally covers large multi-layered aquifer systems, where the interaction between the layers need to be considered. The projects have provided a framework for managing the groundwater and protect the surface water environment. In some of the studies, a proper allocation of pumping sites needed to be addressed. In broader environmental impact assessments, the regional models have in some cases also provided boundary conditions for local detailed models to examine e.g. the threat of contamination of water supply wells from waste deposits. Styczen and Storm (1993, 1995), presented a methodology for using MIKE SHE in connection with non-point agricultural pollution.

Recently, MIKE SHE has been introduced in irrigation and salinity planning and management projects. Mudgeway and Nathan (1993) used it to simulate flow and salt transport processes, over a three months period, between shallow groundwater and surface waters in a 8,9 ha set of irrigation bays in the Tragowel Plains in Victoria Australia. The model was tested against drain flow, groundwater table, soil moisture and salt concentrations. The simulations were used to study the effect of deepening drains on discharge of salt loads.

Storm et al. (1996) repeated the simulations using a much longer time series (eighteen months) of observations. The model was calibrated against drain flow (two stations), groundwater levels (five piezometers) and soil moisture at eight sampling points. The simulation exercise demonstrated an interesting point, namely that calibration against selected state-variables such as drain flow and/or groundwater levels alone, may not necessarily lead to a correct predictions of other state-variables, e.g. soil moisture conditions even though the data availability for parameter assessment may be quite comprehensive.

In a similar study, but on regional scale (approx. 2700 km^2), Storm and Punthakey (1995) used MIKE SHE to evaluate a number of options proposed in a Land and Water Management Plan for the Wakool Irrigation District (WID) in NSW, Australia. The WID experiences, as many other large irrigation districts in semi-arid regions in Australia, an increase in the area with shallow saline groundwater. MIKE SHE provided a complete description of the complex hydrological regime in WID involving temporal and spatial variations in the exchange of water between the ground surface, drainage and supply systems, and the groundwater aquifers within the area. Management options to be analyzed for a time frame of 30 years and included scenarios which focused on the surface water regime (extension of the drainage system and/or sealing of supply channels where seepage losses were observed) as well as the subsurface water regime (groundwater pumping schemes in shallow and deep aquifers).

Figs. 4 a-c show the estimated progress of shallow water table areas for two scenarios, a so-called 'no plan' option, where no additional action is taking place and one where additional 48 pumps are installed in the shallow aquifer system.

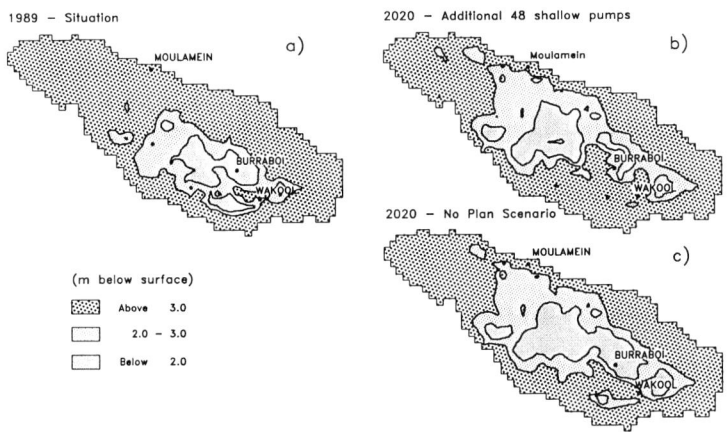

Figure 4   Areas of shallow water table in 1989 and predicted for year for two scenarios

Refsgaard et al. (1995a) used MIKE SHE to evaluate the efficiency of a pump-and-treat system for remediating a severe Chlorinated Carbon Solvent contamination. Model results showed that the pump-and-treat system would not lead to significantly improvements of conditions in the groundwater aquifer and certainly not prevent contamination of the recipient located about 2 kilometres down gradient from the pollution source. The conclusions lead to a close down of the remediation pumping and establishment of a monitoring system to ensure reasonable conditions in the recipient in the future.

5. Problem areas using physically-based distributed models

In most catchment studies carried out so far, it was not possible to use spatial resolution sufficiently detailed to claim that the developed models were physically-based (not the same as claiming the model code to be physically-based). In fact, it was realised from early test applications that considerable lumping is involved at the scale of a grid square. This is not a new experience, but have been recognised by modellers working with groundwater modelling. They have successfully been using transmissivities and storage coefficients derived from e.g. pumping tests. It is evident to everyone that such parameter values are gross simplifications of actual aquifer conditions, but even then many professionals have found great benefits from using groundwater models for different types of problems.

It is the authors experience that the spatial resolutions and variations in properties used, can provide a good representation of the conditions of the area, and account for significant differences in e.g. recharge pattern across the model areas. In practise, spatial variations in catchment characteristics are often obtained from maps describing topography, soil and land-use pattern and interpreted geological conditions, combined with information about representative properties for different map units.

Parameters values obtained from the above information are only initial estimates, which are modified during the model calibration, where simulated state-variables are matching observed conditions at discrete points. The final parameter set for a specific model, is subjective and the outcome of several factors including the modellers perception of the catchment and flow processes, the spatial resolution used, the degree of matching of observed and simulated variables obtained etc.

It is important to emphasize that there is a number of fundamental scale problems which needs to be carefully considered in all model applications involving distributed models. This becomes particular important when considering interactions between the surface flow and the subsurface flows. A few areas where we have encountered scale problems include:

* The flow exchange between groundwater aquifer and river system. Since the flow is based on Darcy's law using the gradient between the river water level and the groundwater heads in the adjacent grid squares, the flow rates and the resultant head changes will depend on the spatial resolution used. This is an important aspect in for example simulating the hydrograph recessions correct.

* In catchments with a dense drainage network it is often not possible to represent entire drainage system (many streams are of ephemeral nature). For such situations sub grid variations in the topography need to be accounted for in order to simulate the hydrograph response in the main streams correctly.

* Simulation of infiltration and vertical unsaturated flow in the soil, the hydraulic parameters used in Richard's equation can be obtained from laboratory measurements at small undisturbed soil samples. However, for grid squares covering large areas

(e.g. 25 hectares) these are seldom representative unless completely homogeneous conditions exist in the horizontal directions. Therefore effective or representative parameters are used, which means that the simulated soil moisture conditions can not be validated directly.

The latter point is often subject to more discussion of MIKE SHE than all the other issues. In most catchment simulations the use of Richard's equation becomes conceptual rather than physically-based and simpler approaches could be chosen. Nevertheless, this equation provides a good routing description, and the capability to simulate capillary rise under shallow water table conditions is important in studies dealing with wetland areas or irrigation in shallow water table area. For situations where we know that Darcy's law is not applicable, empirical formulas describing flow in preferential flow paths is included in the simulation. The similar problem concerning infiltration in cracking soils was demonstrated by Storm et al. (1996), where an empirical cracking option was necessary to describe high infiltration rates in clayey soils during early parts of irrigation events.

Because representative parameter values are used, the certainty in the model results depends on the data available to compare the simulated spatial and temporal variations with observations. Again, this is well-known from groundwater model applications, where the model parameters (conductivities or transmissivities) are estimated from the calibration against observed head variations in discrete points.

For regional catchment studies the model performance is usually evaluated based on comparisons against river discharges and groundwater heads. Very seldom measured soil moisture data are available, and if they are such comparisons will require that the site specific properties are known.

It is often stated that distributed models requires a large number of data and therefore very time consuming and complicated to set up and calibrate. In fact, a number of short-term screening evaluation projects have been carried out with MIKE SHE for example in connection with studying the contamination risks from waste disposals. In these studies only sparse existing information about the hydrogeological conditions was available. The model was used to obtain an improved knowledge about the possible flow patterns around the waste disposal site based on the existing geological interpretations. These applications could also be used to identify where existing knowledge is lacking and assist in defining appropriate monitoring programmes.

Another common argument against fully distributed models is the risk of over-parameterization. This risk is of course always there. However, the general experience is that if the data are lacking to describe the spatial variations in the catchment it is too time-consuming and not worthwhile to modify a large number of parameter values in order to improve e.g. hydrograph predictions. In such cases very few parameters are modified during the calibration and the reliability of the results are evaluated with this in mind.

6. Conclusions

The recent experience with the use of MIKE SHE in a range of water resources projects show that physically-based and fully distributed model codes such as MIKE SHE are important tools for assessing different types of water planning and management issues. Although it can be argued that the governing equations used in this type of modelling systems not always are fully valid on the scale they are applied, they provide an important flexibility to investigate problems on different scales.

MIKE SHE has been used in small-scale research studies (where the equations are directly valid) as well as large scale catchment projects, where considerable lumping and conceptualisation is imposed. In the latter type of studies, the models are calibrated in accordance with the common practice as made in for example groundwater studies. It is important in this type of application to recognise the general limitations of the developed model depending of the various assumptions used in the conceptualisation, calibration and prediction phases.

7. References

Abbott, M.B. and J.A. Cunge (1982) *Engineering applications of computational hydraulics*, Vol. 1, Pitman Advanced Publishing Program, London.

Abbott, M.B., J.C. Bathurst, J.A. Cunge, P.E. O'Connell and J. Rasmussen (1986a) An Introduction to the European Hydrological System - Système Hydrologique Européen, "SHE", 1: History and philosophy of a physically-based, distributed modelling system. *Journal of Hydrology*, 87, 45-59.

Abbott, M.B., J.C. Bathurst, J.A. Cunge, P.E. O'Connell and J. Rasmussen (1986b) An Introduction to the European Hydrological System - Système Hydrologique Européen, "SHE", 2: Structure of a physically-based, distributed modelling system. *Journal of Hydrology*, 87, 61-77.

Bathurst, J.C. (1986) Physically-based distributed modelling of an upland catchment using the Système Hydrologique Européen. *Journal of Hydrology*, 87, 79-102.

Bathurst, J.C. and O'Connell, P.E. (1992) Future of distributed modelling: Système Hydrologique Européen, *Hydrological Processes*, Vol. 6, 265-277.

Bear, J. and Verruijt, A. (1987) *Modeling Groundwater Flow and Transport*. D. Reidel Pub. Com., Dordrecht, Holland.

Beven, K.J. (1985) Distributed Models, in M.G. Anderson and T.P. Burt (Eds.) *Hydrological Forecasting*, Wiley, Chichester.

Beven, K.J. and Binley, A.M. (1992) The future role of distributed models: model calibration and predictive uncertainty. *Hydrological Processes*, 6, 279-298.

Brettensen, K., K.H. Jensen and R. Jacobsen (1993) Tracer test in fractured chalk. 2, Numerical analysis. *Nordic Hydrology*, 24, 275-296.

Calver, A. (1988) Calibration, sensitivity and validation of a physically-based rainfall runoff model, *J. Hydrology*, 103, 103-115.

Crawford, N.H. and Linsley, R.K. (1966) Digital simulation in hydrology, Stanford watershed model IV. Departmant of Civil Engineering, Stanford University, Technical Report 39.

Feddes, R.A., Kabat, P., van Bakel, P.J.T., Bronswijk, J.J.B., and Halbertsma, J. (1988) Modelling Soil Water Dynamics in the Unsaturated Zone - State of the Art. *Journal of Hydrology*, 100, 69-112.

Grayson, R.B, Moore, I.D. and McMahon, T.A. (1992) Physically-based hydrological modelling, 2, Is the concept realistic? *Water Resources Research*, 28(10), 2639-2658.

Hasholt, B. and Styczen, M. (1993) Measurement of sediment transport components in a drainage basin and comparison with sediment delivery computed by a soil erosion model. In: *Sediment problems: Strategies for monitoring, prediction and control*, IAHS publ. 217, 1993, pp 147-159.

Jain, S.K., Storm, B., Bathurst, J.C., Refsgaard, J.C. and Singh, R.D. (1992) Application of the SHE to catchments in India - Part 2: Field experiments and simulation studies on the Kolar Subcatchment of the Narmada River. *Journal of Hydrology*, 140, 25-47.

Jensen, K.H. (1983) *Simulation of water flow in the unsaturated zone including the root zone*. Series paper No. 33. Institute of Hydrodynamics and Hydraulic Engineering. Technical University of Denmark.

Kristensen, K.J. and Jensen, S.E. (1975) A Model for Estimating Actual Evapotranspiration from Potential Evapotranspiration. *Nordic Hydrology*, Vol. 6, 70-88.

Leonard, B.P. (1979) Simple high-accuracy resolution program for convective modelling of discontinuities. *International Journal for Numerical Methods in Fluids*, 8, 1291-1318.

McDonald, M.G. and Harbaugh, A.W. (1988) *A modular three-dimensional finite-difference ground-water flow model*. Techniques of Water Resources Investigations 06-A1, United States Geological Survey.

Monteith, J.L. (1965) Evaporation and environment, *Symp. Soc. Ex Biology*, 19, 205-234.

Mudgeway, L.B. and Nathan, R.J. (1993) *Process modelling of flow and salt transport between shallow groundwater and surface drainage in the Tragowell Plains*, Rural Water Corporation, Investigations Branch, Rep. No. 1993/13.

Nielsen, S.A. and Hansen, E. (1973) Numerical simulation of rainfall-runoff process on a daily basis. *Nordic Hydrology* 4, 171-190.

Prathapar, S.A., Bailey, M.A., Poulton, D., Barrs, H.D. (1995) Evaluating water table control options using a soil water and groundwater simulation model (SWAGSIM). *Proceeding of the international congress on Modelling and Simulation*, Vol. 3 p. 18-23, Newcastle, Australia.

Preissmann, A. and Zaoui, J. (1979) Le module "Ecoulement de surface" due Système Hydrologique Européen (SHE). *Proceedings of the 18th IAHR Congress*, Cagliari.

Refsgaard, A., Refsgaard, J.C. and Høst-Madsen, J. (1994) A hydrological modelling system for joint analyses of regional groundwater resources and local contaminant transport. *Interamerical Congress of Sanitary and Environmental Engineering*, Buenos Aires, Oct. 31 - Nov. 3, 1994.

Refsgaard, A. (1994) The influence on groundwater recharge and surface water discharge of groundwater abstraction - an example from Odense, Denmark (in Danish). *Danish Academy of Technical Science*, Copenhagen, May 4, 1994.

Refsgaard, A., Nilsson, B. and Flyvbjerg, J. (1995a) Skrydstrup waste disposal site - a case study. Proceedings of the 68th international conference WEFTEC'95, Vol 2: *Residuals & biosolids/Remediation of soil & groundwater*, Miami Beach, Florida, Oct. 21 - 25, 1995.

Refsgaard, J.C., Christensen, T.H. and Ammentorp, H.C. (1991) A Model for oxygen transport and consumption in the unsaturated zone. *Journal of Hydrology*, 129, 349-369.

Refsgaard, J.C., Seth, S.M., Bathurst, J.C., Erlich, M., Storm, B., Jørgensen, G.H. and Chandra, S. (1992) Application of the SHE to catchment in India - Part 1: General results. *Journal of Hydrology*, 140, 1-23.

Refsgaard, J.C. and Sørensen, H.R. (1994) Modelling the influence of the Gabcikovo hydro power plant on the hydrology and the ecology of the Danubian Lowland. *Conference on Modelling, Testing & Monitoring for Hydro Powerplants, Budapest*, July 11-13, 1994.

Refsgaard, J.C., Storm, B. and Refsgaard, A. (1995b) Validation and applicability of distributed hydrological models, Modelling and Management of sustainable basin-scale Water resources systems, *IAHS Publ. no. 231*, 387-397.

Refsgaard, J.C. and Storm, B. (1995) MIKE SHE, in V.P. Singh (Ed), *Computer Models of Watershed Hydrology*, Water Resources Publications, 809-846.

Smith, R.E., Goodrich, D.R., Woolhiser, D.A. and Simanton, J.R. (1994) Comment on 'Physically-based hydrologic modelling, 2, Is the concept realistic?' by R.B. Grayson, I.D. Moore, and T.A. Mchahon, *Water Resources Research*, 30, 3, 851-854.

Storm, B. (1991) Modelling of saturated flow and the coupling of the surface and subsurface flow, in D.S. Bowles and P.E. O'Connell (Eds.) *Recent advances in the modelling of hydrologic system*. Kluwer Academic Publishers, Dordrecht.

Storm, B, Jayatilaka, C.J., Mudgeway, L.B. (1996) Simulation of water and salt transport on irrogation-bay scale with MIKE SHE. Submitted to *Journal of Hydrology*.

Storm, B. and Punthakey, J.F. (1995) Modelling of environmental changes in the Wakool Irrigation District. *International Conference, MODSIM 95*, Newcastle, Australia.

Styczen, M. and Nielsen, S.A. (1989) A view of soil erosion theory, process, research and model building: Possible interactions and future developments. *Quaderni di Scienza del Suolo*, Vol. II, Firenze.

Styczen, M. and Storm, B. (1993) Modelling of N-movements on catchment scale - a tool for analysis and decision making. 1. Model description and 2. A case study. *Fertilizer Research* 36: 1-17.

Styczen, M. and Storm, B. (1995) Modelling the effects of Management Practices on Nitrogen in Soils and Groundwater, in P.E. Bacon and M. Dekker (Eds.), *Nitrogen Fertilization in the environment*, Inc. New York.

Thomas, R.G. (1973) Groundwater models. *Irrigation and drainage*. Spec. Pap. Food Agricultural Organis. No. 21, U.N., Rome.

Vested, H.J, Justesen, P. and Ekebjærg, L. (1992) Advection-diffusion modelling in three dimensions. *Applied Mathematical Modelling*, 16, 506-519.

CHAPTER 5
MULTI-SPECIES REACTIVE TRANSPORT MODELLING

Peter Engesgaard
*Water Quality Institute &
Department of Hydrodynamics and Water Resources,
Technical University of Denmark*

1. Introduction

A species transported through the subsurface is either reactive or non-reactive (conservative). Transport codes for non-reactive species were discussed in Chapter 3. When a reactive species is transported through the subsurface it may undergo chemical and microbiological processes that will retard and transform the species. In short, these types of processes will be called *reactions*.

Reactive transport codes are used in many facets of groundwater and environmental investigations such as for simulating regional groundwater chemistry, transport of metals, inorganic and organic pollution plumes at spills and landfills, bioremediation, and artificial recharge. The development and application of reactive transport codes is progressing at an increasingly faster pace justified by a concern for the quality of the water resources. In the early 1980's there existed only a few and relatively simple reactive transport codes but today they are too numerous to count, at least if all types of codes are included. For example, Grove and Stollenwerk (1987) listed 156 references dealing with reactive transport codes and many more have appeared since. Other reviews of reactive transport codes have appeared in the literature (Abriola, 1987; Engesgaard and Christensen, 1988; and Mangold and Tsang, 1991). Consequently, to keep this presentation clear, it is necessary to restrict the chapter to presenting just certain aspects of reactive transport modelling.

The chapter will only include deterministic modelling. Stochastic reactive transport modelling is a growing field, mainly in research, but will not be discussed here. The chapter is restricted to describe multi-species reactive transport codes. Single-species reactive transport codes assume that transport of the constituent is independent of other reactive species. Brusseau (1994) has recently reviewed some of the literature on this aspect. Only codes applicable to the groundwater zone are discussed. The soil science literature reveal an abundance of codes developed for unsaturated flow, perhaps the most prominent code being that of Šimůnek and Suarez (1994). Often it is possible to neglect transient flow as was demonstrated by Christiansen (1994), and if the gas phase can be treated as a homogeneous phase, then saturated reactive transport codes may also be applicable to unsaturated flow. Examples of this approach were given early by Grove and Wood (1979) and lately by Hansen and Postma (1995). Finally, the chapter focuses on codes applicable to field conditions. Over the years many reactive transport codes have been developed, but only a few have been applied and tested at the field scale. Our modelling capabilities certainly exceed our capability to collect the data in the field that are needed to support the formulation of the processes that we would like to include in the model.

2. A Historical Perspective

Anderson (1979) summarized a decade of research in groundwater problems. A small section was devoted to reactive transport modelling. The approach to simulate reactive transport was traditionally based on the constant K_d approach. All reactive processes were lumped into just one parameter, the K_d, describing the equilibrium partitioning between sorbed and dissolved solute concentrations. However, Reardon (1981) clearly demonstrated that the K_d approach was inadequate even for very simple systems. Anderson (1979) concluded that more research was needed to boost our modelling technology to get an improved understanding of how reactive contaminants are transported in the subsurface. From 1980 and to now there has been a sizeable increase in the number and types of codes developed to simulate multi-species reactive transport in groundwater.

Foremost was this expansion in the number of codes able to simulate inorganic reactions (geochemistry). The codes almost exclusively assumed that the reactions are governed by equilibrium. The basis for some of the early codes for simulating ion exchange problems (e.g. Valocchi et al. (1981)) was already given much earlier by Rubin and James (1973). Focus was now on adding more and more complex chemistry to one-dimensional transport codes (Jennings et al., 1982; Miller and Benson, 1983; Dance and Reardon, 1983; Walsh et al., 1984; Kirkner et al., 1984 and Förstner, 1986). These codes used self-developed transport and chemistry codes. Eventually, with the increase in computational power, it became advantageous to couple existing transport and chemical modules, thereby extending the simulation capabilities to two and three dimensions and/or to include the whole suite of geochemical reactions (Cederberg et al., 1986; Narasimhan et al., 1986; Appelo and Willemsen, 1987; Lewis et al., 1987; Engesgaard and Kipp, 1992; Walter et al., 1994; Parkhurst et al., 1995; Stollenwerk, 1995; and Engesgaard and Traberg 1995). Other codes still utilized specifically designed chemistry modules (Yeh and Tripathi, 1991 and Bjerg et al., 1993) which can be more computational efficient. Few codes have incorporated inorganic kinetics (Kirkner et al., 1985; Noorishad, 1987, Liu and Narasimhan, 1989, and Friedly and Rubin, 1992).

The striking feature of this group of codes is that only two of them have been used for simulating real field scale contamination problems. The majority has been used for simulating reactive transport in hypothetical aquifers, column experiments, and small to large-scale controlled field experiments. Only Narasimhan et al. (1986) and

Engesgaard and Traberg (1995) were able to compare model simulations with observations at two accidentally contaminated sites.

Codes to simulate transport of organic constituents were also launched in the 1980's. The development was, and is, hampered by a lack of a general formulation of the microbiological system (see Section 4.2) and by a lack of reliable reaction constants. Most codes were tailored specifically to a certain system, for example a laboratory or field experiment (Sykes et al., 1982; Borden and Bedient, 1986ab; Molz et al., 1986; Bouwer and Cobb, 1987; Widdowson et al., 1988; Celia et al., 1989; McQuarrie et al., 1990; Frind et al., 1990; Kinzelbach et al., 1991; Chen et al., 1992; Zysset et al., 1994b, and Wood et al., 1994). A list of these models describing some of their features can be found in Sturman et al. (1995). For this reason, and possibly also because of a greater concern for organic contamination of aquifers, these codes have found more use at the field scale involving calibration of physical and microbiological parameters.

3. Classification of Reactive Processes

Figure 1 shows a cartoon-like presentation of the myriad of reactive processes that can take place in a pore section of a groundwater system. As a starting point we may conveniently group them into *inorganic* and *organic* processes. This classification follows the modelling tradition discussed above. Clearly, the system is of a multi-species and multi-process nature.

Some of the processes in Figure 1 take place strictly in the aqueous phase (homogeneous processes), while others involve the aqueous and solid phases (heterogeneous processes). The inorganic processes may include aqueous complexation, mineral precipitation/dissolution, surface complexation, ion exchange, and oxidation/reduction. The organic processes may include sorption, degradation, and growth and decay of biomass.

Rubin (1983), in a landmark paper, proposed a more generic classification by dividing all reactions into three levels. The first level, level A, distinguishes between the time scale of reaction, grouping reactions into those of 'sufficiently fast' and reversible and those of 'insufficiently fast' and/or irreversible. The two groups are both divided into homogeneous and heterogeneous processes at level B. Level C subdivides all heterogeneous reactions into surface and classical reactions. Rubin (1983) showed that this classification is essential for developing a mathematical framework for solving reactive transport problems. Any of the processes in Figure 1 can be classified according to this scheme. The definition of what is 'sufficiently fast' relies on the so-called Local Equilibrium Assumption (LEA; Rubin, 1983). For the LEA to be valid, the rates of reactions must be high in comparison with those of other processes of the system that change solute concentration. The other processes could be advection, dispersion, and changes in boundary conditions. The LEA is a pragmatic statement and the question of the validity of the LEA is a question of the error that can be tolerated

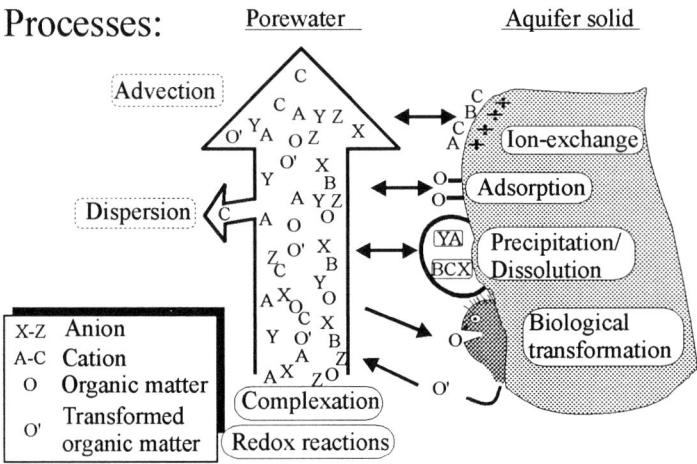

Figure 1 Reactive processes at the pore level.

in solving the reactive transport problem. Several groups of researchers have attempted to derive meaningful criteria that define conditions for the validity of the LEA (Valocchi, 1985, 1986, 1988, 1989, Jennings and Kirkner, 1984; Jennings, 1987; Bahr and Rubin, 1987; Bahr, 1990).

The level A classication is more generic than the classification of reactions as being either inorganic or organic. The distinction between inorganic or organic processes is difficult to maintain, because, for example, oxidation of organic compounds or Fe (II) to Fe-oxide can be biotic or abiotic. However, in the past, inorganic processes have for the most part been described as 'fast' reactions and organic processes as 'slow' reactions. For the remainder of the chapter this distinction will be maintained.

The LEA has traditionally been used when simulating inorganic reactions. The term *geochemistry* is used here to represent equilibrium-controlled inorganic reactions. A reaction between two reactants (A and B) and two products (C and D) can be written (Appelo and Postma, 1993)

$$aA + bB \rightleftarrows cC + dD \qquad (1)$$

Using thermodynamics we may describe the reaction at equilibrium using the mass action equations

$$K = \frac{[C]^c[D]^d}{[A]^a[B]^b} \tag{2}$$

where K is the equilibrium constant and [i] is the activity of species i. It is the easy access to reasonably well-defined constants like K and the fact that many inorganic reactions (aqueous and surface complexation, ion exchange) are quite rapid that make the LEA so appealing. Other reactions such as weathering or abiotically nitrate reduction of pyrite can be very slow, and will thus require a kinetic formulation similar to what is presented below, but where the action of a microbial population is left out.

Generally, the LEA can not be invoked for organic reactions. The term *biodegradation* will be used here to represent microbial, kinetically-controlled degradation of organic components. Bacteria in aquifers can catalyze certain redox processes, whereby organic components are oxidized through the reduction of oxidized species. Recent investigations have shown that, even in pristine aquifers, the number of bacteria can be quite high, and that once contaminated with for example leachate from landfills, the bacteria are able to fuel growth (Christensen et al., 1994). There are traditionally three different ways of idealizing how the bacteria grow in a porous medium; (1) 'the strictly macroscopic', (2) 'the microcolony', and (3) 'the biofilm' models, see Baveye and Valocchi (1989). The first type of model is characterized by making no assumptions concerning the microscopic configuration and distribution of the bacteria in the pores, while the two last types of models involve parameters, like biofilm thickness, that, on the order of μm, specify how the bacteria occupy a pore.

Dependent on the flux of substrate through the porous medium, the bacteria may grow to form more or less continuous biofilms attached to the soil grains (Rittman, 1993). Close to a landfill, a continuous biofilm may develop, while more downstream, the bacteria may be attached in patchy aggregates called colonies. Recently, Zysset et al. (1994b) derived a 'macroscopic biofilm' model by first posing the problem at the microscopic scale and subsequently integrating over a REV (Representative Elemental Volume (Bear, 1972)), to obtain a macroscopic model. The growth of a biomass can affect porosity, hydraulic conductivity, and dispersivity (Taylor and Jaffé, 1990abcd; Vandevivere et al., 1995). The rate of growth may be so high that the porous medium gets plugged. In these situations the flow and reactive transport equations are coupled.

The microbial population can be very diverse, each microbial species having its own capabilities of degradation. It is generally not possible to simulate the endeavor of all microbes due to a lack of information about the mechanisms of growth and decay of each microbe.

The organic components are sometimes referred to as the substrates or electron donors and the oxidizing species as the electron acceptors. The substrate itself may be attached to the soil, Soil Organic Carbon (SOC), or be present as Dissolved Organic Carbon (DOC). SOC and DOC are bulk parameters covering a variety of organic components (Christensen et al., 1994). The electron acceptors can also be numerous and quite often biodegradation of DOC takes place in a redox sequence (Christensen et al., 1994). The electron acceptors may be associated with the aqueous phase (e.g. O_2, NO_3, and SO_4) and the solid phase (Fe- and Mn oxides).

Information about degradation parameters of the various biomasses, organic components, and electron acceptors are frequently non-existent. Modelling attempts thus focus on simulating the behaviour of the bulk parameters employing some kind of effective degradation parameters. Biodegradation is often a slow reaction and must be described by a kinetic-type rate equation. Several empirical equations have been proposed including zero-order, first-order, Michaelis-Menten, and Monod, see Alexander and Scow (1989) for a review on the kinetics of biodegradation. The general Monod equation is

$$\mu = \mu_{max}(\frac{S}{K_s + S}) \tag{3}$$

where the specific growth rate μ is a function of the maximum growth rate μ_{max}, the substrate concentration S, and a half-saturation constant K_s. The Michaelis-Menten and Monod equations are the most widely used rate expressions because both include, for low values of K_s, the zero-order, and, for high values of K_s, the first-order rate equation.   The inorganic and organic processes depicted in Figure 1 can be intimately dependent on each other. As an example natural decomposition of organic matter may trigger important inorganic reactions such as the reduction of iron oxides, sulfate and nitrate or the production of CO_2 can influence reactions involving carbonate minerals (Appelo and Postma, 1993). More complicated systems at landfills (Christensen et al., 1994) and spills of oil (Baedecker et al., 1993) have shown that in order to understand the transport behaviour of inorganic and organic constituents, one need to take into account that the two groups of constituents will interact. Only a few codes have been developed that can assist in this task (McNab and Narasimhan 1994,1995; Brun et al. 1994).

4. Macroscopic Reactive Transport Equations: MIKE SHE as an Example

Examples of the macroscopic reactive transport equations will be presented for the two major classification of reactive processes; geochemistry and biodegradation. The two types of processes have been coupled to the MIKE-SHE code (VKI, 1994, 1995). This modelling framework is used as an example.

4.1 GEOCHEMISTRY
Derivation of the basic reactive transport equations for a geochemical transport system have been described by e.g. Rubin (1983), Kirkner and Reeves (1988), Yeh and Tripathi (1989), and Zysset et al. (1994a). Thus, the governing equations will only

briefly be presented below.

The geochemical processes that will be considered are aqueous complexation, ion exchange, and mineral precipitation/dissolution. Redox reactions can be considered as a complexation or mineral reaction and at the end of this section it will be demonstrated how redox processes can be incorporated.

Box 1 shows a concise formulation of the three types of geochemical processes, see Section 9 for an explanation of the notation. These are all similar to equations (1) and (2). In very concentrated solutions it is necessary to use the activity of the species instead of concentrations (Appelo and Postma, 1993). Generally, it is assumed that the hydraulic properties do not change with the advent of geochemical processes. Mineral dissolution may change porosity and hydraulic conductivity, however, these changes normally occur over geologic time scales (thousands of years) for most natural and contaminated environments (Sanford and Konikow, 1989), and this effect can therefore normally be neglected.

In order to describe the chemical reactions mathematically, a subset of the species must be chosen as components. All other ions (c_j), complexes (x_i), exchanged species (s_j), and minerals (p_k) can then be formed from the components. Typically, these components are selected as the free species, for example Ca^{2+}, Mg^{2+}, and SO_4^{2-}. The components are building blocks from which the other species can be constructed, for example a complex $CaSO_4^0$, a mineral $CaCO_3$, or an exchanged species $Ca-X_2$.

The basic reactive transport equations are

$$\frac{\partial}{\partial t}(c_j + \sum_{i=1}^{N_x} A_{ij} x_i) - L(c_j + \sum_{i=1}^{N_x} A_{ij} x_i) \qquad (4)$$
$$+ \frac{\rho}{\theta}(\frac{\partial s_j}{\partial t} + \sum_{k=1}^{N_p} B_{kj} \frac{\partial p_k}{\partial t}) = 0 \qquad j=1,\ldots,N_c$$

Box 1 Mathematical representation of equilibrium processes

Process	Formula	Equilibrium	
Complexation:	$\sum_{j=1}^{N_c} A_{ij}\hat{c}_j \rightleftarrows \hat{x}_i$	$K_i = x_i / \prod_{j=1}^{N_c} c_j^{A_{ij}}$	$i=1,\ldots,N_x$
Ion Exchange:	$z_1\hat{s}_j + z_j\hat{c}_1^{z_1} \rightleftarrows z_j\hat{s}_1 + z_1\hat{c}_j^{z_j}$	$K_{GT} = \left[\frac{f_1}{c_1}\right]^{z_j}\left[\frac{c_j}{f_j}\right]^{z_1}$	$j=2,\ldots,N_s$
Minerals :	$\hat{p}_k \rightleftarrows \sum_{j=1}^{N_c} B_{kj}\hat{c}_j$	$K_k^s \geq \prod_{j=1}^{N_c} c_j^{B_{kj}}$	$k=1,\ldots,N_p$

where L() is the advection-dispersion operator given by

$$L(c) = \nabla \cdot \mathbf{D} \cdot \nabla c - \mathbf{v} \cdot \nabla c \qquad (5)$$

At this time we can introduce the total aqueous component concentration defined by

$$u_j = c_j + \sum_{i=1}^{N_x} A_{ij} X_i \qquad (6)$$

i.e. the mass of each component in the aqueous phase. We can insert (6) into (4) to obtain

$$\frac{\partial u_j}{\partial t} + L(u_j) = -\frac{\rho}{\theta}\left(\frac{\partial s}{\partial t} + \sum_{k=1}^{N_p} B_{kj} \frac{\partial p}{\partial t}\right) \qquad (7)$$

The three temporal derivatives are capacitance terms, i.e. they reflect the changes in storage of mass in the two phases. The capacitance terms for the exchanged and mineral components are often referred to as source-sink terms, because they add or remove mass to or from the aqueous phase. The total component concentration (the total mass of each component in both the aqueous and solid phase) can be used instead of u_j by appropriately modifying (7), see Kirkner and Reeves (1988) and Yeh and Tripathi (1989).

Redox processes pose a problem because these reactions involve oxygen. In principle, we could proceed as above and define O_2 as a component and formulate a reactive transport equation like (7). Water (H_2O) would then be considered as a complex. However, since the mass of water is very high, and nearly constant, any changes in the oxygen concentration due to geochemistry will barely be noticeable, when solving (7) for oxygen. Numerical errors could then have a great impact on the chemical solution. Instead, a number of other approaches can be taken as described by Engesgaard and Kipp (1992). All of these will not be discussed here, but one approach called the internal approach (Liu and Narasimhan 1989) involves the conservation of valence. Redox reactions involve changes in valence (e.g. oxidation of Fe^{2+} to Fe^{3+}), but no electrons or valence can be lost in the system as a whole. The total operational valence or redox state of a system can be defined as

$$R_s = \sum_{i=1}^{N_c+N_x} V_i m_i \qquad (8)$$

where each species have been given an operational valence V_i. The redox state R_s is therefore the 'mass of valence' in the aqueous phase. This property must be conserved when losses to and from the solid phase is accounted for. In addition to the introduction of the mass balance on R_s, we must also define the electron (e^-) as a component and assume that the concentration of water is constant, so that any redox reaction can be defined on the basis of this component. The redox state is a transportable quantity and a reactive transport equation similar to (7) can be written

$$\frac{\partial R_s}{\partial t} - L(R_s) + \frac{\rho}{\theta}(\sum_{j=1}^{N_s} V_{s,j}\frac{\partial s_j}{\partial t} + \sum_{k=1}^{N_p} V_{p,k}\frac{\partial p_k}{\partial t}) = 0 \qquad (9)$$

Through the definition of the initial and boundary conditions of R_s, changes in reducing or oxidizing solutions can be specified.

4.2 BIODEGRADATION

The derivation of the basic reactive transport equations for biodegradation have been given by e.g. Baveye and Valocchi (1989), Kindred and Celia (1989), Kinzelbach et al. (1991), and Zysset et al. (1994b).

The main questions to ask when conceptualizing a transport system with biodegradation are; 1) How many organic substrates, how many electron acceptors, and how many biomasses should be included?, 2) How should the microscale biomass distribution be accounted for?, and 3) What kind of kinetic formulation should be used to describe biodegradation?

The MIKE-SHE BIODEGRADATION code considers one substrate, typically as a bulk parameter, either in the dissolved or solid phase. It can handle two dissolved electron acceptors (e.g. O_2 and NO_3) and one type of biomass (bulk parameter). The code is based on the 'strictly macroscopic' conceptualization of biomass in a porous media, and thus requires no data on the microscopic configuration of the biomass. The multiplicative Michaelis-Menten equation is used to account for biodegradation, i.e.

$$\frac{dS}{dt} = -k\,B\,\frac{A}{K_A + A}\,\frac{S}{K_s + S} \qquad (10)$$

With these assumptions a biodegradation model can be formulated as shown in Box 2. The net rate of biomass (B) growth is given as the difference between growth (proportional to the cumulative rate of change of the substrate times the yield coefficient) and first order rate of death.

The reactive transport equations are formulated in the same manner as before but now defining an equation for each species. The capacitance or rate terms for biodegradation that appear in these equations are then defined by those in box 2.

4.3 COUPLED GEOCHEMISTRY AND BIODEGRADATION

Modelling coupled geochemistry and biodegradation is very much a pursuit still in its infancy. McNab and Narasimhan (1994,1995), Brun et al. (1994), and Griffioen et al. (1995) use the same overall simulation approach. Geochemistry and biodegradation

> Box 2 Mathematical representation of biodegradation processes
>
> $$\frac{dS}{dt} = \sum_{i=1}^{2} R_i = \sum_{i=1}^{2} -k_i B \left[\frac{A_i}{K_i + A_i}\right]\left[\frac{S}{K_{s,i} + S}\right]$$
>
> $$\frac{dA_i}{dt} = R_i F_i \quad i=1,2$$
>
> $$\frac{dB}{dt} = \left[\sum_{i=1}^{2} -Y_i R_i\right] - \mu_d B$$

are formulated separately and the link between the two are obtained by specifying which organic processes occur. The example from Griffioen et al. (1995) is illustrated in Section 6.

5. Numerical Solution Procedures

It is not possible here to give an exhaustive description of commonly applied numerical procedures for solving reactive transport problems. Kirkner and Reeves (1998), Yeh and Tripathi (1989), and Rubin (1990) have reviewed various methods for coupling transport and reactions (mainly geochemistry). Reeves and Kirkner (1988), Herzer and Kinzelbach (1989), Valocchi and Malmstead (1992), Kaluarachchi and Morshed (1995), Morshed and Kaluarachchi (1995), and Engesgaard et al. (1995) have investigated numerical errors associated with some of the more common methods.

For two and three dimensional problems with multiple species and processes it is recognized that the only feasible way to solve the system of equations is to divide the solution procedure into two steps; (1) solve transport equations and (2) solve reaction equations. This procedure is known as the Operator Splitting (OS) method. The OS method can be of the iterative or non-iterative type. The OS method will work for transport coupled to geochemistry or biodegradation.

The non-iterative OS method is especially interesting because it avoids the need for iterating between solving for transport and reactions and has thus received increased attention recently (Herzer and Kinzelbach, 1989; Valocchi and Malmstead, 1992; Kaluarachchi and Morshed, 1995; Morshed and Kaluarachchi, 1995; Engesgaard et al.

1995). However, the numerical solution is affected in two ways; for geochemistry, the method introduces numerical dispersion, and for biodegradation, the method introduces mass balance problems.

For biodegradation (or kinetic reactions), Valocchi and Malmstead (1992), Kaluarachchi and Morshed, (1995), and Morshed and Kaluarachchi (1995) showed that the non-iterative OS method will have an inherent mass balance problem which is a function of $\mu\Delta t$. They proposed an alternating scheme, where, in one time step, transport is solved first then reaction, in the next time step the order of calculation is reversed, and so on. This scheme will reduce mass balance errors.

Aside from the fact that the OS method is the only feasible way of coupling transport and reactions, the method is also very flexible. It is possible to couple existing transport and reaction codes. As described in Section 1, this approach has been widely used in recent years. The appropriate numerical technique can then be applied to each step.

6. Two examples of field applications

6.1 GROUNDWATER CONTAMINATION AT A WASTE DEPOSIT

The example is taken from Engesgaard and Traberg (1995). Figure 2 shows three plumes of contamination downstream to a waste residue deposit. The high concentrations of Cl and K is due to contamination by leachate from the deposit. The K plume seem to travel at a retarded velocity when compared to the Cl plume. The Ca concentration in groundwater is higher than in the leachate and a source of Ca must therefore exist in the aquifer. Ion exchange was proposed as a mechanism to explain these observations. This was confirmed by soil column experiments using leachate and soil from the site (Kjeldsen, 1983). It was concluded that the contamination possibly occurred as overflow from the deposit during a four year period 1977-1980. Remediation took place in early 1981. A geochemical transport model was developed and used in one dimension to simulate the soil column experiment (to obtain unknown

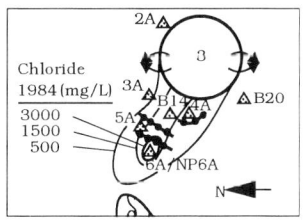 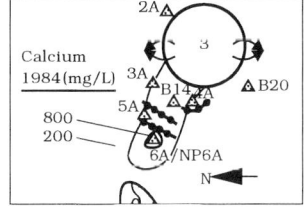 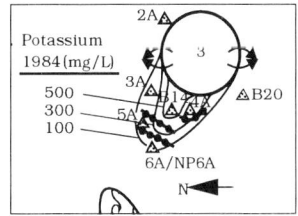

Figure 2 Cl, K, and Ca plumes downstream deposit 3 at the waste site.

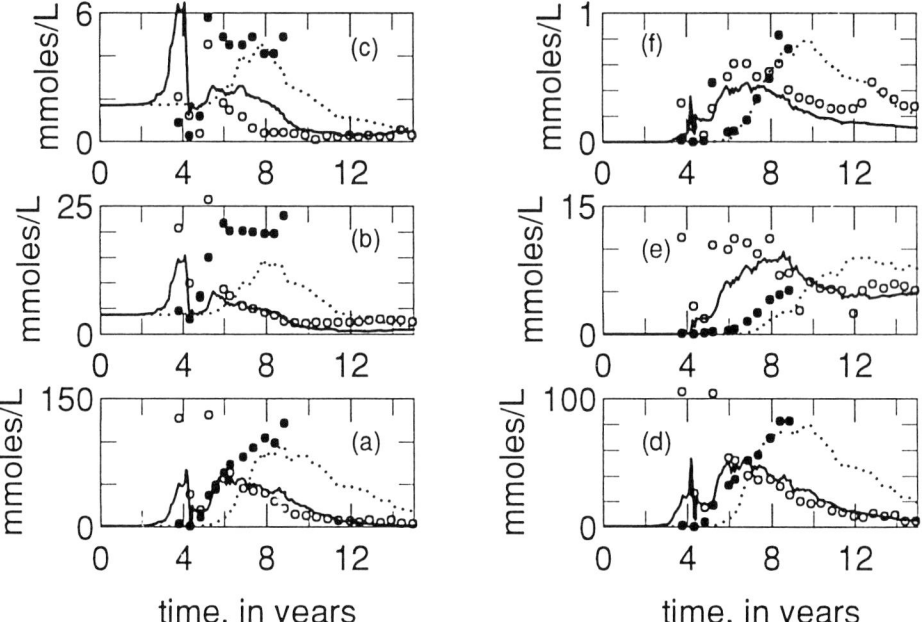

Figure 3 Breakthrough curves in well 5A and 6A for Cl (a), Ca (b), Mg (c), Na (d), K (e), and NH_4 (f). Open/solid circles and solid/dashed lines are data and simulations for 5A/6A.

geochemical constants through a calibration) and in two dimensions to simulate field scale transport of contaminants. Only the field scale application will be shown here.

Wells 5A and 6A are both located approximately along the longitudinal directions of the plumes, and breakthrough curves (BTCs) of Cl and some of the reactive components are shown in Figure 3. Time is relative to the start of the overflow in year 1977. At well 5A there is a good agreement between observations and simulation for all components, except at the time of remediation, at approximately year 4. At well 6A the agreement with the observations varies according to the component. The simulated Cl BTC is shifted in time arriving almost one year later than observed. From the BTCs in 5A it can be seen that Na, K, and NH_4 are not retarded significantly when compared with Cl. This is due to the high inlet concentrations of the cationic components. On a molar basis the inlet concentrations of Na, K, and NH_4 were much higher than the Cation Exchange Capacity (CEC). The mass of these components exchanged from the groundwater to the solid phase is negligible up to the time of remediation. Remediation removed approximately half of the contamination causing a lowering of the concentrations in the aquifer, and at the same time the overflow from the deposit was stopped. The resulting lower concentrations make the loss of mass to the solid phase important and the plumes of reactive contaminants moving from 5A to 6A are affected by ion exchange processes. This is most clearly seen in the observations for K but also to some degree for NH_4.

It was concluded that reactive transport at the site initially is controlled entirely by transport, later by ion exchange and transport. Such observations can be decisive for the design of an effective remediation scheme.

6.2 GROUND WATER QUALITY IN A RIVERINE AREA

The groundwater quality in the vicinity of the Danube river just south of Bratislava, Slovakia has been investigated by groundwater sampling in a transect of multi-level wells and subsequent development and application of a coupled organic-inorganic reactive transport model. The model is based on the MIKE SHE GEOCHEMISTRY and BIODEGRADATION codes described in Section 4.1 and 4.2.

Figure 4 shows schematically the conditions in a 2 km profile. The Danube river recharges the aquifer and high fluxes of Dissolved Organic Carbon (DOC) and nitrate (NO_3) infiltrates to the aquifer. Oxygen has already disappeared within 10 meters from the Danube. Multi-level sampling in 11 wells in the profile shows consumption of DOC and NO_3 and the production of nitrite (NO_2). *Denitrification* is therefore a possible reaction in the profile. A microbial population is responsible for this reaction. Soil Organic Carbon (SOC) played a minor role in denitrification. There were also high manganese (Mn) concentrations in the same area as where NO_2 is produced. It was

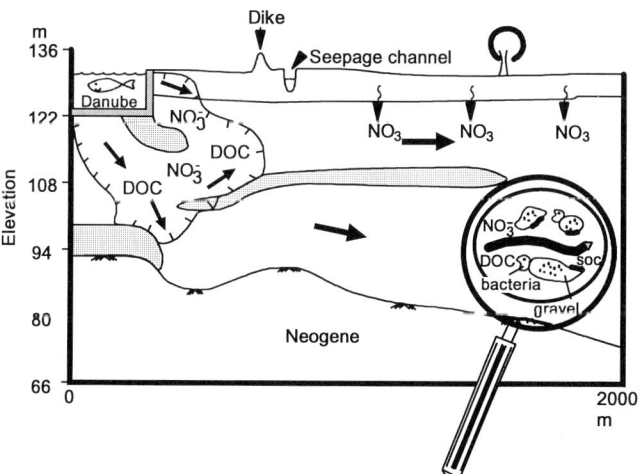

Figure 4 Schematic presentation of cross-section at the Danube river.

therefore postulated that the denitrification reaction, or *carbon/nitrogen redox cycle*, was coupled to a second *manganese/nitrogen redox cycle*, where reduced nitrogen species were oxidized by reducing the mineral manganite (Mn(III)OOH$_s$).

The MIKE SHE GEOCHEMISTRY and BIODEGRADATION codes were linked through the definition of a nitrogen redox capacity, which is part of the total redox state defined in equation (8). Consumption of the nitrogen redox capacity (calculated by the BIODEGRADATION code) will create dis-equilibrium with manganite minerals and subsequently result in the dissolution of manganite and oxidation of reduced nitrogen species, e.g. NO_2 (calculated by the GEOCHEMISTRY code). The consumption of the nitrogen redox capacity does not give actual nitrogen species concentrations, only the changes in the valences within the nitrogen system. The appearance of NO_2 is concocted by allowing the GEOCHEMISTRY code to calculate a pseudo-equilibrium between the nitrogen species. This is done by manipulating the thermodynamic equilibrium constant in the nitrogen speciation.

Figure 5 shows an example of the simulations that were carried out to test the hypothesis about the coupled redox cycles. The denitrification rate was calibrated by matching the observed DOC plumes (coming from the river). Nitrate is consumed in the vicinity of the river and small amounts of NO_2 are produced as a result of the incomplete denitrification reaction. Manganite is dissolved which leads to increased Mn concentrations in the profile. The simulated NO_2 and Mn concentrations agree with the observations and the model could therefore support the hypothesis about two interconnected redox cycles.

7. Future needs in multi-species reactive transport modelling

Some future needs within the area of reactive transport modelling can be identified; (1) Application to the field scale, (2) Reactive heterogeneity, (3) Identification and scaling of processes, (4) Coupled inorganic-organic codes, (5) Inverse modelling, and (6) Numerical methods.

The ultimate goal for developing reactive transport codes is to use them at the field scale. However, the attempts at doing so have been very few. This is especially true for the geochemical transport codes, but increased awareness of the intimate coupling of inorganic-organic systems necessitates that all types of codes be tested at the field scale.

Spatial heterogeneity in reactive parameters will possibly play just as important a role as spatial heterogeneity in physical parameters has on the spreading of contaminants. Stochastic reactive transport codes have been developed to study simple reactions in heterogeneous systems but the simplified representation of the reactions make the codes less applicable to field conditions. These codes are either based on solving stochastic reactive transport equations, whereby the spatial variability in the

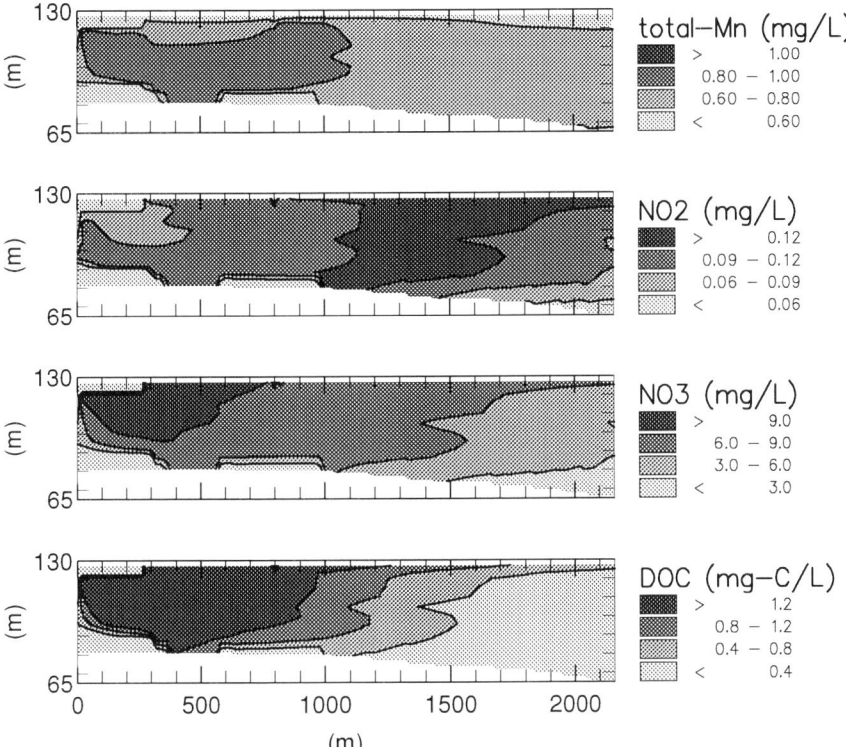

Figure 5 Simulated concentration distributions of DOC, NO_3, NO_2, and total-Mn.

reactive parameters is incorporated directly, or based on a Monte Carlo approach, whereby a determinstic solution to the reactive transport equations is obtained for several specified fields of the reactive parameters.

In a deterministic field scale model, the reactive transport equations are generally macroscopic representing a scale of centimeters to meters. We are unable to represent pore-scale variations in velocity components because of an imperfect knowledge of the porestructure and because of computational constraints. The predicted velocity field predicted is macroscopic and any unknown solute spreading is lumped into the dispersion coefficient. If this is the modelling framework, then the reactive processes should be macroscopic as well. However, quite often the reactive processes are formulated on a microscopic basis. For example, models of degradation of organics in aquifers have borrowed the formulation from biofilm technology and we really need to answer the question of how we can upscale this and other processes from the pore to the macroscopic level.

Recently a new type of code has emerged; coupled organic-inorganic transport codes. These codes are still very immature but they do recognize that it is not always possible to look separately at either the organic or the inorganic system.

Inverse reactive transport modelling should find increased use in order to parameterize not only the physical but also the multi-species reactive processes. Codes exist that are able to parameterize single-species reactive parameters but only a few have been developed for multi-species reactions (Rainwater, 1987).

To accomplish some of these tasks there is a need for developing more efficient and robust schemes for solving reactive transport problems. The classical approach of coupling existing transport and reaction modules probably has to be abandoned. More sophisticated codes are needed that are specifically designed to handle transport of reactive species. Kalatzis et al. (1993) and Wolfsberg and Freyberg (1994) give examples of how the transport solution can be modified to accommodate the need for accuracy and efficiency.

8. Acknowledgement

I would like to express my gratitude to Jasper Griffioen, TNO Institute of Applied Geoscience, The Netherlands, and Karsten Høgh Jensen, Water Quality Institute/Institute of Hydrodynamics and Hydraulic Engineering, Denmark, for observant reviews of this chapter. I would also like to thank Adam Brun, Water Quality Institute/Institute of Hydrodynamics and Hydraulic Engineering, Denmark, for being creative with drafting Figure 1.

9. Notation

SYMBOL	EXPLANATION	UNITS
A_{ij}	stoichiometric coefficient of component j in complex i	
A_i	electron acceptor concentration	ML^{-3}
B_{kj}	stoichiometric coefficient of component j in mineral k	
B	biomass concentration	ML^{-3}
$\hat{c}_j$	chemical formula for component j	
c_j	component concentration	ML^{-3}
$\mathbf{D}$	dispersion tensor coefficient	L^2T^{-1}
D	dispersion coefficient	L^2T^{-1}
F_i	stoichiometric coefficient in biodegradation	
f_j	equivalent fraction of exchanged component	
k	substrate utilization rate	T^{-1}
K_d	distribution coefficient	L^3M^{-1}

K_{GT}	Gaines-Thomas selectivity coefficient	(variable)
K_i	equilibrium constant for complexation	(variable)
K_k^s	solubility constant	(variable)
K_s, K_A	half-saturation constant for electron donor or acceptor	ML^{-3}
$L()$	advection-dispersion operator	
m_i	concentration of aqueous species	ML^{-3}
n	time level	
N_c	number of components	
N_p	number of minerals	
N_s	number of exchange components	
N_x	number of complexes	
$\hat{p}_k$	chemical formula for mineral	
p_k	mineral concentration	ML^{-3}
R_i	substrate utilization	MT^{-1}
R_s	redox state of aqueous solution	ML^{-3}
S	substrate concentration	ML^{-3}
$\hat{s}_j$	chemical formula for exchanged component j	
s_j	exchanged concentration	ML^{-3}
t	time	T
u_j	total aqueous component concentration	ML^{-3}
$\mathbf{v}$	pore water velocity vector	LT^{-1}
v	pore water velocity	LT^{-1}
V_i	operational valence of aqueous species	
$V_{s,j}$	operational valence of exchanged component j	
$V_{p,k}$	operational valence of mineral k	
$\hat{x}_i$	chemical formula for complex i	
x_i	complex concentration	ML^{-3}
Y	yield coefficient	MM^{-1}
z_i	charge of species i	
ρ	soil bulk density	ML^{-3}
θ	porosity	[-]
μ	specific growth rate	T^{-1}
μ_{max}	maximum growth rate	T^{-1}
μ_d	biomass death rate	T^{-1}

10. References

Abriola, L.M., Modeling contaminant transport in the subsurface: an interdisciplinary challenge, *Reviews of Geophysics*, Vol. 25(2), pp. 125-134, 1987.

Alexander M, and K.M. Scow, Kinetics of biodegradation in soil, *Reactions and movement of organic chemicals in soils*, Soil Science Society of America Special Publication no. 22, 243-267, 1989.

Anderson, M.P., Using models to simulate the movement of contaminants through groundwater flow systems, *CRC Critical Reviews in Environmental Control*, Vol 9(2), 97-156, 1979.

Appelo, C.A.J. and D. Postma, *Geochemistry, Groundwater and Pollution*, A.A. Balkema, Rotterdam, 536 pp, 1993.

Appelo, C.A.J., and A. Willemsen, Geochemical calculations and observations on salt water intrusions,

I. A combined geochemical/mixing cell model, *J. Hydrol.*, *94*, 313-330, l987.
Baedecker, M.J., I.M. Cozzarelli, R.P. Eganhouse, D.I. Siegel, and P.C. Bennett, Crude oil in a shallow sand and gravel aquifer-III. Biogeochemical reactions and mass balance modeling in anoxic groundwater, *Applied Geochemicstry*, 8, 569-586, 1993.
Bahr, J.M., and J. Rubin, Direct comparison of kinetic and local equilibrium formulations for solute transport affected by surface reactions, *Water Resour. Res.*, 23, 438-452, 1987.
Bahr, J.M, Kinetically influenced terms for solute transport affected by heterogeneous and homogeneous classical reactions, *Water Resour. Res.*, 2, 21-34, 1990.
Baveye P, and A. Valocchi, An evaluation of mathematical models of the transport of biologically reacting solutes in saturated soils and aquifers, *Water Resour. Res.*, 25(6), 1413-1421, 1989.
Bear, J., *Dynamics of Fluids in Porous Media*, Elsevier, New York, 1972.
Bjerg, P.L., H.C. Ammentorp, and T.H. Christensen, Model simulations of a field experiment on cation exchange-affected multicomponent solute transport in a sandy aquifer, *J. Contam. Hydrol.*, *12*, 291-311, 1993.
Borden, R.C., and P.B. Bedient, Transport of dissolved hydrocarbons influenced by oxygen-limited biodegradation 1. Theoretical development, *Water Resour. Res.*, 22(13), 1973-1982, 1986a.
Borden, R.C., and P.B. Bedient, Transport of dissolved hydrocarbons influenced by oxygen-limited biodegradation 2. Field application, *Water Resour. Res.*, 22(13), 1983-1990, 1986b.
Bouwer, E.J., and G.D. Cobb, Modeling of biological processes in the subsurface, *Wat. Sci. Tech.*, 19, 769-779, 1987.
Brun, A., E.O. Frind, and P. Engesgaard, A coupled microbiology-geochemistry model for three-dimensional groundwater flow, *Proceedings of the IAHR/AIRH symposium on transport and reactive processes in aquifers*, Zurich, Switzerland, 11-15 April, 1994.
Brusseau, M.L., Transport of reactive contaminants in heterogeneous porous media, *Reviews of Geophysics*, 32(3), 285-313, 1994.
Cederberg, G.A., R.L. Street, J.O. Leckie, A groundwater mass transport and equilibrium chemistry model for multi-component systems, *Water Resour. Res.*, 21, 1095-1104, l985.
Celia, M.A., J.S. Kindred, and I. Herrera, Contaminant transport and biodegradation 1. A numerical model for reactive transport in porous media, *Water Resour. Res.*, 25(6), 1141-1148, 1989.
Chen, Y-M, L.M. Abriola, P.J.J. Alvarez, P.J. Anid, and T.M. Vogel, Modeling transport and biodegradation of benzene and toluene in sandy aquifer material: comparisons with experimental measurements, *Water Resour. Res.*, 28(7), 1833-1847, 1992.
Christensen, T.H., P. Kjeldsen, H-J. Albrecthsen, G. Heron, P.H. Nielsen, P.L. Bjerg, and P.E. Holm, Attenuation of landfill leachate pollutants in aquifers, *Critical Reviews in Environmental Science and Technology*, 24(2), 119-202, 1994.
Christiansen, J.S, Modelling reactive transport in the unsaturated zone, M.Sc. Thesis, Institute of Hydrodynamics and Hydraulic Engineering, Technical University of Denmark, 1994.
Dance, J.T., and E.J. Reardon, Migration of contaminants in groundwater at a landfill: A case study. 5. Cation migration in the dispersion test, *J. Hydrol.*, 63, 109-130, 1983.
Engesgaard, P., and Th.H. Christensen, A review of chemical solute transport models, *Nordic Hydrology, 19*, 183-216, 1988.
Engesgaard, P., and K.L. Kipp, A geochemical model for redox-controlled movement of mineral fronts in ground-water flow systems: A case of nitrate removal by oxidation of pyrite, *Water Resources Research,* 28, 10, 2829-2843, 1992.
Engesgaard P., and R. Traberg, Contaminant transport at a waste residue deposit: 2. Geochemical transport modelling, (Accepted for publication in Water Resources Research 1995).
Engesgaard, P., K.L. Kipp, and T. Russell, Analysis of the operator splitting method for modelling reactive transport problems, (in preparation), 1995.

Förstner, R.A., A multicomponent transport model, *Geoderma*, 38, 1-4, 261-278, 1986.

Frind, E.O., W.H.M. Duyinsveld, O. Strebel, and J. Boettcher, Modeling of multicomponent transport with microbial transformation in groundwater: The Fuhrberg case, *Water Resour. Res.*, 26(8), 1707-1719, 1990.

Friedly, J.C. and J. Rubin, Solute Transport with multiple equilibrium-controlled or kinetically controlled chemical reactions, *Water Resour. Res.*, 28(6), 1935-1953, 1992.

Griffioen J, P. Engesgaard, A. Brun, D. Rodak, I. Mucha, and J.C. Refsgaard, Nitrate and Mn-chemistry in the alluvial Danubian Lowland aquifer, Slovakia, *Groundwater Quality: Remediation and Protection*, (Proceedings of the Praque Conference, May 1995), IAHS Publ. no. 225, 1995.

Grove, D.B., and W.W. Wood, Prediction and field verification of subsurface-water quality changes during artificial recharge, Lubbock, Texas, *Ground Water*, 17, 250-257, 1979.

Grove, D.B., and K.G. Stollenwerk, Chemical reactions simulated by ground-water-quality models, *Water Resour. Bull.*, 23(4), 601-615, 1987.

Hansen, B.K, and D. Postma, Acidification, buffering and salt effects in the unsaturated zone of a sandy aquifer, Klosterhede, Denmark, *Water Resour. Res.*, 31, 2795-2809, 1995.

Herzer, J., and W. Kinzelbach, Coupling of transport and chemical processes in numerical transport models, *Geoderma*, 44, 115-127, 1989.

Jennings, A. A., and D.J. Kirkner, T.L. Theis, Multicomponent equilibrium chemistry in groundwater quality models, *Water Resour. Res.*, 18, 1089-1096, 1982.

Jennings, A.A., and D.J. Kirkner, Instantaneous equilibrium approximation analysis, *J. Hyd. Eng.*, ASCE 110, 1700-1717, 1984.

Jennings, A.A., Critical chemical reaction rates for multicomponent groundwater contamination models, *Water Resour. Res.*, 23, 1775-1784, 1987.

Kalatzis, A., R.A. García-Delgado, T-K Pang, A.D. Koussis, and A.R. Bowers, Two-dimensional groundwater transport of reactive solutes with competitive adsorption, *Water Resour. Res.*, 29(7), 2241-2248, 1993.

Kaluarachchi, J.J., and J. Morshed, Critical assesment of the operator-splitting technique in solving the advection-dispersion-reaction equation: 1. First-order reaction, *Advances in Water Resources*, 18, 2, 89-100, 1995

Kinzelbach, W., W. Schäfer, and J. Herzer, Numerical modeling of natural and enhanced denitrification in aquifers, *Water Resour. Res.*, 27(6), 1123-1135, 1991.

Kindred, J.C., and M. Celia, Contaminant transport and biodegradation 2. Conceptual model and test simulations, *Water Resour. Res.*, 25(6), 1149-1159, 1989.

Kirkner, D.J., T.L. Theis, A.A. Jennings, Multicomponent solute transport with sorption and soluble complexation, *Adv. Water Resour.*, 7, 120-125, 1984.

Kirkner, D.J., A.A. Jennings, T.L. Theis, Multicomponent mass transport with chemical interaction kinetics, *J. Hydrol.*, 76, 107-117, 1985

Kirkner D.J., H. Reeves, Multicomponent mass transport with homogeneous and heterogeneous chemical reactions: Effect of the chemistry on the choice of numerical algorithm, 1. Theory, *Water Resour. Res.*, 24, 1719-1729, 1988.

Kjeldsen, P., Th.H. Christensen, and O. Hjelmar, Selection of parameters for groundwater quality monitoring at waste incinerator residue disposal sites, *Environ. Tech. Letters*, 5, 333-344, 1984.

Lewis, F.M, C.I. Voss, and J. Rubin, Solute transport with equilibrium aqueous complexation and either sorption or ion exchange, *J. Hydrol.*, 90, 81-115, 1987.

Liu, C.W., T.N. Narasimhan, Redox-controlled multiple-species reactive chemical transport: 1. Model development, *Water Resour. Res.*, 25, 869-882, 1989a.

Mangold D.C, and C-F Tsang, A summary of subsurface hydrological and hydrochemical models, *Reviews of Geophysics*, 29(1), 51-79, 1991.

McNab, W.W. Jr., and T.N. Narasimhan, Modeling reactive transport of organic compounds in groundwater using a partial redox disequilibrium approach, *Water Resour. Res.*, 30(9), 2619-2635, 1994.

McNab, W.W. Jr., and T.N. Narasimhan, Reactive transport of petroleum hydrocarbon constituents in a shallow aquifer: Modeling geochemical interactions between organic and inorganic speices, *Water Resour. Res.*, 31(8), 2027-2033, 1995.

McQuarrie, K.T.B., E.A. Sudicky, and E.O. Frind, Simulation of biodegradable organic contaminants in groundwater, 1, Numerical formulation and model calibration, *Water Resour. Res.*, 26(2), 207-222, 1990.

Miller, C.W., L.V. Benson, Simulation of solute transport in a chemically reactive heterogeneous system: Model development and application, *Water Resour. Res.*, 19, 381-391, 1983.

Molz, F.J., M.A. Widdowson, and L.D. Benefield, Simulation of microbial growth dynamics coupled to nutrient and oxygen transport in porous media, *Water Resour. Res.*, 22(8), 1207-1216, 1986.

Morshed J., and J.J. Kaluarachchi, Critical assesment of the operator-splitting technique in solving the advection-dispersion-reaction equation: 1. Monod kinetics and coupled transport, *Advances in Water Resources*, 18, 2, 101-110, 1995

Narasimhan, T.N., A.F. White, and T. Tokunaga, Groundwater contamination from an inactive uranium mill tailings pile, 2, Application of a dynamic mixing model, *Water Resour. Res.*, 22, 1820-1834, 1986.

Noorishad, J., C.L. Carnahan, L.V. Benson, Development of the non-equilibrium reactive chemical transport code CHEMTRNS, *LBL-22361*, Lawrence Berkeley Laboratory, University of California, Berkeley, 1987.

Parkhurst, D.L., P. Engesgaard, and K.L. Kipp, Coupling the geochemical model PHREEQC with a 3D multi-component solute-transport model, abstract, V.M. Goldschmidt Conference, May 24-26, The Pennsylvania State University, 1995.

Rainwater, K.A, W.R. Wise, and R.J. Charbeneau, Parameter estimation through groundwater tracer tests, *Water Resour. Res.*, 23(10), 1901-1910, 1987.

Reardon, E.J., K_d's - Can they be used to describe reversible ion sorption reactions in contaminant migration?, *Ground water*, 19(3), 279-286.

Reeves, H., D.J. Kirkner, Multicomponent mass transport with homogeneous and heterogeneous chemical reactions: Effect of the chemistry on the choice of numerical algorithm, 2. Numerical results, *Water Resour. Res.*, 24, 1730-1739, 1988.

Rittmann, B.E., The significance of biofilms in porous media, *Water Resour. Res.*, 29(7), 2195-2202, 1993

Rubin, J., and R.V. James, Dispersion-affected transport of reacting solutes in saturated porous media: Galerkin method applied to equilibrium-controlled exchange in unidirectional steady water flow, *Water Resour. Res.*, 9, 1332-1356, 1973.

Rubin, J., Transport of reacting solutes in porous media: Relation between mathematical nature of problem formulation and chemical nature of reactions, *Water Resour. Res.*, 19, 1231-1252, 1983.

Rubin, J., Solute transport with multisegment, equilibrium-controlled reactions: A feedforward simulation method, *Water Resour. Res.*, 26, 2029-2055, 1990.

Sanford, W.E., L.F. Konikow, Simulation of calcite dissolution and porosity changes in saltwater mixing zones in coastal aquifers, *Water Resour. Res.*, 25(4), 655-667, 1989.

Šimůnek, J., and D.L. Suraez, Two-dimensional transport model for variably saturated porous media with major ion chemistry, *Water Resour. Res., 30*, 1115-1133, 1994

Stollenwerk, K.G., Modeling the effects of variable groundwater chemistry on adsorption of molybdate, *Water Resour. Res.*, 31, 347-357, 1995.

Sturman, P.J., P.S. Stewart, A.B. Cunningham, E.J. Bouwer, and J.H. Wolfram, Engineering scale-up of in situ bioremediation processes: a review, *J. of Contam. Hydrol.*, 19, 171-203, 1995

Sykes, J.F., S. Soyupak, and G.J. Farquhar, Modeling of leachate organic migration and attenuation in groundwaters below sanitary landfills, *Water Resour. Res.*, 18(1), 135-145, 1982.

Taylor, S.W., and P.R. Jaffé, Biofilm growth and the related changes in the physical properties of a porous medium 1. Experimental investigation, *Water Resour. Res.*, 26(9), 2153-2159, 1990a.

Taylor, S.W., and P.R. Jaffé, Biofilm growth and the related changes in the physical properties of a porous medium 2. Permeability, *Water Resour. Res.*, 26(9), 2161-2169, 1990b.

Taylor, S.W., and P.R. Jaffé, Biofilm growth and the related changes in the physical properties of a porous medium 3. Dispersivity and model verification, *Water Resour. Res.*, 26(9), 2171-2180, 1990c.

Taylor, S.W., and P.R. Jaffé, Substrate and biomass transport in a porous medium, *Water Resour. Res.*, 26(9), 2181-2194, 1990d.

Valocchi, A.J., R.L. Street, P.V. Roberts, Transport of ion exchanging solutes in groundwater: Chromatographic theory and field simulation, *Water Resour. Res.*, 17, 1517-1527, 1981.

Valocchi, A.J., Validity of the local equilibrium assumption for modeling sorbing solute transport through homogeneous soils, *Water Resour. Res.*, 21, 808-820, 1985.

Valocchi, A.J., Effect of radial flow on deviations from local equilibrium during sorbing solute transport through homogeneous soils, *Water Resour. Res.*, 22, 1693-1701, 1986.

Valocchi, A.J., Theoretical analysis of deviations from local equilibrium during sorbing solute transport through idealized stratified aquifers, *J. Contam. Hydrol.*, 2, 191-207, 1988.

Valocchi, A.J., Spatial moment analysis of the transport of kinetically adsorbing solutes through stratified aquifers, *Water Resour. Res.*, 25, 273-279, 1989.

Valocchi, A.J., and M. Malmstead, Accuracy of operator splitting for advection-dispersion-reaction problems, *Water Resour. Res.*, 28, 1471-1476, 1992.

Vandevivere, P., P. Baveye, D.S. de Lozada, and P. DeLeo, Microbial clogging of saturated soils and aquifer materials: Evaluation of mathematical models, *Water Resour. Res.*, 31(9), 2173-2180, 1995.

VKI, *BIODEGRATION, Technical Reference Manual*, Vandkvalitetsinstituttet, Hørsholm, 1995.

VKI, *GEOCHEMISTRY, Technical Reference Manual*, Vandkvalitetsinstituttet, Hørsholm, 1994.

Walsh, M.P, S.L. Bryant, R.S. Schechter, L.W. Lake, Precipitation and dissolution of solids attending flow through porous media, *Am. Inst. Chem. Eng. J.*, 30, 317-328, 1984.

Walter, A.L., E.O. Frind, D.W. Blowes, C.J. Ptacek, and J.W. Molson, Modelling of multicomponent reactive transport in groundwater 1. Model development and evaluation, *Water Resour. Res.*, 30(11), 3137-3148, 1994.

Widdowson, M.A., F.J. Molz, and L.D. Benefield, A numerical transport model for oxygen- and nitrate based respiration linked to substrate and nutrient availability in porous media, *Water Resour. Res.*, 24(9), 1553-1565, 1988.

Wolfsberg, A.V., and D.L. Freyberg, Efficient simulation of single species and multispecies transport in groundwater with local adaptative grid refinement, *Water Resour. Res.*, 30(11), 2979-2991, 1994.

Wood, B.D., C.N. Dawson, J.E. Szecsody, and G.P. Streile, Modeling contaminant transport and biodegradation in a layered porous media system, *Water Resour. Res.*, 30(6), 1833-1845, 1994.

Yeh, G.T., and V.S. Tripathi, A critical evaluation of recent developments in hydrogeochemical transport models of reactive multichemical components, *Water Resour. Res.*, 25, 93-108, 1989.

Yeh, G.T., and V.S. Tripathi, A model for simulating transport of reactive multispecies components: Model development and demonstration, *Water Resour. Res.*, 27(12), 3075-3094, 1991.

Zysset, A., F. Stauffer, and T. Dracos, Modeling of chemically reactive groundwater transport, *Water Resour. Res.*, 30(7), 2217-2228, 1994a.

Zysset, A., F. Stauffer, and T. Dracos, Modeling of reactive groundwater transport governed by biodegradation, *Water Resour. Res.*, 30(8), 2423-2434, 1994b.

CHAPTER 6
SOIL EROSION MODELLING

J.K. LØRUP[1,2] & M. STYCZEN[1]
[1] *Danish Hydraulic Institute, Hørsholm, Denmark*
[2] *Department of Hydrodynamics and Water Resources, Technical University of Denmark*

1. Introduction

Increasing rates of soil erosion in developing countries have been given attention for a long time. The increase in population pressure, inequality in societies, and sometimes also legislation have resulted in cultivation of areas unsuitable for crop production or in unsustainable farming which, together with overgrazing, are major reasons for soil erosion. Also erratic rainfall results in ecosystems prone to erosion, in particular in the semi-arid regions where the amount of rainfall impedes the establishment of good ground cover. Whitlow (1988) has estimated that average soil losses on croplands and grazing areas on Communal Lands in Zimbabwe are 50 and 75 t/ha/year, respectively, whereas the rates of soil formation are very slow, e.g. 400 kg/ha/year.

Moreover, soil erosion is increasingly being recognized as a hazard in European countries, in particular in the Mediterranean area and on the loamy, sandy loamy, and sandy soils of northern Europe. In Belgium and England measured erosion rates from bare ground were in the range 7-82 t/ha/year (Bollinne, 1978) and 10-45 t/ha/year (Morgan, 1985), which is far above the soil loss tolerance of 1 t/ha/year for northern Europe as suggested by Evans (1981).

Oldeman (1992) estimated that worldwide 24 percent of the inhabited land area are affected by man-induced soil degradation ranging from 12 percent i North America to 27 percent in Africa and 31 percent in Asia.

The harms of erosion are twofold. At the location where erosion takes place, infiltration rates, crop production and often the waterholding capacity as well are reduced through the removal of organic matter and plant nutrients. Furthermore, the transported material causes decreasing water quality, increasing eutrophication, and reduced life time of reservoirs due to siltation.

Much work has been put into development of soil erosion models over the last years to obtain a good tool for evaluation of soil erosion problems. Models are expected to assist in the following fields:
(a) Assessment of the extent of soil and nutrient losses and sediment transport in various environments.
(b) Land use planning as they can provide important information on the effects of changes in land use and of implementation of different soil conservation measures on soil losses and sediment yields.

(c) A better understanding of the erosion processes, the dynamic and relative importance of the single processes and their interactions. Thus, just as model development relies on the research on erosion processes, models are important tools to test new findings in soil erosion research.

While the first soil erosion models were empirically-based, much of the recent work is now concentrated on the development of more physically-based descriptions of processes and their interactions. Simultaneously, there is a trend towards greater interaction between researchers involved in experimental work, theory development and modelling. This is a good development, because model building is probably the strongest tool available for evaluation of the relative significance of different processes, for evaluation of the sensitivity of the system to different interventions, and for discovering new angles to the given problem.

The present chapter will be restricted to erosion caused by water before it reaches a river or stream. Furthermore, the text will be confined to modelling of rill and interrill (sheet) erosion, and therefore does not include soil loss due to gully erosion and landslides.

Section 2 gives a short discussion of the various types of soil erosion models; from the earlier empirically-based, and mathematically simpler versions to the distributed physically-based models based on the recent research. The data input requirements and advantages and limitations of the various types of models in relation to different types of applications are briefly discussed.

The rest of the chapter focuses on physically-based distributed soil erosion modelling. The model development is discussed in Section 3 emphasizing the soil erosion processes and variables to be included. The requirements to the associated hydrological models as well as the linkages between the erosion and hydrological models are briefly discussed, too.

The construction, calibration, and validation of soil erosion models are shortly discussed in Section 4, while Section 5 contains a case study on the application of EUROSEM/MIKE SHE, a distributed physically-based soil erosion model.

Finally, constraints in soil erosion modelling, the possible application of physically-based models on a catchment scale and future research needs are summarized in Section 6.

2. Classification of Soil Erosion Models

A number of models, of various complexity, has been developed in the past. Like hydrological models soil erosion modelling has moved from empirically-based and simple mathematical models, e.g. the Universal Soil Loss Equation, USLE (Wischmeier and Smith, 1965) towards physically-based and mathematically much more complicated models like the European Soil Erosion Model, EUROSEM (Morgan et al., 1995).

Basically, there are three categories of soil erosion models: 1) Empirical, 2) Conceptual (or partly empirical/mixed), and 3) Physically-based. In Table 1 some of the most used models within each of the three categories are listed.

TABLE 1. List and key characteristics of a number of the most used soil erosion models.

Model name(s)	Type of model	Scale of application	Temporal resolution	Spatial resolution	Separate rill/inter-rill components	Event-based/continuous	References
USLE and RUSLE	Empirical	Hillslope	Yearly soil loss	No	No	-	Wischmeier & Smith, 1978; Renard et al., 1994
SLEMSA	Empirical	Between ridges	Yearly soil loss	No	No	-	Elwell, 1978
ANSWERS	Conceptual	Catchment	Distributed	Distributed(2-D)	No	Event-based	Beasley et al., 1980
CREAMS	Conceptual	Field-scale	Total storm loss	No	Yes	Event-based	USDA, 1980
Calvin Rose	Physically-based	Plane element, e.g. uniform slopes	Distributed	Distributed(1-D)	No	Event-based	Rose et al., 1983a,b
SEM	Physically-based	Catchment	Distributed	Distributed(2-D)	Yes (for hillslope)	Event-based (continuous)	Nielsen and Styczen, 1986; DHI and IoG, 1992
WEPP	Physically-based	Hillslope version Catchment version	Distributed	Distributed(1-D) Distributed(2-D)	Yes	Continuous	Lane and Nearing, 1989
EUROSEM/ KINEROS	Physically-based	Individual fields and small sub-catchments	Distributed	Distributed(2-D)	Yes	Event-based	Morgan et al., 1995
EUROSEM/ MIKE SHE	Physically-based	Hillslopes and small catchments	Distributed	Distributed(2-D)	Yes	Continuous	DHI (1994)
SHESED-UK	Physically-based	Small sub-catchments	Distributed	Distributed(2-D)	No	Continuous	Wicks et al. (1992)

2.1. EMPIRICAL MODELS

Most of the *empirical models* are based on data from field observations, mostly standard runoff plots on uniform slopes, and are usually statistical in nature. The first empirical model to be developed was the Universal Soil Loss Equation, USLE, (Wischmeier and Smith, 1965, 1978), which also is the most well-known and most widely used empirical model. Although highly criticized - and for good reasons - it is still in use and has undergone a number of modifications. Thus a Revised USLE, RUSLE, to be run on a computer has been developed recently. The USLE predicts the annual soil loss from small areas on a slope, and the RUSLE maintains the basic structure of the USLE, namely

$$A = R\,K\,L\,S\,C\,P \qquad (1)$$

where A is the computed annual soil loss, R is the rainfall-runoff erosivity factor, K is the soil erodibility factor, L is the slope length factor, S is the slope steepness factor, C is the cover-management factor, and P is the supporting practices factor (Renard et al., 1994).

A similar model, the Soil Loss Estimation Model for Southern Africa, SLEMSA, primarily based of field plot erosion studies in Zimbabwe, has been developed by Elwell (1977).

The main limitation of USLE/RUSLE and empirical models in general, is their limited applicability outside the range of conditions for which they have been developed. Adaption of e.g. the USLE to a new environment requires a major investment of resources and time to develop the database required to drive the model (Nearing et al., 1994).

Although the empirical models may give reasonable estimates of annual soil loss from a field, they are not adequate for catchment scale estimations. For example, they do not take into account deposition at the lower parts of a hillslope, which is relevant in relation to transport of sediment and pollutants to rivers and reservoirs. The models estimate the annual soil loss and can not therefore be used to study the temporal dynamics of erosion. The empirical models only provide a limited insight in the relative importance of the various variables and their sensitivity in different environments. USLE suffers from the conceptual defect that rainfall and soil factors (among others) cannot simply be multiplied because of the subtractive effect of soil infiltration capacity in generating erosive runoff from a given rainfall (Kirkby, 1980).

A modified version of the USLE, the MUSLE (Williams, 1975) has tried to overcome a number of the above-mentioned problems by introducing an empirical runoff energy factor instead of the rainfall factor, and the model is able to estimate sediment yield from single storms.

2.2. CONCEPTUAL MODELS

Realisation of the insufficiency in the application of the USLE led to the development of a number of *conceptual models* in the 1970s, including CREAMS (USDA, 1980),

ANSWERS (Beasley et al., 1980) and the modified ANSWERS, MODANSW (Park et al., 1982).

These models lie somewhere between the empirically and physically-based models. The main step forward by the development of these models was the introduction of the laws of conservation of mass and energy, i.e. the continuity equation and the grouping of the area of concern into a number of elements/grids in order to describe the spatial variations in erosion and deposition. The detachment and transport of sediment from each element/grid follow the model proposed by Meyer and Wischmeier (1969). The outflux of sediment from a given grid/element is determined by the influx of sediment plus the net detachment of sediment by runoff and rainfall within the element/grid with the maximum limit that the outflux never exceeds the total transport capacity. Thus, regarding these basic concepts these models resemble the physically-based distributed models.

On the other hand, for a number of processes described in the models USLE factors are used, and their physical validity is in some cases questionable. Both the C and K factors developed for USLE refer to total soil loss, and one cannot expect these factors to represent the individual processes in a single storm as is done in ANSWERS and MODANSW. Likewise, some of the processes in CREAMS are questionable. For example, the model calculates detachment on a storm basis whereas sediment transport is calculated on an instantaneous rate basis.

Thus, the main limitations of the conceptual models lie in the poor physical description of the processes which, among other things, results in distortion of parameter values determined by calibration (Elliot et al., 1994).

2.3. PHYSICALLY-BASED MODELS

During the last 10-15 years most of the work on soil erosion modelling has been concentrated on the development of *physically-based erosion models*. The physically-based models are intended to represent the essential mechanisms controlling erosion. The models include most of the factors affecting erosion and their spatial and temporal variability, and the subprocesses and their complex interactions are described as well.

Three physically-based models will be mentioned here: 1) WEPP (USDA Water Erosion Prediction Project) which mainly is based on research results from the USA (Lane and Nearing, 1989; Nearing et al., 1989), 2) EUROSEM (European Soil Erosion Model) which mainly is based on recent soil erosion research in Europe (Morgan, et al., 1995), and 3) the soil erosion model developed by a group of Australian scientists (Rose et al., 1983a,b).

The basic erosion concepts of most of the physically-based models are rather similar, whereas the way they are linked to a hydrological model and the use of equations to model the individual processes vary. All of the above-mentioned models have separate interrill and rill components. Rill erosion is described as a function of the flow's ability to detach sediment, of the sediment transport capacity, and of the existing sediment load in the flow. The models use different transport capacity equations as well as different thresholds for rill initiation. Also, the detachment and transport in interrill areas are modelled in a different way.

It should be stressed that the most important basis for the physically-based soil erosion models is an adequately distributed simulation of the driving variable in the soil erosion and transport processes - namely the overland flow. This is in particular important when it comes to a more precise description of rill initiation and development. Thus, the linkage of the erosion model to a distributed hydrological model with a comprehensive overland flow component is a prerequisite.

The EUROSEM model is the most recent as well as the most comprehensive physically-based distributed model. It includes effects of plant cover on interception and rainfall energy; rock fragment (stoniness) effects on infiltration, flow velocity and splash erosion, and the changes in the shape and size of rill channels as a result of erosion and deposition (Morgan et al., 1995).

The physically-based distributed soil erosion models have a number of advantages as compared to the empirical and conceptual models. Due to a physically-based description of the processes research scientists can use the models to test new theories, and sensitivity analysis can help identifying which factors or erosion processes are the most important to the overall erosion process and therefore should be given more attention in research. Due to the calculation of the spatial as well as the temporal variations of sediment concentration this type of models provides e.g. a much better possibility for identifying areas with high erosion risk and to extrapolate soil erosion rates from plot to catchment scale. Thus, such models are *potentially* much more useful as planning tools.

However, presently the large requirement to input data and computer power as well as the complexity of the models, restrict a wider application of the models.

3. Soil Erosion Processes in Physically-Based Models

3.1. MODELLING THE VARIOUS EROSION PROCESSES

A comprehensive outline of the state-of-knowledge of soil erosion processes used in modelling is outside the scope of this chapter. However, as physically-based soil erosion modelling generally is still at an early stage of development as compared to physically-based hydrological modelling and many of the processes are not yet well understood, the most crucial soil erosion processes, and in particular processes which need more attention, are discussed below.

Key soil erosion processes and their interactions as included in the European Soil Erosion Model, EUROSEM (Morgan et al., 1995) are shown in Fig. 1.

3.1.1. Soil Detachment By Raindrop Impact (Splash Erosion)
The two major variables in modelling the detachment by raindrop impact are the ability of the raindrop to detach the soil (rainfall erosivity) and the ability of the soil to resist the raindrop impact (soil erodibility). At least two other variables have to be considered, namely the vegetation canopy, which influences the raindrop diameter and velocity and the flow depth that may reduce/(dissipate) the erosivity of the rainfall.

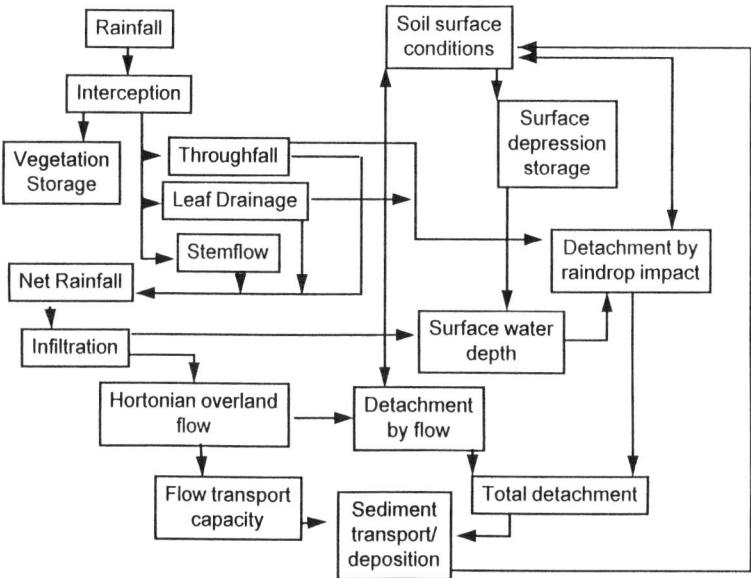

Figure 1. Flow chart of the European Soil Erosion Model, EUROSEM (Morgan et al., 1995).

It is generally accepted that rainfall is the main detaching agent in interrill areas (Foster et al.; 1982, Gilley et al., 1985), mainly because most rains strike the surface at velocities between 5 to 9 m/s, while runoff velocities usually are less than 1 m/s in sheet flow (Meyer, 1981). Most splash erosion models have been correlated to the energy, intensity or momentum of the rain. Morgan et al. (1995) prefers the expression:

$$DET = k\, KE\, e^{bh} \qquad (2)$$

where DET is the soil detachment by raindrop impact (g m^{-2}), k is an index of the detachability of the soil (g J^{-1}), KE is the total kinetic energy of the rainfall at the ground surface (J m^{-2}), h is the depth of the surface water layer and b is an exponent varying between 0.9 and 3.1 (Torri et al., 1987b). As the soil parameters are poorly known, they are often used as a calibration factor.

If the soil is covered by vegetation, the estimation of splash erosion becomes more complicated. Different correction factors have been used, among these the C-factor from the USLE (Beasley et al., 1980), or the C_I, C_{II} and C_{III} factors, defined by Wischmeier (1975). However, these correction factors are always smaller than or equal to 1 and therefore fail to describe situations found by e.g. Mosley (1982), Morgan (1982) and Morgan et al. (1985) where more splash is measured under vegetation than on bare soil.

The theoretically based equation for splash erosion developed by Styczen and Høgh-Schmidt (1988) for cohesive soils can include effects of a canopy, as it considers splash erosion to be proportional to the sum of the squared momentum of each drop hitting the ground:

$$\text{Splash} = \frac{A \, Pr}{2 \, \hat{e}} \sum_{i=0}^{D} N_D \, P_D \qquad (3)$$

where Splash is the amount of splash erosion (kg m^{-2} s^{-1}), ê is the energy needed to break the bonds between two microaggregates (J), Pr is the probability of energy excess (energy left for lifting the particles), A is a soil parameter which is a function of the above probability function, N_D is the number of drops (m^{-2} s^{-1}) with the diameter D, and P_D is the squared momentum of a drop with the diameter D (m). Basically the factors before the summation sign are soil parameters and after the summation sign rainfall parameter (compare with eq. 2). In case of vegetation, the summation will include separate expressions for the throughfall and the leaf drip. For non-cohesive soils (ê = 0), the rainfall parameter becomes a kinetic energy expression instead of a squared momentum. Studies by Styczen and Høgh-Schmidt (1988) indicated that the use of squared momentum is in better agreement with experimental results with and without vegetation than when models based on energy or intensity are used.

The soil erodibility term influencing splash erosion has to be specifically related to forces that bind the soil mass together as in eq. 3. The soil shear strength, τ_s, is probably the best physical measure to represent these forces - e.g. Al-Durrah and Bradford (1981, 1982a,b) and Torri et al. (1987b) found good correlation between shear strength and amount of detached soil. Good correlation has also been found between splash erosion and aggregate stability (Bryan, 1976). Due to the lack of data the soil shear strength has also been related to the soil texture (e.g. Morgan et al., 1991). However, this does not allow for treating the erodibility as a temporal variable, and it does not include the importance of the organic matter content. On the other hand, much more research is needed before the variability of soil shear strength can be properly modelled (see also section 3.1.4).

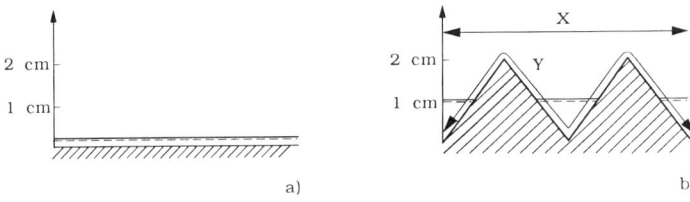

Figure 2a,b. Two soil surfaces having the same average flow depth but the effects of the water depth on the splash erosion are likely to be different.

The splash erosion decreases when the runoff depth increases. Most studies have shown that the splash erosion decreases exponentially with increasing depth, as shown in eq. 2.

However, eq. 2 applies to smooth surfaces, whereas on most surfaces in the field the water seldom is equally distributed. This is illustrated in Fig. 2a,b where two soils are having the same average water depth but are likely to experience different rates of detachment. Also here, a better description of the microtopography would help to estimate the effect of flow depths on splash erosion

3.1.2. Infiltration Conditions and Generation of Overland Flow
As overland flow is the major transporting agent and in some cases also the main detaching agent, a proper description in space and time of the generation and routing of overland flow is crucial.

Regarding the generation of overland flow the *infiltration rate* is the most sensitive variable. This is in particular true in cases where the rainfall intensity and the infiltration rate are of the same order of magnitude as this may give rise to simulation rates which -relatively - deviate considerably. To simulate the infiltration properly the model must include a sub-model for the unsaturated zone, and detailed input data on saturated hydraulic conductivity and soil moisture retention curves are therefore needed. The infiltration rate may be influenced by several other factors, such as frost/thaw, presence of stones and crusting.

Particularly in tropical environments, Hortonian type of overland flow may be an important generator of overland flow. In other environments, and in particular in the temperate regions, the major source of overland flow is saturated overland flow. In humid vegetated areas soil moisture levels tend to build up downslope, especially close to streams, and near-saturated areas generate a disproportionate amount of overland flow runoff (Kirkby, 1980). Thus, information on the spatial and temporal variation of depth to groundwater tables (primary or secondary) and description of existing semi-impermeable or impermeable layers are valuable in order to simulate the generation of saturated overland flow.

In the colder temperate regions erosion, and in particular rill erosion, is often caused by rainfall or snow melt on partly frozen soils. This is a process that is very difficult to model in details, e.g. because the depth of the snow can vary considerably due to local differences in wind and shelter conditions. Apart from snow melt, the issues to be modelled are the depth of thawed surface soil and estimation of shear strength which changes through the frost-thaw cycles.

Rock fragments within the soil will reduce the effective porosity and thereby lead to faster saturation of the soil. Moreover, rock fragments on the soil surface affect the infiltration rates. When modelling on catchment scale, rock outcrops and areas with impermeable hardpan (e.g. around settlements) have to be considered as such areas may result in increased localized erosion.

Areas susceptible to crusting pose special problems, as the crusting/sealing can cause quick changes of the hydraulic conductivity. From the experiments by Bryan and Poesen (1989) it seems important to operate with separate infiltration rates for interrill areas and within the rills, particularly in areas subject to crusting. Rocks that are embedded in a surface seal reduce infiltration while rocks on the surface protect the soil structure and promote infiltration (Poesen and Ingelmo-Sanchez, 1992; Poesen et al., 1994). An improved description of soil erosion on crusting soils must include a feed-back system between the soil erosion model and the hydrological model - e.g. the effect of raindrop detachment on crust development and thereby the hydraulic conductivity which in turn affect the amount of overland flow.

The surface storage capacity influences both infiltration and the time till runoff occurs. It is, among other things, determined by soil type, slope steepness, type and in particular orientation of tillage. Depression storage may be negligible on smooth seedbed tilled up and down the slope, whereas a ridged potato field with the ridges following the contours will have a considerable surface storage depth. Based on laboratory data Morgan et al. (1995) related the surface storage depth, D (mm), to the ratio of the straight line distance between two points on the ground (X) to the actual distance measured over all the micro-topographic irregularities (Y), (see figure 2b):

$$D = \exp(-6.6 + 27(\frac{Y - X}{Y})) \tag{4}$$

This equation does not consider the depths of the depressions and the slope on which they are measured. Although guide values for depression storage may be obtained for various combinations of soil type, slope steepness, tillage methods, and orientation of tillage, the issue of surface storage depression is complicated, e.g. by the fact that roughness element break-down takes place over time.

3.1.3. Soil Surface Conditions and Runoff Processes

Surface flows are influenced by irregularities of micro-topography, caused by management practice, vegetation, and soil clods/aggregates. This results in an uneven distribution of the flow over the surface and influences surface roughness. The process is further complicated because the micro-topography varies considerably during the year, due to management practice, vegetation, etc.

The basic equations for describing surface runoff, Q, as well as sediment discharge, q_s are the conservation of mass equations for flow and sediment, respectively. In the EUROSEM (Morgan et al., 1995) the computation of runoff and sediment is based on a numerical solution of the dynamic mass balance equation (Bennett, 1974; Kirkby, 1980; Woolhiser et al., 1990):

$$\frac{\partial A}{\partial t} + \frac{\partial Q}{\partial x} = r(t) - i(t) \tag{5}$$

$$\frac{\partial AC}{\partial t} + \frac{\partial QC}{\partial x} - e(x,t) = q_s(x,t) \tag{6}$$

where A is the cross-sectional area of the flow (m^2), Q is the discharge (m^3 s^{-1}), r(t) is the rainfall less the interception for interrill flow and the unit discharge into the rills (from interrill areas) in rill flow (m^2 s^{-1}), i(t) is the local infiltration rate (m^2 s^{-1}), x is the horizontal distance (m), t is the time (s), C is the sediment concentration (m^3 m^{-3}), e is the net detachment rate per unit length of the flow (m^3 s^{-1} m^{-1}), and q_s is the external input of sediment per unit length of flow (m³ s⁻¹ m⁻¹). For interrill flow q_s becomes zero.

By using the kinematic wave assumption (and that Q = A v) equations (5) and (6) can be solved by using the Manning equation for flow velocity. Although comprehensive guideline values for Manning's roughness coefficient exist (e.g. Engman, 1986), the use of equations (5) and (6) is associated with many difficulties. Due to the variety of surface conditions in the field one may question how representative these guideline values are. Most data available are still based on laboratory experiments, and Emmett (1978) found that field data indicated a ten-fold increase in resistance on the natural field plots compared to the laboratory surfaces. The large temporal variation in surface roughness over the year as a result of tillage, soil consolidation and rainfall further complicates the situation. Savat (1980) found that the Manning equation underestimates the friction coefficient for thin sheet flow.

Most existing attempts to describe surface roughness, e.g. the MIF-index (Römkens and Wang, 1986), only provide qualitative results. No procedure has yet been discovered allowing field measurements of surface roughness to predict accurately hydraulic roughness coefficients independently of flow measurements. The need for such linkage between physical surface roughness and hydraulic roughness is obvious in soil erosion modelling. This necessarily also has to include a method to estimate the temporal changes in soil surface conditions.

Even with proper hydraulic formulas for thin overland flow, the uneven distribution of the water over the soil surface complicates the description. The uneven distribution has a major impact on the transporting and detaching capacity of the flow and thereby the initiation, development, and spatial location of rills, and the distribution between rill and interrill areas.

The two most important factors regarding the flow distribution is the topography and the tillage orientation. On an experimental plot tilled up-and-down, situated on a uniform slope, the flow direction and distribution are relatively easy to predict, but when it comes to small catchments sloping in more than one direction and tilled along the contours or at an angle to the slope, the routing soon becomes complicated. In such case it requires both detailed input data on micro-topography and macro-topography and a model able to route the water in all directions and not only down the prevailing slope direction.

In cases where the area is tilled up and down the slope, a hypsometric curve (Styczen and Nielsen, 1989), showing the frequency distribution of heights over a unit length perpendicular to the flow direction, can be used to quantify the distribution of flow depths in the rills/depressions.

Concentration of water into rills will greatly increase the detachment as well as the transport capacity of the flow and therefore a correct routing of water as rill flow and interrill flow is important. A good correspondence between observed and simulated hydrographs does not automatically mean that the description of overland flow is correct as the routing between rills and interrill areas may not be correct. To obtain this a hydrological model with a detailed description of the overland flow routing is needed (see Section 3.2).

3.1.4. Soil Detachment and Transport by Overland Flow

It is generally agreed that without occurrence of surface runoff, erosion rates are small. Erosion becomes really severe when the overland flow gains enough power to detach the soil, and rill and/or gully erosion occurs.

The detachment and transport of soil by flow are two issues still under discussion. And although interrelated it is necessary to distinguish between the two. The concept of transport capacity is originally developed for non-cohesive materials, meaning that the shear strength of the material is rather low. Most soils, however, are cohesive materials, with a somewhat higher shear strength. In addition, most soils consist, not of single particles, but of more or less water-stable aggregates, which, however, to some extent may break down under influence of water and physical action.

Looking solely at flow detachment, the concentration in the flowing water is determined as a balance between detachment (which is influenced by the mean flow velocity, the shear stress of the water and the shear strength of the soil, as well as the surface contact area), and deposition. As the final concentration is influenced by the shear strength of the soil, it may be different from what it would have been, had the material been noncohesive. However, in addition to the material detached by flow there may be an addition of material through splash, and the two types of material confuse the discussion.

Meyer and Wischmeier (1969) proposed that the detachment capacity at each point should be compared with the transporting capacity and the actual transport rate at that point. The actual net detachment rate is then taken as the lesser:

$$\frac{\partial C}{\partial x} = D_c \text{ if } C < TC, \quad C = TC \text{ if } C \geq TC \tag{7}$$

TC is the transporting capacity, C is the actual sediment load, and D_c is the detachment capacity. In the alternative approach by Foster and Meyer (1972) the net detachment rate is related to the deficit between the actual sediment load and the transport capacity load:

$$\frac{\partial C}{\partial x} = \frac{TC - C}{h} \tag{8}$$

TC, C and D_c are defined as above, and h may depend on other variables. As the net detachment rate reflects a balance between detachment and deposition processes, the basic approach of equation (8) is used in most physically-based erosion models (e.g. Lane and Nearing, 1989; Morgan et al., 1995). However, without a proper description of the detachment and deposition processes their physical background is still not properly understood.

The most convincing physical description of sediment concentration being a balance between detachment and deposition is made by Torri and Borselli (1991) who base their description on the following assumptions:
(a) hydraulic roughness decreases with increasing sediment load;

(b) overland flow detachment is proportional to the part of boundary shear stress due to the water fraction of total (water+sediment) fluid discharge;
(c) sediment deposition follows Stokes law of motion (and final velocity of the particles is not reached in the thin flows in question).

The algorithm proved to be a good approximation of the physics behind the empirically derived transport equation by Govers (1990):

$$TC = c(\omega - \omega_{cr})^\eta \tag{9}$$

where S is the slope, ω is the stream power (u S), u is the mean flow velocity (cm s^{-1}), S is the slope, ω_{cr} is the critical value of unit stream power (= 0.4 cm s^{-1}) and c and η are experimentally derived coefficients depending on particle size. The algorithm derived by Torri and Borselli (1991) is also able to reproduce the empirical equations for incipient rilling which link shear velocity with soil shear strength (e.g. Rauws and Govers, 1988). The theoretical considerations by Torri & Borselli (1991) aiming at combining physical descriptions and empirical results seem to form a good basis for improving the understanding of the interactions between flow detachment and transport processes of thin overland flow.

Until the recent development of transport capacity equations for thin overland flow a variety of "classical" transport capacity equations for streamflow have been used including the Engelund-Hansen (Engelund and Hansen, 1967) and the Yalin (Yalin, 1963) equations.

Following the logic of the balance approach, material added through splash will increase the concentration in the flow, but the material will deposit following the same rules as the flow-detached material. Models using the approach of eq. 7 have a tendency to deposit material in lumps rather than gradually because the sediment concentration is supposed not to exceed the transport capacity determined for non-cohesive materials. However, the transport capacity is determined from flow detachment alone and does not take into account that material may be added through other processes, without energy expenditure of the flow.

Settling velocities should be calculated separately for groups of particles or aggregate sizes. The use of one d_{50}-value may cause instabilities during modelling as decreasing transport capacity may result in a very abrupt increase in sedimentation rates. Some of the questionable issues may still be calculation of settling velocities using Stoke's law in thin overland flow with turbulence caused by irregularities, and how to treat flows and sediment transport for aggregates that may have a mean diameter larger than the flow depth. Torri and Borselli (1991) calculate effective shear stress as a function of the hydraulic radius minus the sediment diameter, and thus the equations only apply for water depths larger than the sediment diameter.

According to Torri and Borselli (1991), the soil particles are kept in place by the cohesive forces and the submersed weight of the particles. In some equations this is described by the detachment rate being reduced by a coefficient ß, which decreases with increasing shear strength, τ_s.

However, the soil shear strength varies through the year as a result of changes in moisture content, tillage operations, impact of plant roots, age hardening, etc. In addition it changes very fast when the soil is wetted. Table 2 clearly illustrate the dynamic nature of the shear strength due to wetting and drying. Effects of tillage, roots etc. will further complicate the picture. The relative changes in soil shear strength during a rainstorm will among other things be influenced by initial soil moisture content and soil type. Sandy soils will show a larger change upon wetting as a relatively large part of their cohesiveness prior to the rain is due to suction (Andersen & Lørup, 1991) (Table 2). Govers et al. (1990) found that shear strength measured on a saturated soil correlates best with erodibility.

The lack of a physical description of the shear strength and its variation is one of the major constraints in soil erosion modelling presently. By using a continuous model it may eventually be possible to simulate the variation of the shear strength by relating it to the changes in soil moisture, root development, time after tillage, etc., but this will require a better understanding of the dynamic nature of the shear strength. In the longer term, model developers will have to rely on continuous models to predict some of these crucial changes of parameter values. Simulations carried out by Wicks et al. (1992) does indicate that is the way forward.

TABLE 2. Measured values of shear strength for 2 different soils with different content of soil organic matter (SOM) as a function of time since overland flow ceased (Andersen & Lørup, 1991).

Soil type	Shear strength (kPa) after 15 minutes with overland flow	Shear strength (kPa) 16 hours after overland flow has ceased	Shear strength (kPa) 7 days after overland flow has ceased
Loamy sand (1.34% SOM)	0.20	1.55	2.49
Sandy loam (2.34% SOM)	1.10	1.95	2.82
Sandy loam (2.34% SOM) sieved (< 2 mm)	0.50	1.52	2.10

Modelling of transport capacity on cohesive soils is still subject to uncertainty as most work has been carried out using non-cohesive material and the conclusions may not hold for cohesive soils. Here the major part of the soil mass is aggregates having dry (and wet) densities below 2 g cm^{-3} and mean diameters many times higher than d_{50} of the soil mass. A decrease in wet bulk density from 2.65 to 2 g cm^{-3} implies that the Shield parameter, θ, will increase 60%.

3.1.5. Description of Rill Initiation and Development
In principle, flow detachment and sediment transport in rills follow the same rules as discussed in subsection 3.1.4. However, due to the concentration of flow, the final erosion rates are generally much greater.

The processes which need special concern are rill initiation, headcut retreat, wall collapse, and rill tail development.

Rills develop when the shear stress of the flow is large enough to remove "all" sizes of soil particles at particular spots along the slope. Rauws and Govers (1988) used the empirical equation:

$$U_{gcr} = 0.89 + 0.56\,C \tag{10}$$

where U_{gcr} is the critical grain shear velocity of the flow (cm/s) and C is apparent cohesion (kPa). Other approaches (e.g. Torri et al., 1987a) suggest τ_0/τ_s to exceed a certain constant value, τ_0 being the flow shear stress and τ_s being the soil shear strength of the soil top layer. In cases where vegetation is present, it may be better to use $\tau_0 * v/\tau_s$ as an indicator, because a larger roughness gives rise to larger water depth, but only a certain part of the resulting stress actually acts on the surface particles.

Headcut retreat has been suggested described as a function of the potential energy released during the drop of the water into the headcut (Styczen and Nielsen, 1989) or being proportional to the total mass of the water at a given time multiplied by the mean flow velocity before the drop and inversely proportional to the stability of the headcut walls (De Ploey, 1989).

A certain part of the shear stress of the flow in rills will act on the walls of the rill. This will cause erosion. If undermined, the upper part of the rill walls is likely to collapse and provide easily detachable material to the flow. A laboratory study on rill initiation and development (Andersen & Lørup, 1991) showed that on sandy soil where the inherent cohesion is low the rill walls easily collapsed and the rills tended to be shallow and wide, whereas rills on clayey soils were deeper but relatively narrow.

The position of the rill tail is determined by the slope of the rill bottom compared to the slope of the hill (Styczen & Nielsen, 1989). As long as the bottom of the rill is flatter than the hill slope, the rill will continue to develop downslope because the transport capacity of the water is larger outside the rill than inside, and this causes continuous erosion. When the two angles/slopes become equal, downward development ceases.

From a rill model based on the principles described above (DHI and IoG, 1992), the following observations were made. The end point of the rill was easily defined, and when reached, it was independent of simulation time. The width of the rill was not independent of time, but the development slowed down as the rill grew wider, because the depth of flow decreased, and exerted less stress on the walls. The rill depth was dependent on the speed of headcut retreat (and wall collapse), because this determined the rate of net detachment from the bottom. The final shape of the simulated rill bore close resemblance to the rill from which input data were generated.

Presently, most models use predefined shallow rills/depressions rather than allow them to be initiated during the simulation. Particularly on agricultural fields this may be defendable, as the flow pattern to a large degree is determined by the tillage operation, and both spacing and direction can be described. However, the critical issue is whether the hydrological model manages to describe the flow pattern and the confluence of water at certain points in the field, triggering incision. It may be attempted to describe rill initiation through the use of a hypsometric curve or a statistical evaluation of probability of rill occurrence, based on the distribution of shear stress and shear strength or similar parameters.

3.2. COUPLING WITH HYDROLOGICAL MODELS

During the early days of developing soil erosion prediction models these models were developed parallel to and independent of hydrological models, and there was little collaboration between agronomist/agricultural engineers and hydrologists. This was mainly due to the fact that the first models were purely empirical and did not need input from hydrological models. Moreover - in particular among agronomists/agricultural engineers - soil erosion was originally considered mainly as a problem in relation to agricultural production. However, in the 1970s the development of conceptual models and the attention to erosion as a potential cause of pollution of water bodies raised the need for hydrological models to be coupled to erosion models and for a closer collaboration between various disciplines of science. With the present development towards physically-based erosion models collaboration and appropriate hydrological models have become even more crucial.

One of the main aims of a physically-based soil erosion model is to describe the various processes as they appear in the natural system. This necessarily requires that such a model is coupled to a hydrological model that fulfil the same aims, i.e. a model which can provide a detailed description of the spatial and temporal changes in the flow of water. A good agreement between simulated and observed amounts of soil loss/ sediment yield does not indicate model-predictive credibility without a similar good agreement between simulated and observed discharge hydrographs.

Due to the need for a detailed description of the overland flow and the need to simulate the geomorphology of the developed rill the model must be able to work with small grids/elements, preferably down to a few metres. The use of small grids/elements requires the model to run with very small time steps.

Furthermore the hydrological model should be a continuous model with a so-called "hot start" facility. This implies that the model can be used as an event model on the basis of initial conditions retrieved from previous model runs. Prior to the "hot starts" the hydrological model may be run for the whole period of concern. Hereby it is possible to identify the major rainfall events where significant overland flow has been generated and soil erosion is likely to have taken place. Furthermore the result file from this run will provide important input variables, such as initial moisture contents, for running the coupled hydrological and soil erosion models for specific storm events - data which otherwise would have to be collected in the field prior to the rainfall events.

As soil erosion modelling calls for a more detailed description of overland flow than hydrological modelling normally does, the use of existing hydrological models for soil erosion modelling purposes may therefore require a revision of the overland flow component.

The separate modelling of rill and interrill erosion requires separate routing of interrill and rill (concentrated) flow including routing of water from interrill areas to the rills. The hydrological model must include a two-dimensional surface description where tillage orientation may be across the slope or at an angle to the slope and not just downslope. Otherwise this will restrict the application of the model seriously.

The distinction between interrill and rill flow and the way they interact are certainly some of the major challenges in the development of hydrological models to be used in soil erosion modelling.

4. Construction, Calibration and Validation of Soil Erosion Models

4.1. GENERAL CONSIDERATIONS

Soil erosion modelling is a complicated issue, which requires a solid understanding of the different hydrological and soil erosion processes in general and specific data for the particular study area. As a guideline for the various steps to consider and a consistent terminology to use the modelling protocol outlined by Refsgaard (Chapter 2) may be applied.

The first step to consider is the definition of the purpose of the model application. For physically-based erosion models two groups of users can - broadly speaking - be identified (Quinton, 1994): 1) Field personnel and policy makers, and 2) Researchers. The different types of purposes of a model application may be described as follows:

(a) To test alternative theoretical process descriptions and in this way improve the physical understanding (researchers).
(b) To test the range of model validity in terms of scale of application (erosion plot, hillslope, catchment), conditions on soil and geology, land use and hydroclimatological regime. This is very important in order to avoid misuse by extrapolating the use of the model beyond its proven range of applicability (both researchers and field personnel/policy makers).
(c) To predict soil losses/sediment yields and the effects of possible management options regarding soil and water conservation measures at particular localities (field personnel/policy makers).

4.2. ESTABLISHMENT OF A CONCEPTUAL MODEL AND SELECTION OF MODEL CODE

The next step is to establish a conceptual model - the modelling framework. This includes identification of the key processes required to be included in the model. The processes to be included and the degree of details to which they need to be modelled depend on assessments of the relative importance of the various processes in the particular study area, the acceptable accuracy limits, the data availability and the specific purpose of the study. At this stage it is important through readily available information to test ('model qualification') whether the defined conceptual model in qualitative terms appears to be a good representation of the physical system.

On the basis of the conceptual model an appropriate model code must be selected. In this connection it is important to select a code that can describe the processes included in the conceptual model. For instance, a model code that is only able to describe Hortonian type of overland flow should be avoided, if the study areas is characterized by significant saturated overland flow; or, similarly, if a model code is unable to describe rill erosion, it should not be applied to an area, where rill erosion is the prevailing type of erosion. Usually, an appropriate code will be selected among existing codes. However, in connection with research projects development of new or modified codes may be important. In such cases the new codes need to be verified, i.e. tested for their ability to provide mathematically accurate approximations to the given process equations.

4.3. MODEL CONSTRUCTION

Model construction involves data collection and processing into appropriate model formats, assessment of parameter values, definitions of model boundaries and internal discretizations. One of the major difficulties in the application of distributed physically-based soil erosion models is the large data requirements. As the ideal requirements of measured field data for all parameters for all model grid points never will be available, the main part of the parameter assessment will have to be done from secondary information, such as mapped soil and vegetation types. The spatial and temporal variability of model input is one of the major challenges to physically-based distributed erosion models. Thus the *parameterization*, i.e. the proces of defining structures of parameter variations, is a crucial part of the model construction. The spatial variation in parameter values should as far as possible reflect the variation found in the field and the availability of data. As vegetation characteristics and tillage operations are major factors in relation to soil erosion, the spatial variation may depend on the number of fields and the topography, unless there are major changes in soil type within the fields. Similarly, information on temporal variation of certain model parameters, such as annual variation of vegetation parameter values, could be incorporated in the parameterization. For continuous models the temporal variability of certain parameters may partly be measured and partly included in the model simulations, as discussed in section 3.1.4. This may reduce the data input requirement considerably.

4.4. MODEL CALIBRATION AND VALIDATION

The first stage in the calibration is calibration of the hydrological model, which in most cases implies calibration of simulated and observed hydrographs. This is followed by calibration of simulated and observed sediment rates. In cases where the insight into the functioning of the erosion component is the main aim, this should preferably be done when good agreement between hydrographs has been obtained - otherwise the calibration of simulated and observed sediment rates may deteriorate the erosion component rather than improve it.

Even if a reasonably good agreement is established between calibrated simulations and observed data, and the parameter values seem physically reasonable, it is not until the model has been validated for a number of rainfall events that a picture emerges of the model's ability to simulate the physical environment. During a validation, the model is used to simulate other events than those used during the calibration phase. Only if the results of the blind simulation and the observed data resemble each other adequately (according to pre-described performance criteria), the model can be considered validated for that particular situation.

For two-dimensional models as the EUROSEM/MIKE SHE that are able to simulate the spatial distribution of the amount of erosion/sedimentation, rill depths, depths of overland and rill flow, etc., qualitative or semi-quantitative comparison with field observations of location and depth of rills, the amount of sedimentation at specific sites, etc. will be valuable to evaluate the performance of the model.

As resuspension, bed and bank erosion can account for a significant part of the sediment leaving small catchments (Hasholt and Styczen, 1993) this should be considered when calibrating and validating on catchment level.

5. Case study: Application of the EUROSEM/MIKE SHE Soil Erosion Model

5.1. THE MODEL, STUDY AREA AND MODEL CONSTRUCTION

EUROSEM/MIKE SHE is the linkage of the codes of the EUROSEM soil erosion model (Morgan et al., 1991) and the MIKE SHE (Refsgaard and Storm, 1995) distributed physically-based hydrological model. Linking the two models required a new overland flow component to MIKE SHE to make separate routing of rill and interrill flow possible.

The EUROSEM/MIKE SHE provides a comprehensive and detailed description of the hydrological processes including generation of saturated overland flow. In contrast to the EUROSEM/KINEROS code the MIKE SHE version gives a continuous simulation of e.g. soil moisture content, and the rills may run at an angle to the slope and not necessarily perpendicular to the contour lines.

The EUROSEM/MIKE SHE model was tested on a 25 x 35 m^2 soil erosion plot at Woburn Experimental Farm, UK (DHI, 1994). The site has been subject to erosion since at least 1950 (Catt, 1992). The plot is non-uniform with a mean slope of 9% in the main sloping direction (see Fig. 4). The soil is mainly the Cottenham series, a dark brown loamy sand. The plot was tilled up and down the slope and sown with sugar beets approximately a month before the studied rainfall events.

The plot was divided into 2.5 x 2.5 m grids with 2 rills (shallow depressions as a result of tillage) pr. metre width in the tillage direction. The vertical discretization in the unsaturated zone was 2 cm for the upper 10 cm increasing gradually to 0.5 m below 2 m depth. Rainfall data were available on a daily basis prior to the storm and on a minute basis during the main part of the storms. Daily values for evaporation were used. The model was parameterized according to EUROSEM input files for the storms.

5.2. CALIBRATION AND VALIDATION OF EUROSEM/MIKE SHE

Two rainfall events on May 29, 1992 were included in the test. The first event was used to calibrate the model, after which the model was validated on the second event using the parameter values obtained during the calibration of the first event.

Only a few of the original parameters in the EUROSEM input file were changed during calibration of the hydrographs, mainly roughness coefficients, saturated hydraulic conductivity, surface detention storage and cohesion. As no information on the groundwater level was available the depth to groundwater was used as one of the main calibration parameters to fit the simulated and observed hydrographs. All the parameter values were kept within physically realistic limits. For example, the value used for saturated hydraulic conductivity (12 mm/h) corresponded well with the mean of the measured values (12.01 mm/h). For comparison, the value used in EUROSEM/KINEROS simulation of the storm was 2.0 mm/h (Quinton, 1993).

TABLE 3. Comparison between simulated and observed values for runoff and soil loss from a 25 x 35 m plot at Woburn Experimental Farm for two rainfall events on May 29, 1992. The first rainfall event (29.5a) was used for calibration, and the parameter values were then used to run the model for the second event (29.5b).

Rainfall events in 1992	Rainfall		Runoff		Soil loss	
	Total rainfall (mm)	Max. intensity (mm/h)	Observed (m^3)	Simulated (m^3)	Observed (kg)	Simulated (kg)
29.5a	3.19	40.8	0.576	0.623	4.16	4.13
29.5b	4.12	26.4	0.208	0.269	1.68	1.40

Simulation results for the calibration event are shown in Fig. 3. The volume of the observed and the simulated hydrographs was reasonably well matched, whereas it was difficult exactly to match the shape of the observed hydrograph, especially the beginning of the hydrograph. This may be due to uneven distribution of surface detention storage over the surface and/or slightly wrong initial soil moisture content.

The matching of the simulated and observed sediment rates resembles that of the hydrographs; while it was difficult to match the exact shape of the curves (Fig. 3), the simulated value for the total soil loss corresponded well with the observed value. Using the parameter value in the EUROSEM input file (except those modified during the calibration of the hydrograph) resulted in a considerable overestimation of the sediment transport. As the majority of the soil loss is caused by rill erosion, the mean diameter and cohesion were used as the main calibration parameters.

It should be noted that for both rainfall events the values for the total runoff and the total soil loss are very small, as the total soil loss from the two events is 0.07 t/ha. Thus, even a small *absolute* error in the simulated values will result in a large *percentage* error. In general, the use of major rainfall events with considerable runoff and erosion would give a better possibility to test the performance of soil erosion models.

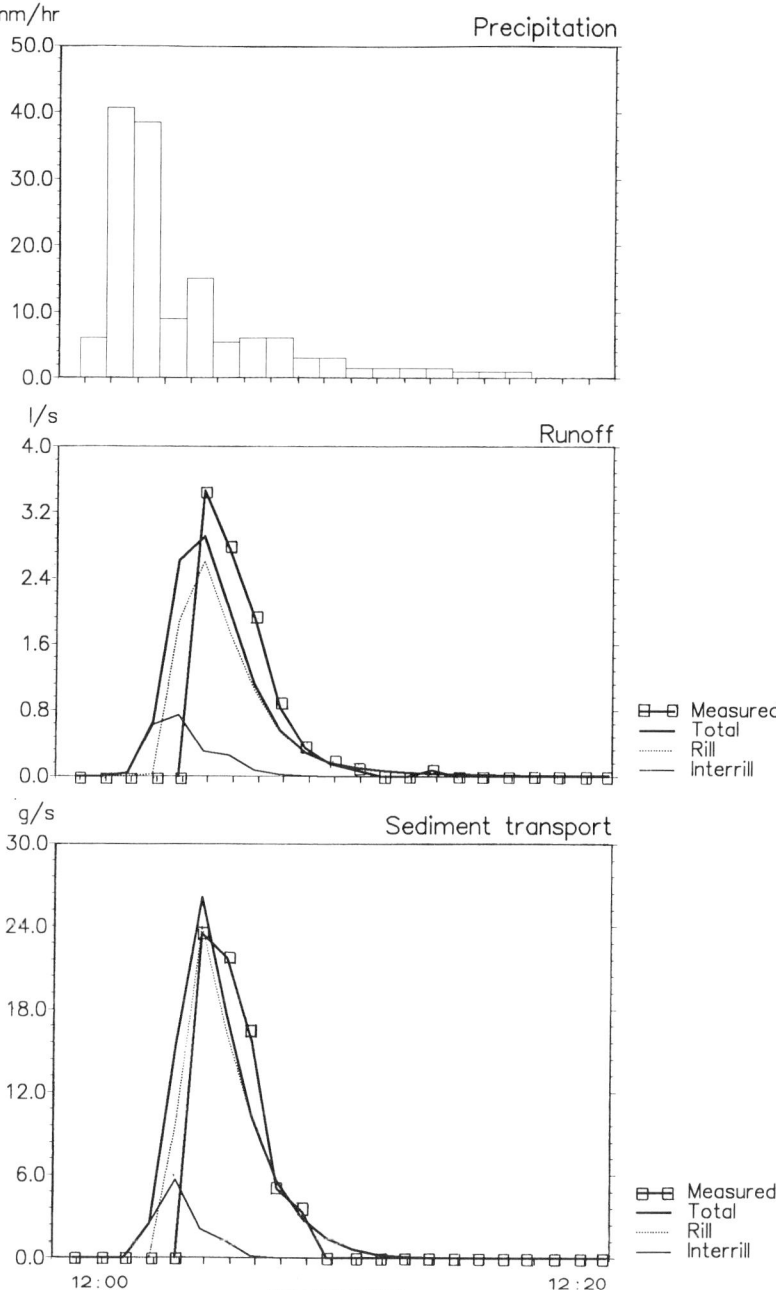

Figure 3. Rainfall, simulated and observed hydrographs, and simulated and observed sediment rates for the first rainfall event on May 29, 1992 on a 25 x 35 m^2 erosion plot at Woburn Experimental Farm, UK.

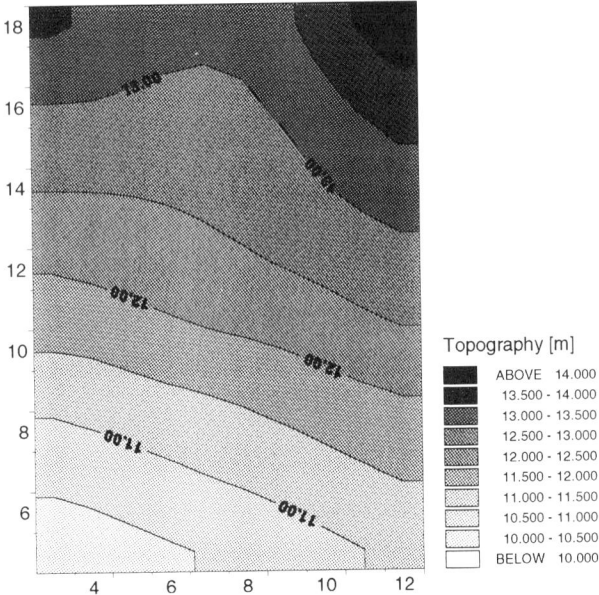

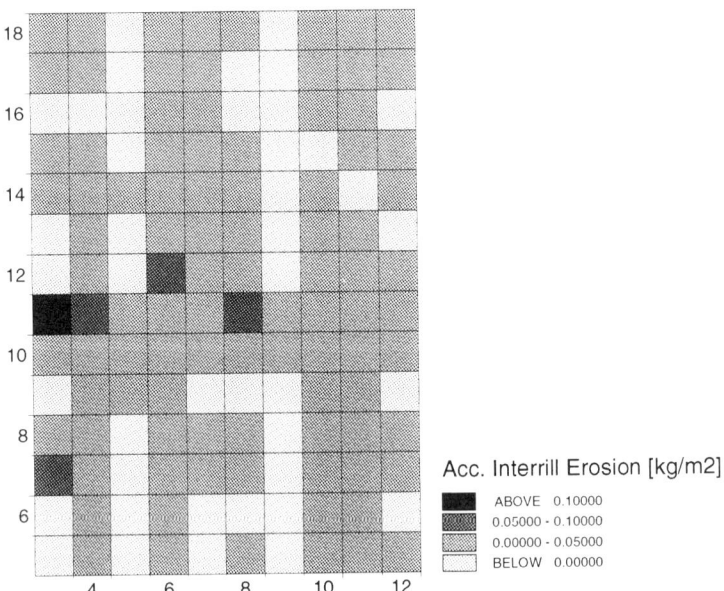

Figure 4. Topography and model grid for the Woburn erosion plot (upper figure) and two-dimensional representation of interrill erosion and sedimentation for the first rainfall event on May 29, 1992 (lower figure). Positive values imply erosion, while negative values correspond to areas with sedimentation.

Some of the spatially distributed features of the model are illustrated in Fig. 4. The topography is shown in the upper part of Fig. 4, and the accumulated simulated interrill erosion and sedimentation are shown in the lower part of the figure for the first rainfall event on 29 May 1992 (same event as shown in Fig. 3).

From Fig. 4 a very significant spatial variability is noticed. Thus, the net erosion rates shown in Fig. 3 turn up to be a result of erosion taking place over the main part of the area minus sedimentation over smaller areas. By comparing the erosion maps with the topography in Fig. 4 it appears that the small topographical 'irregularities' generate differences in overland flows (not shown here) which again generate significant spatial variations in the erosion pattern.

5.3. SENSITIVITY ANALYSES

Although a rather good prediction of runoff as well as of soil loss was obtained, it became obvious during the calibration that especially the hydrological part of the model was very sensitive to changes in parameter values. Thus, for the first storm (29.5a) a number of sensitivity tests was carried out for the most sensitive parameters (Table 4).

The extremely high sensitivity to the hydraulic conductivity is mainly due to the fact that the rainfall intensity for a major part of the storm was in the same order of magnitude as the hydraulic conductivity. For storms with large differences between the rainfall intensity and the hydraulic conductivity the model will be less sensitive to the hydraulic conductivity. The effect of the number of rills shows the effect of the concentration of the overland flow and illustrates the importance of a good surface description. The effect of splash erosion will vary from storm to storm depending on the amount of runoff generation during the rainfall event.

TABLE 4. Sensitivity analyses of EUROSEM/MIKE SHE on a 25 m x 35 m plot at Woburn, UK on May 29, 1992 (DHI, 1994). The observed runoff and soil loss values were 0.576 m³ and 4.16 kg.

Parameter	Parameter change			Simulated runoff		Simulated soil loss	
	From... to		%	m³	Change %	kg	Change %
Original parameters				0.622	-	4.16	-
Manning n ($m^{1/3}$ s)	0.033->0.028		- 21	0.639	+ 3	4.93	+19.4
Surface detention storage (m)	0.0004->0.0005		+ 20	0.529	15	3.22	- 22
Saturated hydraulic conductivity (mm/hr)	12->11		- 8	0.692	+ 11	4.68	+ 13
Splash erosion included	Yes->No			0.622	0	3.64	- 11.9
Number of rills per meter width	2->1			0.678	+ 9	4.92	+ 19
	2->3			0.566	- 9	3.58	- 13

6. Discussion on Limitation, Applicability and Research Needs

While the last decade has seen substantial progress in the process descriptions for soil erosion, and considerably increased the understanding of the interaction between the different processes, there are still basic issues that complicate the modelling.

Sediment transport is highly dependent on the pattern of overland flow. Outside the laboratories, description of flow patterns, roughness elements, flow velocities, and the division of shear stresses between the surface soil and other roughness elements have turned out to be very difficult. Spatial differences in infiltration also play an important role.

An often mentioned limitation of the physically-based distributed models is the amount of data needed to run the models. The issue is dual, because the detailed process descriptions and the extensive data collections have generated much more detailed knowledge about how, where, and under which conditions erosion takes place. It is true at present that the data requirements appear comprehensive, but as more understanding of sensitivity in different environments is generated, it also becomes possible to target the data collection to a higher degree than is presently done.

Thus, a direct application of soil erosion models on medium to large size catchment scales is not yet feasible in practice. The empirical models can be used together with GIS's to prepare qualitative information such as erodibility indexes, but they can not be expected to provide reliable quantitative predictions, and they are not well suited to assess the impacts of alternative soil conservation management options. The physically-based models, on the other hand, are not yet ready for application at such scale. This does not imply, however, that soil erosion models are not useful tools for soil and water management. When used with great care by experienced modellers model results may be very useful for practical purposes.

Examples of possible approaches for applications of distributed physically-based models at different scales include the following:

(a) *Plot and hill slope scale.* Here it is feasible directly to apply a physically-based soil erosion model.

(b) *Small size catchment* (up to a few km^2). Here it is possible to use distributed physically-based models, but maybe some of the process descriptions requiring the finest spatial and temporal resolution, such as rill erosion, may have to be described rather coarsely. Such simplification implies that the models may not be able to simulate all key processes to the same degree of reliability; however, still the results may be very useful.

(c) *Ordinary size catchment* (up to several hundreds or thousands of km^2). In this case there are different ways of utilizing model results, such as:
   * If model parameters for a particular typical site are known, it may be possible - for given rainfall events - to tabulate runoff rates and sediment rates as a function of different interventions. Such tables could aid extension workers in choosing conservation methods according to the sensitivity of that particular type of site. It should be noticed that the traditional empirical

models do not allow extension workers or farmers to judge, for instance, whether, at a particular site, infiltration is the most efficient parameter to manipulate, or whether surface runoff cannot be avoided, so the conservation methods therefore must focus on removing excess water. Potential applications of a physically-based soil erosion model (EUROSEM) for evaluating the effects of soil conservation measures is reviewed by Rickson (1994).

* Soil erosion modelling is carried out at on a number of representative plots or hillslopes, so-called 'soil erosion response units'. The entire catchment is then divided into sub-units and each sub-unit classified as being represented by one of the erosion response units. These response units are characterized by common slope, soil type, land use, climatic regime and possibly other factors. This approach is very suitable for a combination of a soil erosion model and a GIS.

Obviously, when moving from modelling at plot scale to modelling at catchment scale the model accuracy becomes less at each point in the area; but experience indicates that the integrated output from the catchment is not necessarily much less accurate than the model output from the plot scale.

Some of the critical issues which need to be addressed in future research include:

(a) The variation of soil shear strength over the year as well as during a rainstorm as a function of soil moisture content, tillage, vegetation and time, so that such changes can be modelled in continuous soil erosion models.
(b) Relations between soil surface roughness (including effects of vegetation) and the hydraulic roughness of the flow.
(c) Transport capacity of thin overland flow where soil material mainly is transported as aggregates rather than single grains.
(d) Interaction between soil and vegetation properties which influence the hydrological processes and parameters, such as hydraulic conductivity.
(e) Effects of tillage on parameter values.

At present, the physically-based soil erosion codes are not much applied outside the group of researchers who have been involved in the development of the codes. A main reason for this, in addition to the problems outlined above, is that these codes are generally not very well documented and not very user friendly. The necessary technological innovations in this regard can be expected to be made gradually, as the research results improve the model applicabilities and as the demands for model use increase.

7. References

Al-Durrah, M.M. and Bradford, J.M. (1981) New methods of studying soil detachment due to water drop impact. *Soil Sci. Soc. Am. J.* 45, 949-953.

Al-Durrah, M.M. and Bradford, J.M. (1982a) Parameters for describing soil detachment due to single water drop impact. *Soil Sci. Soc. Am. J.* 46, 836-840.

Al-Durrah, M.M. and Bradford, J.M. (1982b) The mechanism of raindrop splash on soil surfaces. *Soil Sci. Soc. Am. J.* 46, 1086-1090.

Andersen, H.E. and Lørup, J.K. (1991) A Theoretical and Experimental Study of Rill Erosion. M.Sc. Thesis, The Royal Veterinary and Agricultural University, Copenhagen.

Beasley, D.B., Huggins, L.F. and Monke, E.J. (1980) ANSWERS: A model for watershed planning. *Transactions of the ASAE* 23, 938-944.

Bennett, J.P. (1974) Concepts of mathematical modelling of sediment yield. *Water Resources Research* 10, 485-492.

Bollinne, A. (1978) Study of the importance of splash and wash on cultivated loamy soils of Hesbaye (Belgium). *Earth Surface Processes* 3, 71-84.

Bryan, R.B. (1976) Considerations on soil erodibility indices and sheetwash. *Catena* 3, 99-111.

Bryan, R.B. and Poesen, J. (1989) Laboratory experiments on the influence of slope length on runoff, percolation and rill development. *Earth Surface Processes and Landforms* 14, 221-231.

Catt, J.A. (1992) Soil erosion on the Lower Greensand at Woburn Experimental Farm, Bedfordshire - evidence, history and causes, in M. Bell and J. Boardman (eds) Past and Present Soil Erosion, Oxbow Monograph 22.

De Ploey, J. (1989) A model for headcut retreat in rills and gullies. *Catena Supplement* 14, 81-86.

DHI (1994) Testing of the MIKE-SHE-EUROSEM code. Unpublished.

DHI and IoG (1992) Modification of and simulation with a soil erosion model, Technical Report Proj. No. 13-4291.

Elliot, W.J., Foster, G.R. and Elliot, A.V. (1994) Soil Erosion: Processes, Impacts and Prediction, in R. Lal and F.J. Pierce (eds), *Soil management for sustainability*, Soil and Water Conservation Society, Iowa, pp. 25-34.

Elwell, H.A. (1977) Soil loss estimation system for southern Africa. Department of Conservation and Extension, Research Bulletin No. 22. Salisbury, Rhodesia.

Emmett, W.W. (1978) *The hydraulics of overland flow on hillslopes*. Geological Survey Professional Paper 662-A.

Engelund, F. and Hansen, E. (1967) A monograph on sediment transport in alluvial streams. Teknisk forlag, Copenhagen.

Engman, E.T. (1986) Roughness coefficients for routing of surface runoff, *Journal of Irrigation and Drainage Engineering* 112, 39-53.

Evans, R. (1981) Potential soil and crop losses by soil erosion. *Proceedings, SAWMA Conference on Soil and Crop Loss: Development in soil erosion control.* National Agriculture Centre, Stoneleigh.

Foster, G.R. and Meyer, L.D. (1972) A closed-form soil erosion equation for upland areas, in H.W. Shen (ed) *Sedimentation: Symposium to Honour Professor H.A. Einstein*, Fort Collins, Colorado, pp. 12.1-12.19.

Foster, G.R., Lombardi, F. and Moldenhauer, W.C. (1982) Evaluation of rainfall-runoff erosivity factors for individual storms. *Transactions of the ASAE* 25, 124-129.

Gilley, J.E., Woolhiser, D.A. and McWorther, D.B. (1985) Interrill erosion. Part I: Development of model equations. *Transactions of the ASAE* 28, 147-153.

Govers, G. (1990) Empirical relationships on the transporting capacity of overland flow. *International Association of Hydrological Sciences Publication* 189, 45-63.

Govers, G., Everaert, W., Poesen, J., Rauws, G., De Ploey, J. and Lautridou, J.P. (1990) A long flume study of the dynamic factors affecting the resistance of a loamy soil to concentrated flow erosion. *Earth Surface Processes and Landforms* 15, 313-328.

Hasholt, B. and Styczen, M. (1993) Measurement of sediment transport components in a drainage basin and comparison with sediment delivery computed by a soil erosion model, in R.F. Hadley and M. Takahisa (eds) *Sediment Problems: Strategies for monitoring, Prediction and Control*, IAHS Publ. no. 217, pp. 147-158.

Kirkby, M.J. (1980) Modelling water erosion processes, in M.J. Kirkby and R.P.C. Morgan (eds), *Soil Erosion*, John Wiley and Sons Ltd., Chichester, pp. 183-216.

Lane, L.J. and Nearing, M.A. (1989) USDA - Water erosion prediction Project: Hillslope profile model documentation. NSERL Report No. 2, USDA-ARS National Soil Erosion Research Laboratory, West Lafayette, Indiana 47907.

Meyer, L.D. (1981) How rain intensity affects interrill erosion. *Transactions of the ASAE* 24, 1472-1475.

Meyer, L.D. and Wischmeier, W.H. (1969) Mathematical simulation of the process of soil erosion by water. *Transactions of the ASAE*, 12, 754-758, 762.

Morgan, R.P.C. (1982) Splash detachment under plant covers: results and implications of a field study. *Transactions of the ASAE* 25, 987-991.

Morgan, R.P.C. (1985) Soil erosion measurement and soil conservation research in cultivated areas of the UK. *Geographical J.* 151, 11-20.

Morgan, R.P.C., Finney, H.J., Lavee, H., Merritt, E. and Noble, C.A. (1985) Plant cover effects on hillslope runoff and erosion: evidence from two laboratory experiments, in A.D. Abrahams (ed) *Hillslope processes*, Allen and Urwin, Winchester, Mass., pp. 77-96.

Morgan, R.P.C., Quinton, J.N. and Rickson, R.J. (1991) Eurosem User Guide - version 1. Silsoe College, United Kingdom.

Morgan, R.P.C., Quinton, J.N., Smith, R.E., Govers, G., Poesen, J.W.A., Auerswald, K., Chisci, G., Torri, D. and Styczen, M.E. (1995) The European Soil Erosion Model (EUROSEM): A Process-based Approach for Predicting Soil Loss from Fields and Small Catchments, Submitted to *Earth Surface Processes and Landforms*.

Mosley, M.P. (1982) The effect of a New Zealand beech forest canopy on the kinetic energy of water drops and on surface erosion. *Earth Surface Processes and Landforms* 8, 569-577.

Nearing, M.A., Lane, L.J. and Lopes, V.L. (1994) Modelling Soil Erosion, in R. Lal (ed) *Soil Erosion Research Methods*, Soil and Water Conservation Society, Ankeny, pp. 127-156.

Nearing, M.A., Foster, G.R., Lane, L.J. and Finkner, S.C. (1989) A process-based soil erosion model for USDA-water erosion prediction project technology. *Trans. ASAE* 32(5), 1587-1593.

Nielsen, S.A. and Styczen, M. (1986) Development of an areally distributed soil erosion model, in *Proceedings of the Nordic Hydrologic Conference*, Aug. 11-13, Reykjavik, pp. 797-808.

Oldeman, L.R. (1992) *Global extent of soil degradation*, Bi-annual report, International Soil Reference and Information Center, Wageningen, The Netherlands, pp. 19-36.

Park, S.W., Mitchell, J.K., Scarborough, J.N. (1982) Soil erosion simulation on small watersheds: A modified ANSWERS model. *Transactions of the ASAE* 25, 1581-1588.

Poesen, J,W. and Ingelmo-Sanchez, F. (1992) Runoff and sediment yield from topsoils with different porosity as affected by rock fragment cover and position, *Catena* 19, 451-474.

Poesen, J.W., Torri, D. and Bunte, K. (1994) Effects of rock fragments on soil erosion by water at different spatial scales: a review, *Catena* 23, 141-166

Quinton, J.N. (1993) Personal communication.

Quinton, J.N. (1994) The validation of physically-based erosion models with particular reference to EUROSEM, in R.J. Rickson (ed) *Conserving Soil Resources: European Perspectives*, CAB International, pp. 300-313.

Rauws, G. and Govers, G. (1988) Hydraulics and soil mechanical aspects of rill generation in agricultural soils, *Journal of Soil Science* 39, 111-124.

Refsgaard, J.C. and Storm, B. (1995) MIKE SHE. In V.J. Singh (Ed) *Computer models in watershed hydrology*. Water Resources Publications.

Renard, K.G., Laflen, J.M., Foster, G.R. and McCool, D.K. (1994) The Revised Universal Soil Loss Equation, in R. Lal (ed) *Soil Erosion Research Methods*, Soil and Water Conservation Society, Ankeny, pp. 105-124.

Rickson, R.J. (1994) Potential application of the European Soil Erosion Model (EUROSEM) for evaluating soil conservation measures, in R.J. Rickson (ed) *Conserving Soil Resources: European Perspectives*, CAB International, pp. 326-355.

Rose, C.W., Williams, J.R., Sander, G.C. and Barry, D.A. (1983a) A mathematical model of soil erosion and deposition processes. I. Theory for a plane element. *Soil Sci. Soc. Am. J.* 47, 991-995.

Rose, C.W., Williams, J.R., Sander, G.C. and Barry, D.A. (1983b) A mathematical model of soil erosion and deposition processes. II. Application of data from an arid-zone catchment. *Soil Sci. Soc. Am. J.* 47, 996-1000.

Römkens, M.J.M. and Wang, J.Y. (1986) Effects of tillage on surface roughness. *Transactions of the ASAE* 29(2), 429-433.

Savat, J. (1980) Resistance to flow in rough supercritical sheet flow. *Earth Surface Processes* 5, 103-122.

Styczen, M. & Nielsen, S.A. (1989) A view of soil erosion theory, process-research and model building: possible interactions and future developments, *Quaderni di Scienza del Suolo* 2, 27-45.

Styczen, M. and Høgh-Schmidt, K. (1988) A new description of splash erosion in relation to raindrop size and vegetation, in R.P.C. Morgan and R.J. Rickson (eds) *Erosion assessment and modelling*. Commission of the European Communities Report No. EUR 10860 EN, pp. 147-184.

Torri, D. and Borselli, L. (1991) Overland flow and soil erosion: Some processes and their interactions, in H.R. Bork, J. de Ploey and A.P. Schick (eds) *Erosion, Transport and Deposition Processes - Theories and Models*, CATENA Supplement 19, Catena Verlag, Cremlingen-Destedt, Germany, pp. 129-137.

Torri, D., Sfalanga, M. and Chisci, G. (1987a) Threshold conditions for incipient rilling. *CATENA Supplement* 8, 97-106.

Torri, D., Sfalanga, M. and Del Sette, M. (1987b) Splash detachment: Runoff depth and soil cohesion. *Catena* 14, 149-155.

USDA (1980) CREAMS - A Field-Scale Model for Chemical, Runoff and Erosion from Agricultural Management Systems. U.S. Department of Agriculture. Conservation Research Report No. 26, 640 pp.

Whitlow, R. (1988) Land Degradation in Zimbabwe. A Geographical Study. Geography Department, University of Zimbabwe, Harare.

Wicks, J.M., Bathurst, J.C. and Johnson, C.W. (1992) Calibrating SHE Soil-Erosion Model for Different Land Covers, *Journal of Irrigation and Drainage Engineering* 118(5), 708-723.

Williams, J.R. (1975) Sediment-yield prediction with universal equation using runoff energy factor, In: *Present and Prospective Technology for Predicting Sediment Yield and Sources*, ARS-S-40, Agr. Res. Serv., U.S. Dept. Agr., Washington D.C., pp. 244-252.

Wischmeier, W.H. (1975) Estimating the soil loss equation's cover and management factor for undisturbed areas, in *Present and Prospective Technology for Predicting Sediment Yields and Sources*, ARS-S-40, USDA Agricultural Research Service Publ., pp. 118-124.

Wischmeier, W.H. and Smith, D.D. (1965): Predicting rainfall erosion losses from cropland East of the Rocky Mountains. Agricultural Handbook No. 282. Agricultural Research Service, United States Department of Agriculture, Purdue Agricultural Experiment Station.

Wischmeier, W.H. and Smith, D.D. (1978): Predicting rainfall erosion losses. USDA Agricultural Handbook No. 537.

Woolhiser, D.A., Smith, R.E. and Goodrich, D.C. (1990) KINEROS: A Kinematic Runoff and Erosion Model: Documentation and user manual. USDA Agricultural Research Service ARS-77.

Yalin, Y.S. (1963) An expression for bed-load transportation. *J. Hyd. Div. ASCE* 89, 221-250.

CHAPTER 7
AGROCHEMICAL MODELLING

M. THORSEN[1], J. FEYEN[2] AND M. STYCZEN[1]
[1] Danish Hydraulic Institute
[2] Katholieke Universiteit Leuven

1. Introduction

1.1. BACKGROUND

During the last decade problems with increased emission of agrochemicals have become more and more obvious. In a number of countries, the stage is now reached, where political decisions regarding control of such emissions have been taken or in the process of being taken and alleviation measures are being implemented. Most of these measures are based on rough estimates of the risk of agrochemical pollution, which not always consider the interaction between climate, crop, soil and hydrology, and additional tools for assessment of the long term effects of the suggested measures of pollution control are lacking. It is therefore very relevant to investigate which tools are available for prediction, how reliable they are, and what are their limitations.

Intensive large scale monitoring programmes are being established in many countries aiming to assess the magnitude of the pollution problems from non-point sources in rural areas. However, such programmes are only able to identify the problems and quantify the results. They are not capable of performing cause and effect analysis in order to identify critical areas with high pollution risk or critical management practices causing higher losses than others. For this purpose understanding of the processes responsible for transport and transformation of the chemicals in the various hydrological compartments is crucial. In this respect, mathematical models describing the relevant processes provide strong tools which can support the interpretation of the monitoring results and provide the possibility to investigate and compare the effect of different management practices on potential losses to the surrounding environment.

1.2. TYPES OF MODELS AVAILABLE

A large number of computer codes have been developed in the past years to describe the transport and fate of agrochemicals in the different parts of the hydrological cycle. They vary in complexity, ranging from simple empirical formulas to comprehensive distributed physically/chemically-based descriptions. Traditionally, there has been a distinction between leaching models and field or catchment models. Leaching models are confined to one dimensional descriptions of the root zone processes, while field or catchment models consider smaller or larger parts of other surface and subsurface processes.

It is outside the scope of this chapter to give a comprehensive review of existing model codes. The main emphasis will be put on presenting a few state-of-the-art descriptions for some of the most promising tools, both from a research and a management point of view, and to describe the principal differences in modelling methodologies.

2. Process Modelling at Point Scale

2.1. GENERAL

A large number of model codes describes the unsaturated zone, including the root zone, in a single profile. The important leaching models are deterministic models, implying that they, as far as possible, describe the physics and chemistry of the processes. The following sections briefly present state-of-the-art leaching model codes describing transport and transformation of nitrogen, phosphorous and pesticides.

When evaluating the features available in leaching models, some general considerations regarding the process requirements are necessary. A prerequisite for describing solute movement in soils is that the description of water flow and the available boundary conditions are adequate for the situation under consideration. The models must be able to handle the hydrological conditions present in the soil. For instance, if shallow groundwater is present, the selected model must be able to handle groundwater fluctuations and capillary rise. Some common approaches and related assumptions are described in Table 1.

Another important part of the water balance which must be considered is the evapotranspiration. Several different approaches exist implying that the actual climatic conditions must be analysed when evaluating the simulated water balances. Additionally, the chemical transformation processes included should reflect the current knowledge regarding processes having significant influence on solute behaviour in the soil.

2.2. NITROGEN MODELLING CODES

The impacts of agricultural crop production on the environment in terms of nitrogen losses to surface water and groundwater is related to the input level of fertilizers as well as the structure of the cropping system. It is well known that eg high application rates of organic manure, and in particular cropping systems without crop cover during periods in which mineral nitrogen is released from organic matter in the soil, may result in increased nitrogen losses in subsequent periods with water discharge. During the recent four decades nitrogen losses from rural areas to the aquatic environment have increased causing deterioration of the water quality and subsequently created great concern on how environmentally and economically sustained agricultural crop production can be developed.

TABLE 1. Examples of approaches and assumptions used in leaching models.

Process	Approach	Assumption
Water flow	Capacity model	Water flow depends on the storage capacity of each layer + an empirical drainage rule. Flow between layers only occurs when the capacity is exceeded. Capillary rise is not taken into account.
	Richards' equation	Water flow depends on the hydraulic gradient and the soil physical properties (hydraulic conductivity and soil water retention curves), and is calculated dynamically for the entire column. Capillary rise is automatically accounted for.
	Preferential flow paths considered	Soil matrix contains macropores or similar preferential flow paths which, when activated, transport water at fast rate from surface layer towards the bottom of the root zone.
Solute transport	Piston displacement	Convective transport with water flow only. Dispersion is set by the user or indirectly accounted for by numerical dispersion.
	Convection-dispersion equation	Convective and dispersive transport assumed. Hydrodynamic dispersion calculated.
	Mobile/immobile water considered	Soil matrix divided in active, mobile fraction where the water movement takes place and an immobile fraction. Diffusion between mobile and immobile phases.

Agricultural crop production as well as losses of nitrogen is determined by a number of physical, chemical and biological processes in the soil-plant-atmosphere continuum which interact simultaneously in a complex way. In the nitrogen cycle in the soil-plant system, the pathway of nitrogen is a complex series of transformation and transport processes all of which are affected by external factors. Thus, it is difficult to predict how changed management practices will effect crop production, nitrogen use efficiency and nitrogen losses. In the conventional scientific approach, field experiments have been used to explore possibilities for appropriate system management practices. This type of research has limitations due to the complexity of the system. Simulation models are therefore increasingly used to support experimental research and, though still at minor scale, to assess the effect of legislation meassures.

A large number of model codes exist aiming to describe the interrelationships between energy, water, carbon and nitrogen cycles of the soil-plant-atmosphere system under various external conditions with different levels of complexity.

14 nitrogen leaching model codes were reviewed by de Willigen (1991). Intercomparative tests of five codes were carried out under the auspices of the CEC (CEC, 1991), Hansen (1992) tested and compared two nitrogen leaching modelling codes and Diekkrüger et al. (1995) compared the simulation results of 19 agroecosystem models of which 8 contained approaches related to the nitrogen cycle. Examples of such model codes are reviewed in Table 2.

Table 2. Review of some of the most comprehensive nitrogen modelling codes.

Process	ANIMO	SOIL-N	DAISY	WAVE	RZWQM	LEACHM
Water flow	External model	External model	Richards' eqn.	Richards' eqn. in matrix	Green & Ampt/Richards' eqn.	Richards' eqn.
Preferential flow	-	-	- (+)	+	+	- (+)
Solute transport	Piston displacement	Piston displacement	Conv./disp.	Conv./disp., mobile/immobile water	Piston displacement + mobile/immobile water	Conv./disp.
Boundary conditions	External model	External model	Free drainage, fixed or fluctuating gwt.	Free drainage, gwt., lysimeter, fluxes, pressure head	Free drainage, pressure head	Free drainage, fixed gwt, zero flux, lysimeter
Mineralization	4 OM pools. 1. order kinetics. f(OM pools, T, θ, pH, C/N, O_2-demand)	3 OM pools. 1. order kinetics. f(OM pools, C/N, T, θ)	4 OM pools + 2 BM pools. 1. order kinetics. f(T, θ, Cl, C/N, OM+BM pools, [NH_4^++NO_3^-])	3 OM pools. 1. order kinetics. f(OM pools, C/N, T, θ)	5 OM pools + 3 BM pools. 1. order kinetics. f(T, pH, θ, BM pool, I)	3 OM pools. 1. order kinetics. f(OM pools, C/N, T, θ)
Nitrification	1. order kinetics. f([NH_4^+], aeration, θ, T, pH)	1. order kinetics. f([NH_4^+], T, θ, NH_4^+/NO_3^--ratio)	Michaelis-Menten kinetics. f([NO_3^-],T, θ)	1. order kinetics. f([NH_4^+],θ, T)	Zero order at low conc. 1. order at high conc. f(T, θ, pH, BM pool, I)	1. order kinetics. f([NH_4^+], θ, T, NH_4^+/NO_3^--ratio)
Denitrification	Zero order reaction. f(air content, T, θ, O_2-demand, pH, [NO_3^-])	Michaelis-Menten kinetics. f([NO_3^-], T, θ, depth)	Denitric. capacity concept based on CO_2-evol. f([NO_3^-], T, θ)	1. order kinetics. f(T, θ)	1. order kinetics. f(T, θ, pH, BM pool, I)	Michaelis-Menten kinetics. f([NO_3^-], θ)
NH_4^+-adsorption	Linear isotherm	Immobile	Non-linear isotherm	Linear isotherm, different partitioning mobile/immobile sites	Linear isotherm	Linear isotherm
Crop uptake	NH_4^+ + NO_3^- uptake. f(transpiration, flux, [NO_3^-], [NH_4^+])	Predefined or separate module	NH_4^+ + NO_3^- uptake. From calc. pot. N-demand + avai': "ility. f(root density, [NH_4^+], [NO_3^-])	From predefined or calculated pot. uptake. f(root density, [NH_4^+], [NO_3^-], diffusive + convective flux)	Pot. crop demand + N-availability. Passive or active transport to root.	2 options. From pot. N-demand + availability or f(root density, [NH_4^+], [NO_3^-])
Volatilization	Given fract. of [NH_4^+] in manure	-	Given fraction of [NH_4^+] in manure	1. order kinetics. f([NH_4^+])		1. order reaction
Crop production	-	Separate module	Calc. of LAI, CAI, Dry matter produc. from Rn.	Input or calc. of LAI and Dry matter produc. from Rn.	Calc. of LAI and Dry matter production from Rn.	Crop cover calc. from empirical functions.
Crop modules	-	-	13 crops	7 crops	2 crops	-
References	Berghuijs et al. (1985) Rijtema et al. (1991)	Johnson et al. (1987) Jansson et al. (1991)	Hansen et al. (1990) Hansen et al. (1993)	Vereecken et al. (1991) Vanclooster et al. (1994, 1995)	DeCoursey et al. (1989, 1992)	Wagenet and Hutson (1987), Hutson and Wagenet (1992)

(+): Planned in future versions, OM: Organic matter, BM: Biomass, C: Carbon content, Cl: Clay content, θ: Soil moisture content, T: Temperature, gwt: Groundwater table, I: Ion strength, LAI: Leaf area index, CAI: Crop area index, Rn: Net radiation.

Two of these codes, ANIMO and SOIL-N, only contain descriptions of the nitrogen cycle and hence require an external water flow model. For the codes containing internal flow descriptions two types of approaches are recognized. The more physically-based codes use Richards' equation, whereas the simpler codes use the capacity approach. The basic difference between the two approaches is that capacity models are not able to calculate fluxes based on pressure head gradients. These models are therefore not suitable for conditions with capillary rise. Preferential flow processes described by dual conductivity and/or mobile/immobile water approaches, are only considered in WAVE and RZWQM.

Solute transport is either described by the convection-dispersion equation, corresponding in complexity to the Richard's equation for water flow, or solely by convective transport calculated by multiplying water flux and solute concentration.

All models describe the nitrate transport and transformation in two steps, by first carrying out the water flow calculations, then the N calculations. This approach may be acceptable where N is not a limiting factor, but in situations with serious N deficiency, it may pose a problem. The plant growth simulated in the first step may be optimal from a water availability point of view, but restricted by N deficiency during the second run. The actual evapotranspiration calculated in step 1 will therefore be overestimated, resulting in unreliable estimates of nitrate concentrations and fluxes.

The major differences between the existing nitrogen model codes arise from the approaches applied for describing the components of the N-balance. Especially the complexity regarding mineralization, nitrification, denitrification and plant uptake varies among the models. As these four processes are of major importance for the overall performance of the models with respect to nitrate leaching, the model review presented in Table 2 focuses on differences and assumptions related to the N-dynamics.

All the models describe the kinetics of the mineralization as a 1. order process, but the number of interacting organic pools range from 3 to 7. Only ANIMO, DAISY and RZWQM take explicit account of one or more pools of biomass.

One of the major processes removing nitrogen from the soil profile is plant uptake. Crop N-uptake may be either simulated directly by a crop module accounting for gross photosynthesis, respiration, dry matter and nitrogen distribution between organs etc., as in DAISY, WAVE and RZWQM or estimated indirectly using predefined curves for potential N-uptake. In some models (eg SOIL-N), the maximum uptake is specified by the user. This may be an advantage in research studies where this component can be assessed, but it hampers the use of the model for predictive purposes. The other models contain or may be combined with a growth module. The N-uptake may be calculated on the basis of transpiration fluxes, assuming uniform concentrations at the root surface and in the bulk soil (ANIMO), or by calculating transport of water and solute from the bulk soil to the root (DAISY, RZWQM, and WAVE)

The plant growth modules differ considerably among the model codes. ANIMO and LEACHM deal with plant processes only in a sketchy way, and SOIL-N hardly includes them. DAISY, WAVE and RZWQM simulate crop production while accounting for gross photosynthesis, respiration, distribution of dry matter and nitrogen between the different organs etc., though the types and number of available crop modules varies.

The general conclusion from the test of 14 model codes conducted by de Willigen

(1991) was that prediction of nitrogen uptake by crops and dry-matter production requires one of the model codes containing a detailed growth module. However, for both soil water and mineralization simulations, the results showed that the detailed mechanistic model codes were not necessarily better than simple models. They require detailed information about soil hydraulic and chemical properties, and are very sensitive to parameter values. On the other hand they apply to a wider range of conditions than more simple models do.

2.3. PHOSPHOROUS MODELLING CODES

Traditionally, losses of phosphorous to groundwater and surface water from non-point sources have been regarded as a minor problem compared to the loads arising from point sources such as urban sewage discharge. However, due to large efforts put into controlling these point sources during recent years the relative load from non-point sources is increasing. Additionally, the magnitude of these loads has periodically been found to be rather large (Culley, 1983; Schjønning et al., 1995). Phosphorous has been identified as being the limiting nutrient for the primary production in many North European lakes, and during spring it may also be the limiting factor in coastal areas. The ability to control losses of phosphorous from non-point sources is therefore crucial for the water quality in these compartments.

Phosphorous is a key plant nutrient which is applied to agricultural areas either in mineral fertilizers or through organic manure. There are two main processes responsible for transport of phosphorous to surface waters and groundwater. The primary process is surface transport of particulate bound phosphorous driven by hydraulic soil erosion. The second process is vertical transport of soluble inorganic or organic and/or particulate phosphorous through preferential pathways in the upper soil to drains and groundwater. The significance of this latter pathway and the relative importance of transport of viz. solute and particulate phosphorous through the unsaturated soil column is generally not well known. However, as Phosphate is expected to be rather mobile under saturated and hereby reducing conditions, leaching of solute phosphorous out of the root zone may under certain conditions, eg local saturation in the upper soil layers, contribute significantly to the total losses.

Generally, the level of development of phosphorous models is lower than for nitrogen and pesticide models, and the main efforts have been put into modelling of the overland flow processes, which primarily involve transport of particulate bound phosphorous during heavy rainfall events causing erosion. Transport of soluble phosphorous is considered to occur to a less extent and is depending on the desorption processes. As the processes responsible for the overland transport of phosphorous are distributed in nature, point/field scale representation of these processes are very simplistic. An example of a field scale model is EPIC (Sharpley and Williams, 1990; Williams, 1995) which uses the SCS-curve method to estimate runoff of water and sediment on a standardised hill slope. An example of a distributed approach for modelling of the transport of phosphorous in surface runoff is the phosphorous module developed for the ANSWERS model code (Storm et al. 1988) which is an event based description of the transport of particulate bound and soluble phosphorous along with surface runoff of water. The solube fraction of phosphorous is calcultated as nonequilibrium desorption from the soil surface to the runoff water.

With respect to the vertical representation of the chemical processes involving phosphorous the existing model codes differ in complexity and in the assumptions made when simulating the complex nature of phosphorous cycling. Examples of model codes aiming to describe the chemical reactions of phosphorous in an unsaturated soil column are EPIC, ANIMO-P (Rijtema et al. 1991) and the approach by van der Zee and Gjaltema (1992).

2.4. PESTICIDE MODELLING CODES

During recent years pesticide losses from rural areas have been recognised as a major problem. Several different pesticides have been detected in groundwater and surface waters in many countries (Fielding et al. 1990), revealing a demand for reassessment of previous and present management practices. Especially, findings located in deep groundwater aquifers usually considered to be protected by impermeable clay layers (Brüsch and Kristiansen, 1994) have cast doubt on the knowledge and assumptions associated with the current procedure for registration and approval of pesticides.

The numerous findings have also revealed a need for development of predictive tools capable of quantifying the transport processes in order to perform risk assessment.
This has converted the objective of model development from a research level to a functional level, and put more focus on reliability and validation.

In Europe, use of models in the pesticide registration has been included in the legislation through EEC-directive (91/414), regarding "uniform principles", stating that use of numerical models shall be incorporated into the registration procedure. At present only two countries, The Netherlands and Germany, have implemented model simulations in the standard procedure. The remaining countries are awaiting the results and recommendations from an EU working group regarding selection and use of pesticide fate models (FOCUS, 1995).

At the moment, many different models exist, claiming to be able to describe the fate of pesticides from application on the soil surface and through the unsaturated zone, hereby predicting the final load to the groundwater. Some models also consider surface runoff, and very few contain descriptions of lateral transport in the saturated zone.

In general, the descriptions of the transformation processes i.e. sorption and degradation differ in complexity and hereby also in parameter requirement. For instance, in some model codes the degradation rate is allowed to be specified differently with depth, phase (solid/liquid), site (matrix/macropores) or reaction type (eg hydrolysis, photolysis, biodeg.). However, the input parameters required for such complex decriptions are usually not available (Styczen and Villholth 1994, Bosch and Boesten, 1994a). This implies a high degree of uncertainty on the simulation results and makes calibration necessary.

Additionally, does the selection of appropriate input parameters and the following interpretation of the simulation results suffer from lack of knowledge regarding the variation in pesticide related parameters in the soil. Large variation in eg sorption coefficients and degradation rates for various pesticides has been observed at different sites, soil types and depths.

In general, testing and validation of models under a variety of conditions still remain to be performed. Some examples of model evaluations and comparisons are described in Pennel et al. 1990, Jarvis et al. 1994, Styczen and Villholth 1994, Bosch and Boesten, 1994b.

A review of selected approaches describing processes known to significantly influence the fate of pesticides in the unsaturated zone are shown in Table 3.

The main differences between the described modelling codes are their ability to mimic various hydrological conditions and their complexity with respect to chemical transformation processes. In order to take part in a standard registration procedure, it is important that the model codes are able to simulate different scenarios known to represent the variation in hydrology in the area under consideration. As an example which is valid for Danish conditions, this implies that the codes should provide various options for selection of the lower boundary conditions such as groundwater present in the root zone, and that special features like subsurface drainage can be included. One of the described model codes (PELMO) only contain the lysimeter boundary as an option while the number of available boundary conditions in the other model codes range from 4 to 9. The option for including subsurface drainage is only provided by MACRO, RZWQM and MIKE SHE.

The most common approach for simulating chemical degradation reactions is to assume one type of reaction described as 1. order decay and allowed to depend on temperature and soil moisture content. Some codes however (RZWQM and LEACHM), consider different degradation reactions, and MACRO allows different reaction rates to be associated with the matrix and the macropores even though the different degradation rates required as input are not commonly available, making parameter assessment difficult. The same problem is identified in the descriptions of the sorption processes. PESTLA, for instance, allows for kinetic sorption which require input of a sorption rate.

The general conclusion from the evaluation of existing pesticide modelling codes is that they provide strong and useful tools for research and comparative risk assessment, but that the present validation status is not adequate for predicting environmental concentrations (PEC) under different hydrological conditions for legislative purposes.

3. Modelling at Field and Catchment Scale

3.1. PROBLEMS AND APPROACHES IN UPSCALING

The modelling approaches presented in the previous sections focused on describing the transport processes in single soil columns, also regarded as point scale approaches.

AGROCHEMICAL MODELLING

Table 3. Review of some of the most comprehensive pesticide modelling codes.

Process	PELMO	PESTLA	MACRO	LEACHM	RZWQM	WAVE	MIKE SHE
Water flow	Capacity model	Richards' eqn.	Richards' eqn. in matrix	Richards' eqn.	Green & Ampt/ Richards' eqn.	Richards' eqn. in matrix	Richards' eqn. in matrix
Pref. flow	-	-	+	-	+	-(+)	+
Lower boundary	Free drainage	Daily groundwater level, daily flux, predescribed flux, daily potential, free drainage.	Free drainage, constant potential, constant hydraulic gradient, groundwater in root zone or zero flux	Constant potential, free drainage, zero flux, lysimeter	Free drainage, constant potential	Free drainage, gwt., lysimeter, fluxes, pressure head	Free drainage, constant potential, predescribed flux, groundwater in root zone, daily ground water level.
Solute transport	Piston displacement	Conv./disp. eqn.	Conv./disp. eqn.	Conv./disp. eqn.	Piston displacement, + mobile/immobile water	Conv./disp. eqn. + mobile/immobile water	Conv./disp. eqn.
Heat	Empirical $f(T_{air}, depth)$	Heat flux calc. $f(T_{air}$, heat conductivity, $\delta T/\delta z)$	Heat flux calc. $f(T_{air}$, heat conductivity, $\delta T/\delta z)$	Heat flux calc. $f(T_{air}, \theta$, heat conductivity, $\delta T/\delta z)$	Heat flux calc. $f(T_{air}$, heat conductivity, $\delta T/\delta z)$	Heat flux calc. $f(T_{air}$, heat conductivity, $\delta T/\delta z)$	Empirical $f(T_{air}, depth)$
Sorption	Linear, Freundlich, kinetic. f(depth)	Linear + Freundlich	Linear, K_d defined in depth.	Linear, Freundlich, two site linear: equilibrium/kinetic	Two sites: viz. linear and kinetic	Two sites, linear + kinetic.	Linear, Freundlich, Langmuir. Kd defined in depth.
Degradation	n. order, $f(\theta, T, depth)$	1. order, $f(\theta, T, depth)$	1. order, $f(\theta, T)$, defined separately in depth, matrix/macropores, solid/liquid phase.	1. order, $f(T, \theta)$. Defined separately in liquid/solid phase.	1. order, $f(T, \theta)$. Different reaction types.	1. order, $f(\theta, T)$.	1. order, $f(\theta, T)$. Defined separately in depth.
Volatil.	+	-	-	+	+	+	-
Plant uptake	Passive, f(transpiration)	Passive, f(transpiration)	Passive, f(transpiration)	Passive, f(transpiration)	Passive, f(transpiration)	Passive, f(transpiration)	Passive, f(transpiration)
Other functions included			Subsurface drainage. Swell/shrink of macropores. Switch between 1 & 2 domains.	Simultaneous simulation of several compounds.	Includes biodegradation, oxidation, complexation, photolysis, hydrolysis, bicarbonate buffering.	Hysteresis	Subsurface drainage. Three-dimensional ground water and surface runoff can be simulated at catchment scale.
References	Klein (1993)	Boesten and van der Linden (1991), Boesten (1993)	Jarvis (1991, 1994)	Wagenet and Hutson (1987), Hutson and Wagenet (1992)	DeCoursey et al. (1989, 1992)	Dust et al. (1994), Vereecken et al. (1994), Vanclooster et al. (1994)	Abbott et al. (1986), Refsgaard and Storm (1995)

(+): Planned in future versions, T: Temperature, θ: Soil moisture content, z: Depth,

These models are valuable research tools for studying transport and transformation processes, but contain a range of limitations when it comes to predicting loads of agrochemicals to streams and aquifers arising from different agricultural systems. The models do not consider spreading processes arising from flux of water and solute in two or three directions in the groundwater zone, nor do they allow for considerations regarding the horizontal variation in hydraulic and chemical properties in the soil or the distributed nature of geology, topography, drainage networks and agricultural management.

For projects focusing on larger geographical areas, such as studies of the impacts of agricultural management practices on solute concentration in groundwater aquifers or streams, extrapolation based on small-scale studies becomes difficult because such studies are likely to represent only a limited selection of the characteristics (soil types, depth of unsaturated zone, vegetation, etc.) found in a larger area. Models covering larger areas may therefore provide a better basis for decision making with regard to management strategies or policies. On the other hand, the problem here is how adequately the study area should be characterized. As the scale increases, the information required for running the models cannot be derived directly (e.g. from measurements), and the results become only approximate due to the simplifications introduced, and the neglected spatial variability of certain features. In addition, model validation is difficult. Site-specific comparisons against observed data cannot be made because representative (or effective) parameter values rather than measured values are used. It is therefore of major importance that model users recognize and report the limitations and uncertainties in the model predictions.

As described in Chapter 4 of this book, numerical groundwater models describing the flow and transport mechanisms of aquifers have been developed since the 1970's and applied in numerous pollution studies. They have mainly described the advection and dispersion of conservative solutes. More recently, geochemical and biochemical reactions have been included to simulate the fate of reactive pollutants from point sources such as industrial and municipal waste-disposal sites (see Chapter 5 of this book). Few attempts have been made to simulate non-point pollution from fertilizers and chemicals used in agriculture. The main problem arises from the need for characterization of physical, chemical and biochemical properties of large areas. An additional problem in connection with groundwater modelling is to provide an estimation of the solute input from the unsaturated zone to the groundwater. If the estimates are not based on results from leaching models, the timing and volume of nitrate fluxes are difficult to assess, because they depend on several factors, such as the depth of the unsaturated zone. This will have an important effect on the simulated concentrations in the groundwater. In areas with a shallow groundwater table, the surface application is reflected in the temporal variation of nitrate concentrations in the groundwater to a higher degree than in areas with large distances to the groundwater table.

Two principally different approaches for upscaling simulations of agricultural managements systems have been developed during the recent years. As these two approaches have similarities with two of the classes of hydrological models described in Chapter 2 of this book they will be denoted lumped conceptual and distributed physically-based approaches, respectively.

In the lumped conceptual approach the area under consideration, typically an agricultural field or a small catchment, is conceptualized as having horizontally homogenous soil properties, uniform rainfall distribution and only one type of landuse and management practice each year. The process descriptions related to the transport and transformation processes in the root zone correspond to those applied in the single column models, whereas the components describing surface run off usually are extended in order to handle overland flow and erosion as functions of areal and topographical data. Percolation from the root zone is routed to the groundwater, but lateral groundwater flow is typically not considered. Similarly, feedback from groundwater zone to unsaturated zone is usually not included.

The distributed physically-based approach allows for horizontal distributions of physical and chemical parameters, rainfall, land use, topography etc. Lateral surface and subsurface flows are included.

In the following two subsections examples of the two model types are briefly introduced and intercompared.

3.2. LUMPED CONCEPTUAL FIELD SCALE MODELLING

Use of the lumped approach for modelling agricultural aspects on larger scales involves conceptual representation of the area in question as being homogenous with respect to climatic conditions, topography, geology, land use, management practice, soil characteristics etc.. This implies that the hydrological unit in this approach is representing a field, hillslope or subcatchment.

Most existing model codes of this type use a simple hydrological model for description of the water balance such as the capacity approach for vertical water flow. The major reason for this is that the spatial resolution and assumptions of the lumped approach are too coarse basis for more complex physically based modelling. The simple water balance approach does not allow direct interaction between groundwater and surface waters. Only fluxes out of the conceptual hydrological unit are accounted for and are routed to rivers or groundwater based on empirical equations. Fluxes between single units and local phenomena such as periodically ponding due to inhomogeneous topography or soil characteristics are not accounted for.

Examples of lumped conceptual models used for assessment of agricultural systems are the family of model codes developed by the U.S. Department of Agriculture having the hydrology in terms of the modified SCS curve number technique in common. These models are CREAMS (Knisel et al., 1980; Knisel and Williams, 1995) which contain fairly simple descriptions of water, nitrate and pesticide transport, GLEAMS (Leonard et al. 1987) which has more chemical focus on pesticides and latest SWRRB (Arnold et al. 1990; Arnold and Williams, 1995), which is described as a basin scale model for simulation of water, nitrate, phosphorous and pesticides. SWRRB include the hydrology part of CREAMS and the pesticide part of GLEAMS and allow for simultaneous calculations of hydrological units representing different fields or subcatchments differing in eg landuse, and agricultural management practice.

Some of the models operate on a continuous simulation basis, while others, such as

ANSWERS (Beasly et al. 1980) and AGNPS (Young et al. 1995) can only simulate single events. For the event models the very difficult estimation of catchments initial conditions are crucial for obtaining reliable model predictions.

3.3. DISTRIBUTED PHYSICALLY-BASED CATCHMENT SCALE MODELLING

Using a distributed physically-based model for agricultural impact assessment provides the possibility of including the distributed nature of different agricultural management practices within an entire catchment, and allow for detailed descriptions of water and solute fluxes within the catchment. The hydrological unit in this approach consists of a large number of internal grids which are defined independently of the catchment structure with a spatial resolution reflecting the complexity of the catchment area in terms of spatial distribution in climatical conditions, topography, geology, land use, agricultural management practice etc,. The choice of grid size is defined by the modeller allowing for simulations varying in detail and complexity, depending on availability of model parameters and the objective of the study. Fluxes of water and solutes are routed between the internal grid elements depending on the spatial representation of catchment, and the physically based nature of the approach also allow for simulation of the interaction with groundwater and surface water.

This type of modelling approach require a large number of parameters describing the spatial variation within an entire catchment. If such parameters are not available application of a lumped conceptual approach may be adequate. However, especially for studies involving detailed assessment of solute fluxes and concentrations, the spatial representation and the physically based nature of the calculations is crucial.

An example of a fully distributed model is the MIKE SHE (Refsgaard and Storm, 1995)

Examples of studies attempting to couple the unsaturated and saturated zone models in order to perform assessments of the impact of agricultural management on the nitrate load to streams and aquifers, are described in Bogardi et al. (1988), Storm et al. (1990), Styczen and Storm (1993).

4. Case study: Modelling of Nitrogen Transport and Transformation on a Catchment Scale

4.1. INTRODUCTION

The present case study is one of the outputs from a comprehensive Danish research and development programme (1986-90), which was carried out with the aim of studying the pollution from nutrients and organic matters in agriculture. The research programme was multidisciplinary and involved a large number of research institutions. It included field investigations, process studies and modelling.

The present case study briefly describes a distributed hydrological modelling of nitrate transport and transformation for the 440 km^2 Karup River catchment. The nitrogen modelling covers the entire land phase of the hydrological cycle - from the source on the soil surface, through the soil zone and the groundwater to the streams. The modelling was based on the MIKE SHE for catchment processes and for the DAISY model (Hansen et al. 1991) for simulation of the nitrogen dynamics in the root zone. The concepts in the coupling of the one-dimensional leaching model (DAISY) and the three-dimensional model (MIKE SHE) is illustrated in Fig. 1. A more detailed description of the present case study is presented by Styczen and Storm (1993).

4.2. MODEL SET-UP

The Karup catchment was represented in a three-dimensional network. The discretization is 500 m in the horizontal directions and varies in the vertical from 5 to 40 cm in the unsaturated zone, and 5 m in the permanently saturated zone. Information on soil and vegetation properties were collected and processed based on information from a number of wells, a three-dimensional geological map was superimposed on the model grid to provide the hydrogeological parameter values. The topography and the river network have been digitized, and all relevant climatological data collected. The overall land use has been identified.

4.3. RESULTS

4.3.1 Discharge and groundwater table hydrographs
The streamflow is simulated for the period 1969 - 1988 at several sites. A comparison with measured discharge at the catchment outlet is shown for four years in Fig. 2. In addition the simulated groundwater table is compared with observations in selected wells (Fig. 3). The comparison indicates that the modelling system simulates the hydrological regime with acceptable accuracy.

4.3.2 Leakage from the root zone
To simulate the trend in the nitrate concentrations in the groundwater and the streams, it is necessary to have information on the history of the fertilizer application in space and time. This information is difficult to obtain in details, for example it is not possible to estimate which type of crop was growing on one particular field in one particular year in the past. The most detailed information one can expect to obtain is a spatial percentage of the various crops, and the types of farming practices that have been carried out in the area. Based on this information a series of 14 crop rotation schemes covering the period of interest was established, and at random distributed over the area.

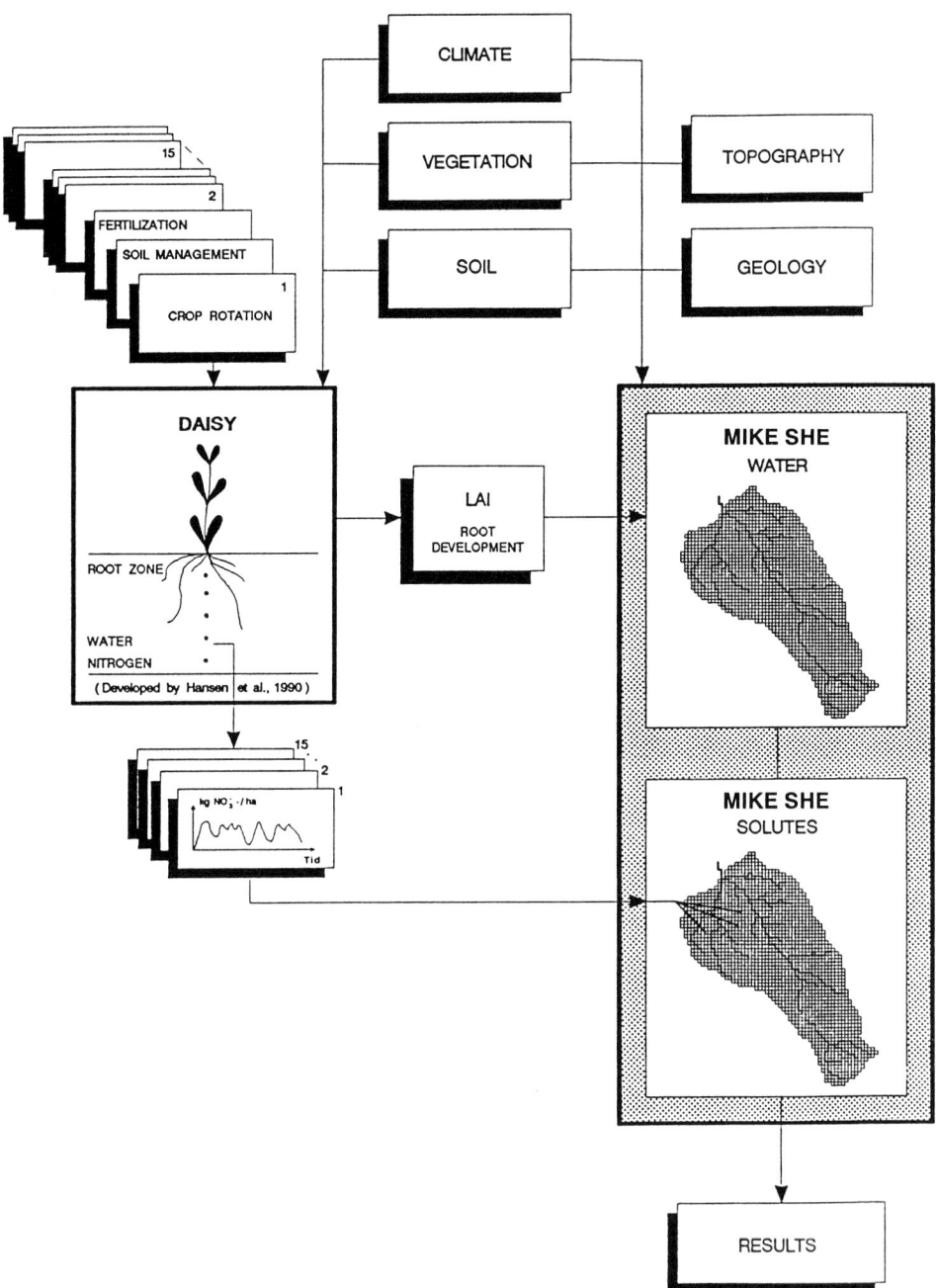

Figure 1. Coupling of the one-dimensional leaching model (DAISY) and the three-dimensional model (MIKE SHE)

AGROCHEMICAL MODELLING

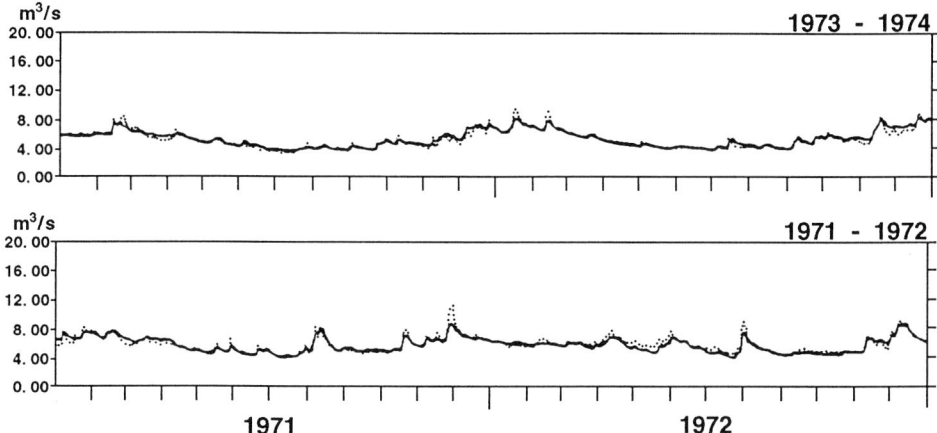

Figure 2. Comparison between simulated and observed river runoff for the period 1971-74

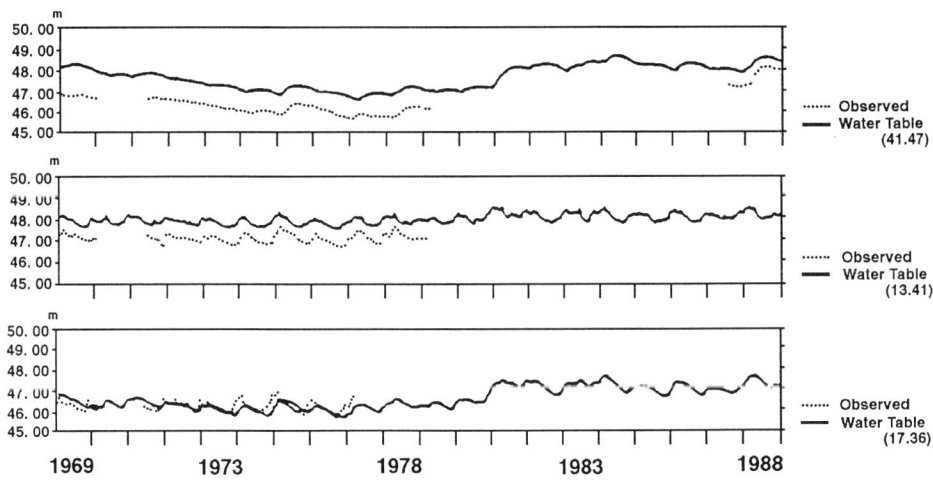

Figure 3. Comparison between simulated and observed groundwater table time series in selected wells

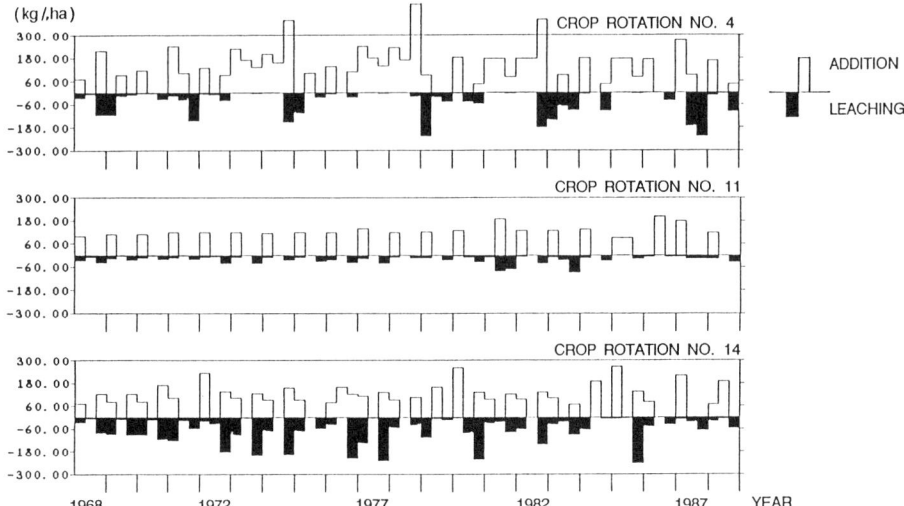

Figure 4. Nitrate leaching (NO$_3^-$-N) from three of the crop rotations calculated by DAISY and summarized over four-months periods. The shown additions of N only include mineral fertilizer and the already mineralized part of manure.

Based on estimated application rates of organic and mineral fertilizer to the individual crops each year, the DAISY model simulates the crop growth, root uptake, mineralization and leakage of nitrate from the root zone. Fig. 4 shows time series of application and leakage for selected crop rotation schemes. On farms which are based on mainly meat production a large amount of organic fertilizer will often be applied on the fields in the autumn. In this period there is a potential risk for significant losses to the groundwater system.

4.3.3 Nitrate concentrations in groundwater

While the root zone model simulates one 'soil column' at a time the total model allows studies of the variations in space and time at regional scale. Fig. 5 illustrates the variation in simulated NO$_3^-$-concentrations in the upper groundwater layer of the Karup catchment below three selected cropping schemes for two points with different depths of the unsaturated zone. A deep unsaturated zone is seen to dampen the influence of a single year.

Fig. 6 shows the spatial variation in simulated NO$_3^-$-concentrations in the upper groundwater layer at a specific time. The very large variation of concentration both in space and time is noticed.

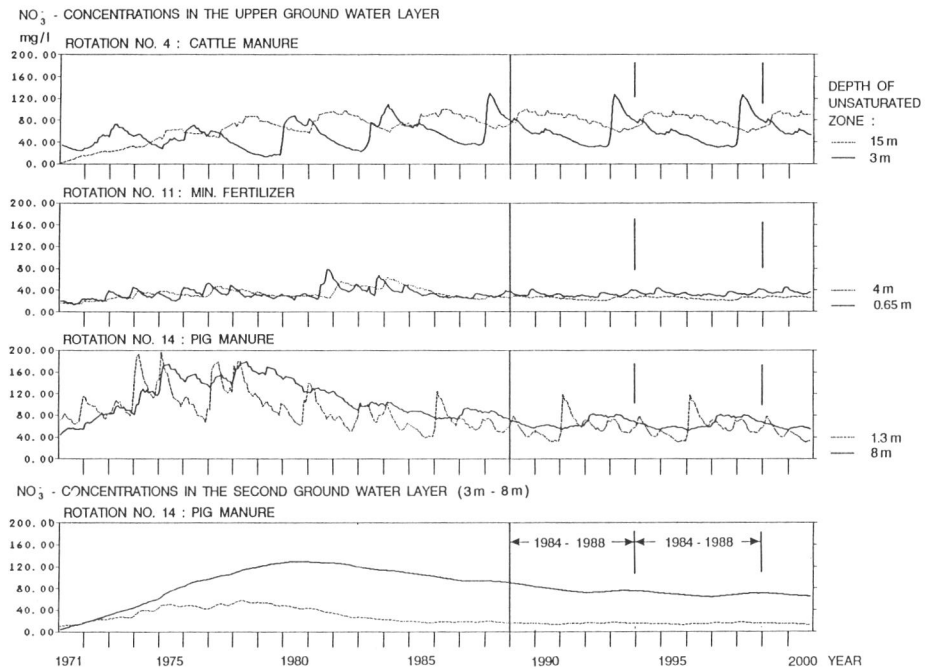

Figure 5. Temporal variation in NO$_3^-$ concentrations in the upper groundwater layer beneath three selected rotation schemes, with two different distances to the groundwater table. The data are extracted from selected grids (not averaged).

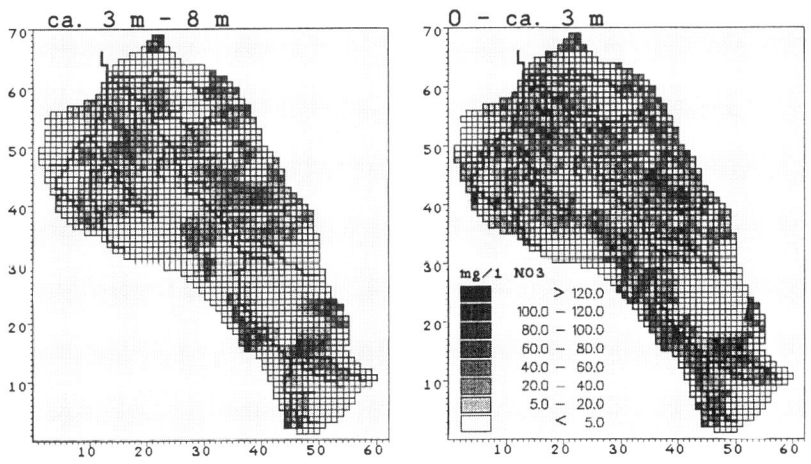

Figure 6. Spatial variation in NO$_3^-$ concentrations in the upper groundwater layer over the entire catchment at a specific time.

5. Discussion on Model Applicability and Limitations

Compared to 'pure' hydrological modelling the agrochemical modelling has naturally not reached the same level of reliability and applicability. The issue is much more complex because it, in addition to all the inherent hydrological problems, comprises agrochemical problems with regard to process understanding, field data availability and modelling methodology. On the basis of the significant progresses made during the past decade and the increasing demand for agrochemical modelling, very significant progress can, however, be expected during the coming years.

The key problems which need to be addressed are insufficient understanding of processes at a local scale, insufficient availability of field data and inadequate methodology for treating the effects of heterogeneity of process parameters and variables.

Many of the process descriptions, even in the so called 'physically-based' model codes, comprise a theoretically based frame, but include empirical equations, which cannot be parameterized further until more knowledge on processes or more particular field data become available. This is similar to the situation on soil erosion modelling, where the importance and difficulties in describing local scale processes are outlined in Chapter 6 of this book.

The most significant progress made in agrochemical modelling during the past decade is related to development of comprehensive physically-based leaching models. In spite of considerable uncertainty involved due to the deficiencies described above, such models appear to have some predictive capability, and can, with professional and cautious use, be very useful tools for management purposes.

For catchment scale modelling the present status is less advanced. A key problem in this respect is the very large spatial (and temporal) variability within a catchment of important parameters, such as soil hydraulic, soil chemical, geological and topographical parameters as well as cropping pattern, fertilization practise, tillage etc. The existing lumped conceptual models are rather easily operational; but have significant theoretical limitations and are, due to their limitations not able to address the effects of many key management options. The distributed physically-based models, based on the advanced (point scale) leaching models in a distributed hydrological modelling framework have significant potentials. The few existing examples of such approaches, such as the one described in Section 4 above, have indicated the usefulness of the approach, but have not highlighted the inherent difficulties. The key fundamental problem in this regard relates to the problem of using the point scale leaching models at grid scales in the distributed models. In the above example for the 440 km^2 catchment the grid size was 25 ha. Such discretization is sufficient for producing a good hydrological simulation of discharges and groundwater levels, but may be problematic for simulation of nitrogen and pesticide leaching, transport and transformation. In connection with many policy questions information at national scale or even above (e.g. EU) is relevant. In such cases the discretization may, in practise, have to be even larger. Fundamental research on these issues is required in the coming years.

6. References

Abbott, M.B., J.C. Bathurst, J.A. Cunge, P.E. O'Connel and J. Rasmussen (1986). An introduction to the European Hydrological system - Systeme Hydrologique Europeen - "SHE", 2:Structure of a physically-based distributed modelling system. *Journal of Hydrology,* 87, 61-77.

Arnold, J.G. and J.R. Williams, A.D. Nicks and N.B. Sammons (1990). SWRRB--A basin scale simulation model for soil and water resources management. Texas A & M University Press, College Station. 241 pp.

Arnold, J.G. and J.R. Williams (1995). SWRRB--A Watershed Scale Model for Soil and Water Resources Management. In: Computer Models of Watershed Hydrology. Ed. V.P. Singh. Water Resources Publication, 847-908.

Beasley, D.B, L.F. Huggins and E.J. Monke (1980). ANWERS: A model for watershed planning. *Trans ASAE*, 23 (4), 938-944.

Berghuijs, J.T. van Dijk, P.E. Rijtema and C.W.J. Roest (1985). ANIMO, Agricultural nitrogen Model. Nota 1671, Institute for Land and Water Management Research, Wageningen, the Netherlands.

Bergström, L. and N. Jarvis (1994). (Eds.) *Journal of Environmental Science and Health*. Special issue on the Evaluation and Comparison of Pesticide Leaching Models for Registration Purposes.

Boesten, J.J.T.I. and A.M.A. van der Linden (1991). Modelling the influence of sorption and trans formation on pesticide leaching and persistence. *Journal of Environmental Quality*, 20, 425-435.

Boesten, J.J.T.I. (1993). Users manual for version 2.3 of PESTLA. Interne medelingen 275. DLO Winand Staring Centre (SC-DLO). Wageningen, Netherlands.

Booltink, H.W.G (1994). Field-scale distributed modelling of bypass flow in a heavily textured clay soil. *Journal of Hydrology*, 163, 65-84.

Bosch, H. van den, and J.J.T.I. Boesten (1994a). Validation of the PESTLA model: Field test for leaching of two pesticides in a humic sandy soil in Vredepeel (The Netherlands). Report nr. 82, DLO Winand Staring Centre, Wageningen, the Netherlands.

Bosch, H. van den, and J.J.T.I. Boesten (1994b). Validation of the PESTLA model: Evaluation of the validation status of the pesticide leaching models PRZM, LEACHMP, GLEAMS, and PELMO. Report nr. 83, DLO Winand Staring Centre, Wageningen, the Netherlands.

Bogardi, I., J.J. Fried, E. Fried, W.E. Kelly and P.E. Rijtema (1988). Groundwater Quality Modelling for Agricultural Non-point sources. In: *Proceedings of the International Symposium on Water Quality Modelling of Agricultural Non-point Sources, June 19-23, Utah State University, Logan, Utah* (ed. D.G. DeCoursey) USDA, ARS81, 307-325.

Brüsch, W. and H. Kristiansen (1994). Fund af pesticider i grundvand. 11. Danske Planteværnskonfe rence 1994, SP report no. 6, 93-103.

CEC (1991). Soil and Groundwater Research Report II. Final Report, Contract nos. EV4V-0098-NL and EV4V-00107-C. EUR 13501.

Culley, J.L.B., E.F. Bolton and V. Bernyk (1983). Suspended solids and phosphorous loads from a clay soil. 1. Plot studies & 2. Watershed studies. J. Env. Qual., 12: 493-503

DeCoursey, D.G., K.W. Rojas and L.R. Ahuja (1989). Potentials for non-point source groundwater contamination analyzed using RZWQM. Paper No. SW892562, presented at the International American Society of Agricultural Engineers' Winter Meeting, New Orleans, Louisiana.

DeCoursey, D.G., L.R. Ahuja, J. Hanson, M. Shaffer, R. Nash, K.W. Rojas, C. Hebson, T. Hodges, Q. Ma, K.E. Johnsen, and F. Ghidey (1992). Root Zone Water Quality Model, Version 1.0, Technical Documentation. United States Department of Agriculture, Agricultural Research Service, Great Plains Systems Research Unit, Fort Collins, Colorado, USA.

Diekkrüger, B., D. Söndergerath, K.C. Kersebaum and C.W. McVoy (1995). Validity of agroecosystem models. A comparison of results of different models applied to the same data set. *Ecological Modelling*, 81, 3-9.

DHI (1994). MIKE SHE WM - A short description. Danish Hydraulic Institute, Hørsholm, Denmark.

Dust, M., H. Vereecken, M. Vanclooster and F. Fuhr (1994). Comparison of model calculations and lysimeter study results on dissipation and leaching behaviour of 14Clopyralid. In Proceedings of the

International Conference on Pesticide Fate, IUPAC, Washington DC, June 1994.

Fielding, M., D. Barcelo, A. Helweg, S. Galassi, S. Torstensson, P. Van Zoonen, R. Wolter, and G. Angeletti (1991). Pesticides in ground and drinking water. Water Pollution Research Report 27, Commission of the European Communities.

FOCUS (1995). Leaching models and EU registration. Final report of the work group - FOrum for the Co-ordination of pesticide fate models and their Use, funded by the European Commission, the European Crop Protection Association and COST Action 66. DOC.4952/VI/95.

Hansen, S. (1992). Comparison of two management-level simulation models of nitrogen dynamics in the crop-soil system DAISY and RZWQM. Dina Research Report no. 13. The Royal Veterinary and Agricultural University, Dep. of Agric. Sci., Copenhagen, Denmark.

Hansen, S., H.E. Jensen, N.E. Nielsen and H. Svendsen (1990). DAISY: A soil plant system model. Danish simulation model for transformation and transport of energy and matter in the soil plant atmosphere system. NPo Research in the NAEP, Report A10. National Agency of Environmental Protection, Copenhagen, Denmark.

Hansen, S., H.E. Jensen, N.E. Nielsen and H. Svendsen (1991): Simulation of nitrogen dynamics and biomass production in winter wheat using the Danish simulation model DAISY, *Fertilizer Research*, 27, 245-259.

Hansen, S., H.E. Jensen, N.E. Nielsen and H. Svendsen (1993). The soil plant system model DAISY. Simulation model for transformation and transport of energy and matter in the soil plant atmosphere system. Basic principles and modelling approach. Jordbrugsforlaget, the Royal Veterinary and Agricultural University, Copenhagen, Denmark.

Hutson, J.L. and R.J. Wagenet (1992). LEACHM. Leaching Estimation and Chemistry Model. Version 3. Department of Soil, Crop and Atmospheric Sciences, Research Series No. 92-3, New York State College of Agriculture and Life Sciences, Cornell University, Ithaca, New York.11 + 112.

Jansson, P.E., H. Eckersten and H. Johnsson (1991). SOILN model - User's manual. Communications 91:6, Swedish University of Agricultural Sciences, Uppsala, Sweden.

Jarvis, N. (1991). MACRO - A model of water movement and solute transport in macroporous soil. Monograph, Reports and dissertations, 9. Dept. of Soil Science, Swedish Univ. of Agric. Sci., Uppsala, Sweden.

Jarvis, N. (1994). The MACRO Model (Version 3.1) - Technical description and sample simulations. Monograph, Reports and dissertations, 19. Dept. of Soil Science, Swedish Univ. of Agric. Sci., Uppsala, Sweden.

Johnsson, H., L. Bergström, P.E. Jansson and K. Paustian (1987). Simulated Nitrogen dynamics and losses in a layered agricultural soil. *Agr. Ecosystems Environ*, 18, 333-356.

Klein, M. (1993). PELMO - Pesticide Leaching Model, version 1.5. Users Manual. Frauenhofer-Institut für Umweltchemie und Ökotoxikolgie, 57392 Schmallenberg.

Knisel, W.G. (ed.), (1980). CREAMS: A Field-Scale Model for Chemicals, Runoff, and Erosion from Agricultural Managements Systems. U.S. Department of Agriculture, Science, and Education Administration. Conservation Research Report No. 26. 643 pp.

Knisel, W.G., R.A. Leonard and F.M. Davis (1989). GLEAMS User's Manual, Southeast Watershed Laboratory, Tifton, Ga.

Knisel, W.G. and J.R. Williams, (1995). Hydrology Component of CREAMS and GLEAMS Models. In: Computer Models of Watershed Hydrology. Ed. V.P. Singh. Water Resources Publication, 1069-1114.

Leonard, R.A., W.G Knisel and D.A Still (1987). GLEAMS: Groundwater Loading Effects of Agricultural Management Systems. *Trans ASAE*, 30, 1403-1418.

Pennel, K.D., A.G. Hornsby, R.E. Jessup and P.S.C. Rao (1990). Evaluation of five simulation models for predicting aldicarb and bromide behaviour under field conditions. *Water Resources Research*, 26, 2679-2693.

Refsgaard, J.C. and B. Storm (1995). MIKE SHE. In: Computer Models of Watershed Hydrology. Ed. V.P. Singh. Water Resources Publication, 809-846.

Rijtema, P.E, C.W.J. Roest and J.G. Kroes (1991). Formulation of the nitrogen and phosphate behaviour in agricultural soils, the ANIMO model. Report no. 30. DLO Winand Staring Centre, Wageningen, the Netherlands.

Schjønning, P., E. Sibbesen, A.C. Hansen, B. Hasholt, T. Heidmann, M.B. Madsen and J.D. Nielsen (1999). Surface runoff, erosion and loss of phosphorous at two agricultural soils in Denmark - plot studies 1989-92. SP report no. 14, Danish Institute of Plant and Soil Science, Foulum, Denmark, 196 pp.

Sharpley, A.N. and J.R. Williams, eds. (1990). EPIC--Erosion/Productivity Impact Calculator. 1. Model documentation. U.S. Dept. Agric. Tech. Bull. No. 1768.

Storm, D.E., T.A. Dillaha III, S. Mostaghimi and V.O. Shanholtz (1988). Modelling phosphorous transport in surface runoff. Transactions of the ASAE, vol. 31, 1, 117-127.

Storm, B., M. Styczen and T. Clausen (1990). Regional Model for Nitrate Transport and Transformation. *NPo Research in the NAEP*, Report B5. National Agency of Environmental Protection, Copenhagen (in Danish).

Styczen, M. and B. Storm (1993). Modelling of N-movements on catchment scale - a tool for analysis and decision making. 1. Model description & 2. A case study. *Fertilizer research*, 36, 1-17.

Styczen, M. and K. Villholth (1994). Pesticide Modelling and Models. Technical Report to the Danish Environmental Protection Agency, Copenhagen. Bekæmpelsesmiddelforskning fra Miljøstyrelsen, Nr. 9 1995.

Vanclooster, M., J. Viaene and K. Christians (1994). WAVE - a mathematical model for simulating agrochemicals in the soil and vadose environment. Reference and user's manual (release 2.0). Inst. for Land and Water Management, Katholieke Universiteit Leuven, Belgium.

Vanclooster, M., P. Viaene, j. Diels and J. Feyen, 1995. A deterministic validation procedure applied to the integrated soil crop model. *Ecological modelling*, 81, 183-195.

Van der Zee, S.E.A.T.M. and A. Gjaltema (1992). Simulation of phosphate transport in soil columns. 1. Model development. Geoderma, 52: 87-132.

Vereecken, H., M. Vanclooster, M. Swerts and J. Diels (1991). Simulating nitrogen behaviour in soil cropped with winter wheat. *Fertilizer Research*, 27, 233-243.

Vereecken, H., M. Dust, Th. Putz, M. Vanclooster and F. Fuhr (1994). Modelling the fate of methabenzthiazuron in arable soil using two year lysimeter study. In Proceedings of the International Conference on Pesticide Fate, IUPAC, Washington DC, June 1994.

Wagenet, R.J. and J.L Hutson (1989). LEACHM - Leaching Estimation and Chemistry Model. A process-based model of water and solute movements, transformations, plant uptake and chemical reactions in the unsaturated zone. Continuum Vol. 2 (Version 2.0). Water Resources Institute, Center for Environmental Research, Cornell University.

Williams, J.R. (1995). The EPIC Model. In: Computer Models of Watershed Hydrology. Ed. V.P. Singh. Water Resources Publication, 909-1000.

de Willigen, P. (1991). Nitrogen turnover in the soil-crop system; Comparison of fourteen simulation models. *Fertilizer research*, 27, 141-149.

Young, R.A, C.A. Onstad and D.D. Bosch (1995). AGNPS: An Agricultural Nonpoint Source Model. In: Computer Models of Watershed Hydrology. Ed. V.P. Singh. Water Resources Publication, 1001-1020.

CHAPTER 8

WEATHER RADAR PRECIPITATION DATA AND THEIR USE IN HYDROLOGICAL MODELLING

C G COLLIER
Telford Institute of Environmental Systems
Department of Civil and Environmental Engineering
University of Salford
M5 4WT
United Kingdom

1. Introduction

Most variables in the hydrological cycle show large and frequent spatial variations, and often exhibit rapid temporal variations. Of particular importance is precipitation, and it is often necessary in many hydrological process studies and applications to monitor precipitation continuously at as many points as possible.

Point measurements are insufficiently representative of sub-catchment scales, and hence it is difficult to understand and model hydrological processes using such data. The basic measurement requirement of hydrology is for areal measurements albeit over sometimes very small, as well as very large, areas.

The hydrological requirement for measurements of precipitation is shown in Table 8.1, where the maximum, minimum and most usual values for resolution, frequency of observations and accuracy (freedom from bias) are given. Clearly which requirement value is appropriate will depend upon the application to which the measurements are put.

The precision with which measurements should be made refers to their reproducibility in space, time and quantity. Precision is expressed quantitatively as the standard deviation of results from repeated trails under identical conditions. Hence a measurement may be quantitively accurate, but imprecise and vice-versa. The reasons for this depend upon the inter-relationship between resolution, frequency and numerical accuracy (see for example, Chatfield, 1983). A measurement from a raingauge is spatially precise, but may be imprecise in time and quantity. Hence in Table 8.1 we have added requirements for precision where this quantity is taken as the standard deviation about the mean requirements for resolution, frequency and accuracy. It is therefore a tolerance within which the requirement stated must be met.

2. Use of precipitation measurements in distributed hydrological models

Lumped models attempt to represent the relationship between the characteristics of the hydrograph and its physiographic factors as a simple relationship between excess rainfall and direct runoff. The main criticism of the lumped models is that they require an assumption that the rainfall input is uniformly distributed over the catchment, or at least, that the spatial distribution is constant. This is an unrealistic assumption. Semi-distributed models consist of a system of interconnected cell units, each cell representing a definite portion, or area unit, of the catchment (see for example Diskin and Simpson, 1978) Figure 8.1 shows how sub watersheds (subcatchments) 8 and 11 of the Walnut Gulch (Arizona) Experimental Catchment are divided into model cells.

All cell units receive rainfall excess input which is variable with time, but assumed to be uniform across each area unit. Two types of cells are recognised, exterior or interior. Exterior cells without any channel inflow, have only one input, namely the rainfall excess input. Interior cells are cells receiving channel inflow from upstream cells in addition to the rainfall excess input. As shown in Figure 8.1 the interconnections between the cells form a branching tree-like structure which approaches the form of the main drainage pattern of the catchment.

This modelling approach may be extended to formulations of many partial differential equations governing various physical processes and equations of continuity for surface and soil water flow. The SHE model (Abbott *et al* 1986a, b) is an example. The equations are applied to grids as small as 250m x 250 m. Unfortunately experience has shown that, even with the adequate calibration data in the form of flow records, these fully distributed models may still fail to provide consistent improvements in forecast accuracy. This is because rainfall inputs need to be provided with a spatial scale close to that of the model grid, and this can only be achieved using radar data (see for example, Anderl *et al*, 1976, Moore, 1987). However, Obled *et al*, (1994) found that, for a rural medium-sized catchment (71 km^{-2}), the sensitivity of a range of model formulations to the spatial variability of rainfall, although important, is not sufficiently organised in time and space to overcome the effects of smoothing and dampening caused by the model formulations. They conceded that this might not be the case for smaller urban catchments or over larger rural catchments. Nevertheless, in general hydrological models may act as a broad band filter to enable a certain level of error in the input data to be tolerated. This is important since, as we shall see, the error characteristics of radar estimates of precipitation maybe variable in time and space. This does not mean that errors in input data are unimportant, but rather that consistency in error characteristics is as important as absolute error.

Table 8.1 Hydrological requirements for measurements of precipitation showing maximum, minimum and most usual values (partly after Herschy *et al*, 1985)

Quantity		Resolution (km)			Frequency			Accuracy (% or as shown)			Precision (% or as shown)		
		max	min	most usual	max	min	most usual	max	min	most usual	max	min	most usual
Rainfall													
(a)	Point	0.1	5	1	5 min	1 month	1 hour	2	10	5	1	5	3
(b)	Area												
	- rural (>20km²)	0.5	5	2	5 min	1 month	1 hour	10	30	20	5	20	15
	- urban/very small rural (<20 km²)	0.25	2	1	1 min	5 min	2 min	5	20	10	1	5	3
Snowfall													
(a)	Depth	0.03	10	0.2	1 hour	1 month	24 hours	2 cm	10 cm	5 cm	1 cm	30 cm	2 cm
(b)	Water equivalent	0.03	10	1	1 hour	1 month	24 hours	1 mm	100 mm	10 mm	2 mm	30 mm	4 cm

Note: The requirement for snowfall measurement refers to both point (max) and areal (min) values

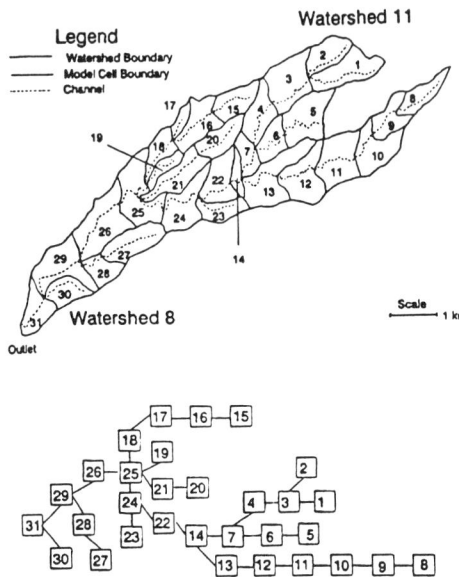

Figure 8.1 Sub watersheds 8 and 11 of the Walnut Gulch (Arizona) experimental catchment divided into model cells (from Karnieli *et al*, **1994**)

3. Single Frequency and Polarisation Radar Measurements

3.1 OUTLINE OF TECHNIQUES

Ground-based radar offers areal measurements of rainfall from a single location, over large areas in near real time. Techniques are based on four types of measurement: (1) the intensity of the back scattered radiation (radar reflectivity), (2) the difference in reflectivity or phase between two orthogonal radiation polarisations, (3) the attenuation of radar energy, and (4), attenuation and reflectivity determined simultaneously at two wavelengths. Of these measurements only the first has been used widely and implemented operationally. Hence we confine most of our discussions to this method, although technically more advanced techniques offer potential for operational development.

As a radar beam rotates about a vertical axis, measurements are made at many ranges out to 100km or more, and at different azimuths, of the energy back scattered from precipitation particles in volumes above the round. Given a radar wavelength λ, and considering a spherical raindrop with diameter D, we may define a back scattering cross section $\sigma_b(D)$ and a total attenuating cross section $\sigma_a(D)$ in proportional to D^6/λ^4. This is the justification for introducing a physical parameter called 'radar reflectivity factor', Z, defined as:

$$Z = \int_0^\infty N(D)D^6 dD \tag{8.1}$$

where N(D) is the drop size distribution (DSD) within the resolution cell (Z is in mm^6 · m^{-3}, D in mm, N(D) in mm^{-1}m^{-3}). In the absence of attenuation along the radar path, and as long as the Rayleigh theory holds, the back scattered power from a resolution cell is proportional to Z. However, when the ratio $\pi D/\lambda$ becomes larger than 0.1, then Mie theory should be used in place of Rayleigh Theory. To take account of this effect, an 'equivalent' radar reflectivity faction Z_e is generally considered:

$$Z_e = \frac{\lambda^4 Z}{\pi^5 |K|_w^2} \tag{8.2}$$

where the subscript w indicates that the value appropriate for water (approximately 0.93 for the usual meteorological radar wavelengths) is used by convention. This convention is adopted because when radar measurements are made, it is often not certain whether the particles are water or ice.

For ice particles:

$$Z_e = \frac{|K|_i^2}{|K|_w^2} \cdot Z \tag{8.3}$$

Smith (1984) discussed the appropriate value of $|K|_i^2$ pointing out that there are two possible "correct" values depending upon how the particle sizes are determined. If the particle sizes used are melted drop diameters then $|K|_i^2$ is 0.208 and

$$Z_e = 0.224 \, Z \tag{8.4}$$

However, if the particle sizes are expressed as equivalent ice sphere diameters then $|K|_i^2$ is 0.176 and,

$$Z_e = 0.189 \, Z \tag{8.5}$$

Normally radars use the "water equivalent" Z_e defined with $|K|_w^2 = 0.93$, and the dielectric factor is not changed when the precipitation form changes from liquid to solid. Table 8.2 compares equivalent radar reflectivity factors calculated for precipitation rates of 1 and 10 mm h^{-1} for rain, using the Marshall-Palmer relationship and for snow, using the Sekhan and Srivastava (1970) relationship (section 3.4). At

$R = 1$ mm h^{-1}, the Z_e value for snow is 3 dB higher than that for rain. Hence in general radar echoes from snow are not weaker than those from rain, although there is a tendency for the precipitation rates to be generally lower in snow than in rain.

	Precipitation rate R (mm h^{-1})	
	1	10
Z_e (rain) - dBz	23	39
Z_3 (snow) - dBz	26	48

Table 8.2 Example values of R and Z_e for rain and snow (from Smith, 1984)

Z is related to the rate of rainfall, R, by

$$Z = A R^B \tag{8.6}$$

where A and B depend on the type of rainfall. Many values have been suggested (Battan, 1973), but values often used for rain are A = 200 and B = 1.6 (Marshall and Palmer, 1948).

Use of R:Z relationships to measure rain, modifying A and B as appropriate, would appear to be straightforward. There are a number of problems, however, arising from the characteristics of both the radar and the precipitation. The importance of these problems will depend upon the particular radar configuration in use and the meteorology of particular situations. Hence they are not listed in order of importance in what follows.

3.2 PROBLEMS ARISING FROM THE CHARACTERISTICS OF THE RADAR AND OF THE RADAR SITE:

Maintenance of a stable radar system is extremely important. Fortunately modern technology allows this to be achieved relatively easily. However, many aspects of a radar site affect the quality of radar measurements. Of particular importance is the pattern of ground clutter and occultation caused when part of the radar beam intersects the ground causing "permanent" echoes and partial beam loss at further ranges (see for examples Collier, 1989). Such effects are detrimental to measurements of surface precipitation.

The radar wavelength governs the amount of beam attenuation through precipitation. Whilst this is negligible at S-band (10cm wavelength), it is not so in heavy rainfall at C-band (5.6 cm wavelength), and can be very large at even quite small rainfall rates at X-band (3 cm wavelength). A similar effect is observed when a film of water resulting from moderate to heavy rainfall occurs on the radome protecting the radar antenna at the radar site.

3.3 PROBLEMS ARISING FROM THE CHARACTERISTICS OF THE PRECIPITATION:

Temporal and spatial variations in rainfall intensity, if large, introduce errors when the radar received power is averaged (Joss and Waldvogel, 1990). Fabry *et al* (1994) investigated the dependence of the accuracy of short-period radar rainfall accumulations on periodic sampling of the rain field. Errors due to sampling can be greater than all the other errors if accumulations are improperly computed. It was found that the best accumulations are obtained with very high time resolution data. For a given time resolution, there is an optimum spatial reduction that minimises rainfall accumulation errors, Figure 8.2.

The radar beam at far ranges from the radar site is at a considerable height about the surface of the Earth. For a beam elevation of $0.5°$ the axis of the beam is at a height of 2 km at 130 km range and 4 km at 200 km range. Reflectivity varies in the vertical because of growth or evaporation of precipitation, vertical air motion, and melting (where melting of snowflakes occurs, there is a layer of enhanced reflectivity known as the bright band).

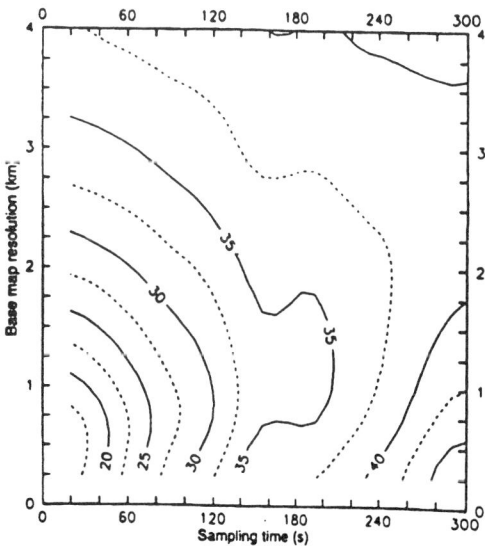

Figure 8.2 Absolute error (%) in 5 min accumulations as a function of the resolution of the reflectivity maps and sampling intervals (from Fabry *et al*, 1994).

Joss and Waldvogel (1990) concluded that errors in the radar measurement of surface precipitation are dominated by the effect of variations in the vertical profile of reflectivity. Such variations occur on the scale of individual pixels (Kitchen and Jackson, 1993), and therefore it would seem that any correction method would have the greatest potential benefit by providing adjustments on a pixel-by-pixel basis. Kitchen *et al* (194) discuss three approaches to adjusting radar estimates of precipitation for both range and bright-band effects namely,

* raingauge adjustment: All such schemes suffer from the problem of random and bias errors introduced by representativeness errors in comparisons between the raingauge and radar values. Recent work by Rosenfeld *et al* (1995) offers a possible approach which is more stable.

* analytical methods based on using radar data alone: An average reflectivity profile is derived from several beam elevations and used to correct the data (see for example, Harrold and Kitchingman, 1975, Andrieu and Creutin, 1995, Andrieu *et al* 1995).

* physically - based methods using independent meteorological data: A parameterised reflectivity profile is derived using surface data or a microphysical model (see for example, Kitchen *et al* 1994, Hardaker *et al*, 1995)

Whilst it remains to be seen which type of approach is most operationally robust, it is likely that physically-based methods will be the most reliable. No method, however, is likely to solve the problem completely.

3.4 Z:R RELATIONSHIPS

Measurements of radar reflectivity (Z) are transformed to measurements of rainfall (R) using the empirical relationship discussed in section 3.1. However, the exact form of this relationship depends upon rainfall type, and Battan (1973) lists 60 published relationships all of which have been found to be applicable in particular circumstances. Fortunately, except for those applicable to orographic rain, most of the relationships do not differ greatly at rainfall rates between about 20 and 200 mm h^{-1}. The most typical relationships for particular rainfall types are shown in Table 8.3, which also illustrates the variability of the published data for the same type of precipitation. The Z:R relationship for orographic rain must be viewed with caution as, since these data were published, an understanding of the growth of orographic rainfall at very low levels has been gained (see for example Browning, 1990). Low level orographic precipitation enhancement may well account for the relationship given in Table 8.3

Table 8.3 Typical empirical relationships between reflectivity factor Z (mm^6 m^{-3}) and precipitation intensity, R (mm h^{-1}) (after Battan, 1973). Discussion of relationships in snow are included here for comparison.

Equation	Precipitation Type	Reference
140 R$^{1.5}$	Drizzle	Joss *et al* (1970)
$\{$ 250R$^{1.5}$	Widespread rain	Joss *et al* (1970)
200R$^{1.6}$	Stratiform rain	Marshall and Palmer (1948)
31R$^{1.71}$	Orographic rain	Blanchard (1953)
$\{$ 500R$^{1.5}$	Thunderstorm rain	Joss *et al* (1970)
486R$^{1.37}$	Thunderstorm rain	Jones (1956)
$\{$ 2000R$^{2.0}$	Aggregate snowflakes	Gunn and Marshall (1958)
1780R$^{2.21}$	Snowflakes	Sakhan and Srivastara (1970)

Following work by Zawadzki (1984), Collier (1986), Joss and Waldvogel (1990) and Rosenfeld *et al* (1992) it has become clear that uncertainties in the dropsize contribution may not be largest sources of errors in radar-based rainfall measurements. Collier *et al* (1983) recognised the need to apply different Z:R relationships derived from raingauge data for different rainfall regimes.

3.5 WINDOW PROBABILITY MATCHING METHOD (WPMM) AND OTHER APPROACHES TO RAINGAUGE ADJUSTMENT

Some recent work on the raingauge adjustment of radar data has sought to use the high correlation between the area average rain intensity, $<R>$, and the fraction of the domain covered with reflectivity greater than a given threshold (τ), F (λ). Doneaud *et al* (1981, 1984) derived the volumetric rainfall, V from

$$V = \int_\tau \int_A R \, da \, dt = R_c \int_\tau \int_A da \, dt = R_c \sum_i A_i \Delta t_i \qquad (8.7)$$

where R is the instantaneous local rain rate
  da and dt are incremental elements of area and time respectively
  R_c is the average rain rate.

The integrals are taken over the entire area A for duration T. The double integral is the area-time integral (ATI).

Following this work Chiu and Kedem (1990) developed a regression model to estimate from the fractional rainy area, the conditional probability that rain rate over an area exceeds a fixed threshold given the values of related covariates. Tests showed that this approach is superior to multiple regression. However, variabilities of meteorological parameters must be accounted for if this technique is to be applied to estimate rain rate reliably.

Atlas *et al* (1990) developed a unified theory for the estimation of both the total rainfall from an individual convective storm over its lifetime, and the area wide instantaneous rain rate from a multiplicity of such storms, by use of measurements of the area coverage of the storms within a threshold rain intensity isopleth or the equivalent threshold radar reflectivity. (8.7) was generalised to

$$V = [\overline{A}(t).T].S(t) \qquad (8.8)$$

where ATI $= A(\tau)\phi.T$
$\tau = $ threshold
$S(\tau) = R_c(\tau)\phi$ and may be defined in terms of the probability density function (pdf) i.e. $\int_{\infty}^{\infty} RP(R)dR / \int_{\tau}^{\infty} P(R)dR$
$\phi = $ the fraction of the total volumetric rate as shown in Figure 8.3

Sauvageot (1994) found that P(r) can be represented by a lognormal distribution, and explained the stability of $S(\tau)$.

Divide (8.8) by total area observed A_0 then V / A_0 is the average area wide rain rate $<R>$ and $A(\tau) / A_0$ is the fractional area $F(\tau)$, covered by rain within the threshold, τ. Thus,

$$<R> = F(\tau).R_c(\tau)\phi \qquad (8.9)$$

Rosenfeld *et al* (1990) examined the generality of the instantaneous area wide method, and developed the method of utilising measurements of storm height to enhance the accuracy of the rain estimates relative to that which is attainable with storm areas only. They defined a parameter E_e, the "effective efficiency" as,

$$E_c = (Q_b - Q_t) / Q_b \qquad (8.10)$$

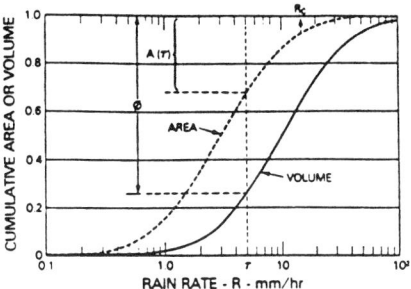

Figure 8.3 Lognormal cumulative distributions of rain area (dashed curve) and volume (solid curve) similar to those observed in GATE, Texas and South Africa as shown by Rosenfeld *et al* (1990). For a rain rate threshold of τ, the fractional rain volume and area encompassed within τ are ϕ and $A(\tau)$, respectively; the average area rain rate within that area is R_c. The curves correspond to a lognormal distribution with mean and standard deviation of log R equal to 1.1 mm h^{-1} (from Atlas *et al*, 1990)

where Q_b and Q_t are the water vapour mixing ratios at the base and top of the storm respectively. Hence E_c is the fraction of the water vapour carried up through the cloud base which is potentially available for precipitation. Q_t is determined by the actual height which is reached by the storm. As E_c increases from one class to another the corresponding pdfs move to larger rain rates in a systematic manner as shown in Figure 8.4. Therefore it is possible to stratify the $<R>$ versus $F(\tau)$ relationships according to both E_c and the corresponding best correlated τ.

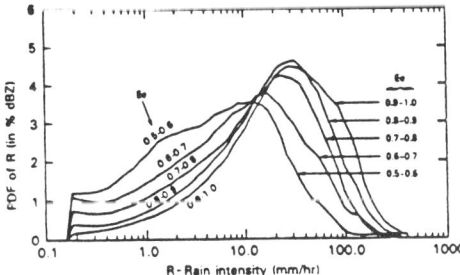

Figure 8.4 The pdfs of the rain intensities as derived from high-resolution radar data between 31 and 90 km, for convective rain situations having different average depth, denoted by E_c intervals of 0.1. These intervals range from 0.5 for the shallowest convection to 1.0 for the deepest convection (from Rosenfeld *et al*, 1990).

This technique is referred to as the window probability matching method (WPMM), and was improved further by Rosenfeld *et al* (1995a, b). Rainfall types are classified using the horizontal radial reflectivity gradients, the cloud depth as indicated by the effective efficiency, the bright band fraction within the radar field window and the height of the freezing level. The results of tests in Australia and Israel are very impressive, and it is felt by the authors that this procedure should result in considerable improvement of both point and areal rainfall measurements with respect to any other method using reflectivity-only radars. If this is so, then the extent to which raingauge adjustment schemes using interpolation of raingauge to radar ratios must be questionable. Also the WPMM does not require that range effects be removed or that the bright-band be corrected in the radar imagery, but rather that its presence and the fraction of enhanced echoes close to the 0^0C level be recognised. This requires only an independent method of defining 0^0C or bright-band altitude. Finally, Rosenfeld *et al* (1995a) define and test successfully procedures for recognising windows contaminated by ground clutter or spurious echoes which negate the need for procedures more complex than the use of a simple clutter map. In spite of the great potential of this approach, further extensive testing is required to verify its operational robustness. Also, it must be remembered that a number of, albeit not very many, raingauges are an essential element in this approach.

If a dense network of raingauges exists then the WPMM approach may not out perform methods based on surface fitting of radar (R) / raingauge (G) ratios (see for example Moore *et al*, 1991). Also, it must be remembered that whilst raingauge adjustment can clearly improve the accuracy of radar measurements of precipitation it should be noted that on any individual case this is not necessarily so. Indeed in convective precipitation raingauge adjustment can actually decrease accuracy. Kitchen and Blackall (1992) note that much of the difference between point raingauge values and radar measurements is due to gauge sampling problems. They conclude that where precipitation is dominated by orographic effects, raingauge adjustment is useful, but in other areas or on other occasions this is not so. Even techniques using data from a very dense network of raingauges suffer from the problem of random and bias errors introduced by representativeness errors in the gauge-radar comparisons. In addition, economic networks cannot completely resolve errors due to the bright band or orographic enhancement. Nevertheless, the advantage of raingauge adjustment is that in relating radar measurements to surface precipitation a range of the errors may be partially dealt with in a single process. Time integrated, G/R values are probably the most appropriate adjustment factors to use. However, point values are not representative of wide area radar measurements and it is important to use schemes involving some type of area integration.

4. Multi-Parameter Radar Techniques

The departure of the shapes of precipitation particles from spherical gives rise to different radar reflectivity properties. Seliga and Bringi (1976) related signals in two orthogonal linear polarisation planes, horizontal (H) and vertical (V), to a two-parametric dropsize distribution. Likewise, circular polarisation has also been used in this way by McCormick and Hendry (1972).

The oblateness of raindrops, when falling at terminal velocity in air, increases with drop volume. Since models for the shape and minor-to-major axis ratio and fall speed data exist, it is possible to relate the dropsize distributions so measured to rainfall rates. If raindrop canting is neglected (mean observed canting angles are in most cases less than 20^0, Brussaard,(1976), then the radar cross-section of the raindrops for horizontal polarisation is expected to be higher than the cross-section for vertical polarisation, that is

$$\sigma_h(D) > \sigma_v(D) \qquad (8.11)$$

where D is the equivolumetric sphere diameter of a raindrop

With a dual polarisation radar, both quantities, Z_H and Z_V can be measured and be used to calculate the so called differential reflectivity $Z_{DR} = Z_H/Z_V$ where, .

$$\bar{Z}_{H,V} = \frac{10^{18}.\lambda^4.N_0}{N^5|K_o|^2} \int_{D=0}^{D_{max}} \sigma_{H,V}(D).\exp(-3.67.D/D_0) \, dD \quad mm^6 m^{-3} \qquad (8.12)$$

It is seen that Z_{DR} depends on D_0 only, and therefore can be used to determine D_0. Having D_0, (8.11) can be used to find the second unknown distribution parameter N_0 (section 3.1)

Z_{DR} has a dynamic range of the order of 5 dB, and a somewhat higher sensitivity with D_0 is evident for C-band. Z_{DR}, therefore, has to be measured in an accuracy of better than 0.2 dB (standard deviation), which is one of the main problems in practical implementations. If, for example, Z_H and Z_V are measured independently as means of 100 integrated independent single-pulse reflectivities, 10% of Z_H and Z_V, are expected to be associated with fluctuation errors above 0.5 dB. This is clearly insufficient for deriving Z_{DR} with the stated accuracy. Improvements can be made by:

(a) increasing the integration number, leading to single cell measurement times in the order of seconds;

(b) acquiring single-pulse echoes for both polarisations nearly simultaneously, so that fluctuation errors in Z_H and Z_V cancel out when calculating $Z_{DR} = Z_H / Z_V$

In practice method (b) is being applied, relying on the empirical relationship for the decorrelation time of single pulse echoes at constant carrier frequency, due to drop rearrangement (Atlas, 1964);

$$Z_{DR} = \frac{\overline{Z}_H}{\overline{Z}_V} = \frac{\int_{D=0}^{Dmax} \sigma_H(D).\exp(-3.67.D/D_0).dD}{\int_{D=0}^{Dmax} \sigma_V(D).\exp(-3.67.D/D_0).dD} \qquad (8.13)$$

It would appear that the Z_{DR} radar technique has the potential for accurately measuring rainfall rate without any need for raingauge adjustment. However, as Jameson *et al* (1981) point out, single-point measurements of Z_{DR} may be associated with significantly diverse rainfall rates. Although two radar variables (Z_H, Z_{DR}) are used, two additional parameters (the maximum drop size and drop shape) are considered, so that there is little net gain in quantitative information and further radar parameters are required (Atlas *et al* (1982), or temporal and areal averaging, perhaps even some form of adjustment in particular meteorological situations. Hence the technique could begin to suffer from the same kinds of problem as the reflectivity-alone technique. Goddard *et al* (1982) have addressed these problems and suggested ways of empirically reducing the errors which result from them. Nevertheless, Hendry and Antar (1984) note difficulties caused by propagation effects, and Herzegh and Conway (1986) and Liu and Herzegh (1986) discuss problems arising from side-lobe effects.

In summary, an over-estimation using Z_{DR} has been attributed to gradients in rainfall rate, dropsize departure from the exponential relationship, departure of drop-oblateness from the model used and drop canting effects due to wind shear and turbulence. The differential reflectivity technique, like other radar techniques, is adversely affected by the presence of reflectivity gradients below the radar beam, which may be significant in cases of isolated thunderstorms or orographic rainfall. In other words, even if the radar measures the rainfall rate accurately aloft within the beam, this measurement may still be unrepresentative of the rainfall rate at the surface. The use of a narrow beamwidth (½ - 1°) helps but does not entirely overcome such problems, particularly where measurements of surface rainfall are required at ranges up to around 100 km from the radar, or in hilly areas of specific interest to hydrologists.

A range of other parameters may be derived from polarisation diversity radar (Doviak and Zrnic, 1994), which may be used to recognise hydrometeor canting effects and so correct Z_{DR}. In addition the value of the cross polar reception lies in the possibility of identifying hydrometeor type. Also Holt (1988) showed that the differential phase parameter (related to mean orientation angle) may be estimated from non-switched circularly polarised systems, opening up new possibilities for more conventional radars. Sachidananda and Zrnic (1987) showed that if accurate measurements of the propagation phase can be made, then this parameter can be used to improve estimates of heavy rainfall. None of these techniques have been tested operationally.

5. Hydrograph Forecast Accuracy Attained Using Radar Estimates of Rainfall

The impact of radar upon the accuracy of hydrograph forecasts will be dependent upon the spatial, temporal and intensity resolution of the data used. Kouwen and Garland (1989) found that a radar resolution of 10 km x 10 km over a 3250 km^2 catchment was sufficient for modelling floods produced by either thunderstorms or frontal systems. However, Ogden and Julien (1994), investigating two catchments of size 32 km^2 and 121 km^2, found that the effect of radar data spatial resolution depended upon the importance of two processes namely "storm smearing" and "watershed smearing".

Storm smearing occurs when the rainfall data length scale approaches or exceeds the rainfall correlation length (~2.3km for convective cells). This tends to decrease rain rates in high intensity regions, and increase rain rates adjacent to low intensity regions thereby effectively reducing rainfall gradients. It is independent of basin size. Watershed smearing occurs when the radar grid size approaches the characteristic catchment size (square root of catchment area). In this case, the uncertainty of the location of rainfall within the catchment boundary is increased.

Hence, in convective rainfall in which large rainfall gradients are present, it is necessary to use a radar grid size of around 1 km x 1 km. For every small urban catchments (sewage systems) it may even be necessary to use a smaller grid size if watershed smearing is to be avoided. However, for large urban or rural catchment grid sizes of 2 km x 2 km or even 5 km x 5 km are quite adequate. Michaud and Sorooshian (1994) found that, for convective rainfall in an arid region spatial averaging of rainfall over 4 km pixels led to consistent reductions in peak flow that, on average, represented 50% of the observed flow.

The temporal variation of rainfall input to a hydrological model also has a significant effect on the predicted hydrograph. Ball (1994) noted that estimation of the time of concentration for a catchment is dependent on the temporal pattern of the rainfall excess, and may be 22% longer or 19% shorter than that predicted using a constant rate of rainfall excess. Hence, the time of concentration for a catchment must be determined taking into account the pattern or magnitude of the rainfall excess. However, the peak discharge was found to be independent of the pattern of rainfall excess. Hence we may conclude that if temporal resolution of the rainfall data prevent a close reproduction of the actual rainfall variations in time, then the resultant predicted hydrograph could be subject to significant timing errors. In practice, for convective rainfall, data must be available at intervals not greater than 5 minutes and even this could be too coarse in some cases as illustrated by Figure 8.5.

Cluckie *et al* (1991) investigated the intensity resolution required of radar data. In this case it was concluded that 8-bit (8 intensity levels) were adequate for most rural catchments and probably most urban catchments. The bulk of the process-describing information is concentrated at the low-frequency end of the spectra, and this portion of the signal is retained by flow forecasting models despite degradation of intensity resolution. Nevertheless, in convective rainfall a reduced intensity resolution may have the same effect as spatial resolution in causing storm smearing, that is to reduce rainfall gradients.

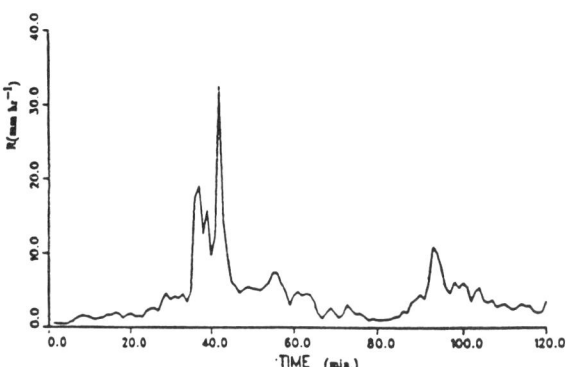

Figure 8.5   Minute by minute variations of rainfall rate derived using a dropsize disdrometer (after Chandrasekar and Gori, 1991)

Early assessments of the impact of radar data input on flow forecasting (Anderl *et al*, 1976, Barge *et al*, 1979) were optimistic, and hydrograph simulations were described showing considerable improvements over hydrographs derived using raingauge data alone. However, others (Gorrie and Kouwen, 1977) noted that generally very little improvement was evident in convective rainfall. The effects of errors in radar measurements of precipitation were highlighted by Collier and Knowles (1986). Similar results were reported by Roberts (1987), who stressed the difficulty of using radar data which were not adjusted accurately in real time by whatever method.

Further problems are encountered when rainfall forecast are input to hydrological models. Schultz (1987) investigated the impact on forecast hydrographs of variations in the forecast rainfall scenario used. A real-time adaptive model minimizing the sum of squares of deviations between the observed and forecast hydrograph was tested, and found to be a promising approach. Likewise, Cluckie and Owens (1987), using an adaptive transfer function model also reported encouraging results, although the occasional grossly incorrect forecast was noted. More recently adaptive distributed models have been developed, for example Chander and Fattorelli (1991), but these will require extensive real-time testing.

6. Summary

Radar data provide good spatial and temporal precipitation measurement coverage from a single location. In particular they have the capability to provide measurements over areas of 1 km^2 or less and at intervals of 1 minute or less.

Unfortunately these data require interpretation and comprehensive quality control before they can be used as input to hydrological models. However, radar does provide a direct measurement of precipitation, albeit in terms of the reflectivity of the hydrometers.

The retrieval algorithms for radar systems introduce uncertainties into the measurements of precipitation which may have a profound impact on flow forecasts produced using these data. Nevertheless, unless raingauge networks are very dense, the measurements provided by them also introduce sampling errors.

The advantages of data collection at a single location, and the spatial and temporal comprehensiveness of the measurements mean that the accuracy limitations of radar data compared with measurements from a dense raingauge network are, in most circumstances, acceptable. This is not to say that work should not be undertaken to improve accuracy, but rather that present limitations of radar data should be no bar to the development of distributed hydrological models for use with these data. It is necessary though to ensure that the modelling approach embodies appropriate error correction feedback mechanisms matched to the error characteristics of the radar data.

Finally, it remains unclear whether the use of distributed rainfall data derived from radar with distributed hydrological models can at present provide more accurate hydrograph simulations than the use of a limped approached. Although this is now being addressed, it does require joint research between hydrological modellers and radar hydrometeorolgists.

7. References

Abbott. M.B., Bathurst, J.C., Cunge, J.A., O'Connell, P.E. and Rasmussen, J. (1986a) "An introduction to the European Hydrological System - Systéme Hydrologique Européen, "SHE" 1. History and philosophy of a physically-based, distributed modelling system", *J Hydrology*, 87, 45-59
(1986b) "2. Structure of a physically-based distributed modelling system", *J Hydrology*, 87, 61-77

Anderl, B., Attmannspacker, W. and Schultz, G.A. "Accuracy of reservoir inflow forecasts based on radar rainfall measurements", *Water Resour. Res.*, 12, No. 2, 217-223

Andrieu, H., Creutin, J.D. (1995) "Identification of vertical profiles of radar reflectivity for hydrological applications using an inverse method. Part I: Formulation," *J. App. Met.*, 34, 225-239.

Andrieu, H., Delrieu, G. and Creutin, J D. (1995) "Identification of vertical profiles of radar reflectivity for hydrological applications using an inverse method. Part II: Sensitivity analysis and case study", *J. App. Met.*, 34, 240-259

Atlas, D. (1964). "Advances in radar meteorology" in *Advances in Geophysics*, 10, 318-478, publ. Academic Press, New York.

Atlas, D., Rosenfeld, D. and Short, D.A. (1990). "The estimation of convective rainfall by area integrals, Part I: The theoretical and empirical basis", *J. Geophysics. Res.*,. 95, 03, 2153 - 2160.

Atlas, D. Ulbrich, C.W. and Meneghini, R. (1982). "The multi-parameter remote measurement of rainfall". Tech. Mem. No. NASA TM 83971, Goddard Space Flight Centre, NASA, Greenbelt, Maryland, 84 pp.

Ball, J.E. (1994). "The influence of storm temporal patterns on catchment response", *J . Hydrology*, 158, 285-303.

Barge, B.L., Humphries, R.G., Mah, S.J. and Kuhnke, W.K. (1979). "Rainfall measurements by weather radar: applications to hydrology", *Weather Resour. Res.*, 15, No. 6, 1380-1386.

Battan, L.J. (1973). "Radar Observations of the Atmosphere". University of Chicago Press, Chicago, 324 pp.

Blanchard, D.C. (1953). "Raindrop size distribution in Hawaiian rains", *J. Met*, 10, 457-473.

Browning, K.A. (1990). "Rain, rainclouds and climate", *Quart. J.R. Met. Soc*, 116, no 495, 1025-1051.

Brussaard, G. (1976) "A meteorological model for rain-induced cross polarization", *IEEE Trans.*, AP-24, 5-11.

Chander, S., Fattorelli. (1991). "Adaptive grid-square based geometrically distributed flood-forecasting model". Paper 38 in *Hydrological App. of Wea. Radar*, ed. I.D. Cluckie and C.G. Collier, publ. Ellis Horwood, Chichester, 424-439.

Chandrasekar, V. and Gori, E.G. (1991). "Multiple distrometer observations of rainfall". *J. App. Met.*, 30, 1514-1520.

Chiu, L.S. and Kedem, B. (1990). "Estimating the excedance probability of rain rate by logistic regression", *J Geophys. Res.*, 95, D3, 2217-2227.

Cluckie, I.D. and Owens, M.D. (1987). "Real-time rainfall - runoff models and use of weather radar information", Chapter 12 in *Weather Radar and Flood Forecasting*, editors V.K. Collinge and C. Kirby, publ. John Wiley & Sons, Chichester, 171-190.

Cluckie, I.D., K.A. Tilford and Shepherd, G.W. (1991). "Radar signal quantization and its influence on rainfall-runoff models", Chapter 39 in *Hydrological applications of Weather Radar*, edits I.D. Cluckie and C.G. Collier, pub. Ellis Horwood, Chichester, 440-451.

Collier, C.G. (1986) "Accuracy of rainfall estimates by radar, Part I: Calibration by telemetering raingauges", *J. Hydrology*, 83, 207.223. Part II: comparison with raingauge network", *J Hydrology*, 83, 225-235.

Collier, C.G. (1989). "Applications of Weather Radar Systems. A guide to uses of radar data in meteorology and hydrology". Published by Ellis Horwood Ltd., 294 pp.

Collier, C.G. and Knowles, J.M. (1986) "Accuracy of rainfall-estimates by radar, Part III: Application for short-term flood forecasting", *J. Hydrology,* 83, 237-249.

Collier, C.G. Larke, P.R. and May, B.R. (1983). "A weather radar correction procedure for real-time estimation of surface rainfall", *Quart. J.R. Met. Soc.* 104, 589-608.

Diskin, M.H. and Simpson, E.S. (1978). "A quasi-linear, spatially distributed cell model for the surface runoff system" *Water Resour. Bull,* 14, 903-918.

Doneaud, A.A., Smith, P.L., Dennis, A.S. and Senggupta, S. (1981). "A simple method for estimating convective rain volume over an area", *Water Resour. Res.,* 17, 1676-1684

Doneaud, A.A., Niscov, S.I., Priegrutz, D.L. and Smith, P.L. (1984). "The area-time integral as an indicator for convective rain volumes", *J. Clim. App. Met,* 23, 555-561.

Doviak, R.J. and Zrnic, D.S. (1984). "Doppler Radar and Weather Observations", *Academic Press,* New York, 458 pp.

Fabry, F., Austin, G.L. and Tees, D. (1992). "The accuracy of rainfall estimates by radar as a function of range", *Quart. J.R. Met. Soc.,* 118, 435-453

Goddard, J.W.F., Cherry, S.M. and Bringi, V.N. (1982). "Comparisons of dual-polarization radar measurements of rain with ground-based distrometer measurements", *J. App. Met,* 21, No. 2, 252-256.

Gorri, J.E. and Kouwen, N. (1977). "Hydrological applications of calibrated radar precipitation measurements". Preprint Volume, 2nd Conf. on Hydromet. Toronto, Ontario, 25-27 October, *Am. Met. Soc.,* Boston, Mass. 272-279.

Gunn, K. and Marshall, J. (1958). "The distribution with size of aggregate snowflakes", *J. Met,* 15, 452-461.

Hardaker, P.J., Holt, A.R. and Collier, C.G. (1995). "A melting layer model and its use in correcting for the bright band in single polarisation radar echoes", *Quart J.R. Met. Soc.* 121, 495-525.

Harrold, T.W. and Kitchingman, P.G. (1975) "Measurement of surface rainfall using radar when the beam intersects the melting layer", Preprints, 16th Radar Met. Conf., Houston, AMS Boston, 473-478.

Hendry, A. and Antar, Y.M.M. (1984). "Precipitation particle identification with centimetre wavelength dual-polarisation radars", *Radio Science,* 19, No. 1, 115-122.

Herschy, R.W. Barrett, E.C. and Roozekrans, J.N. (1985). Remote Sensing in Hydrology and Water Management, Final Report, ESA, Contract No. 5769/A84/D/JS(Sc), EARSeL, Strasbourg, 268 pp.

Herzegh, P.H. and Conway, J.W. (1986). "On the morphology of dual-polarization radar measurements: distinguishing meteorological effects from radar system effects", Preprints 23rd Conf. on Radar Met., Vol. 1, Snowmass, Colorado, Am. Met. Soc., Boston, R55-R58.

Holt, A.R. (1988) "Extraction of differential propagation phase shift from data from S-band circularly polarised radars", *Electronics Letters*, 24, 1241-1242.

Jameson, A.R., Beard, K.V. and Bresch, J. (1981). "Complications in deducing rain parameters from polarization measurements", Preprints, 20th Conf. on Radar Met., 30 Nov. - 3 Dec., Boston, Mass., Am. Met. Soc., Boston, 586-589.

Jones, D.M.A. (1956). "Raindrop size distribution in Hawaiian rains", *J. Met.*, 10, 457-473.

Joss, J. and Waldvogel, A. (1990). "Precipitation measurement and hydrology: A review", in *Battan Memorial Volume, Radar in Meteorology*, editor D. Atlas, Am. Met. Soc., Boston, Chapter 29a, 577-606.

Joss, J. Schram, K., Thoms, J.C. and Waldvogel, A. (1970). "On the quantitative determination of precipitation by radar", Wissenschaftliche Mitteilung Nr. 63, Zurich, Eidenossische Kommission Zum Studium der Hagelbildung und dert Haglabwehr.

Karnieli, A.M. Diskin, M. Hy., Lane, L.J. (1994). "CELMOD5 - a semi-distributed cell model for conversion of rainfall into runoff in semi-distributed watersheds", *J. Hydrology*, 157, 61-86.

Kitchen, M. and Blackall, R.M. (1992). "Representativeness errors in comparisons between radar and gauge measurements of rainfall", *J. Hydrology*, 134, 13-33.

Kitchen, M, Brown, R and Davies, A.G. (1994). "Real-time correction of weather radar data for the effects of bright-band, range and orographic growth in widespread precipitation", *Quart. J.R. Met. Soc.* 120, 1231-1254.

Kouwen, N. and Garland, G. (1989). "Resolution considerations in using radar rainfall data for flood forecasting", *Can. J. Civ. Eng*, 16, 279-289.

Liu, J. and Herzegh, P.H. (1986). "Differential reflectivity signatures in ice-phase precipitation: radar-aircraft comparisons", Preprints, 23rd Conf. on Radar Met., Vol. 1, Snowmass, Colorado, Am. Met. Soc., Boston, R59-R61.

Marshall, J.S. and Palmer, W. (1948). "The distribution of raindrops with size", *J. Met.*, 5, 165-166.

McCormick, G.C. and Hendry, A. (1972). "Results of precipitation back-scatter measurements at 1.8cm with a polarization diversity radar", Preprints 15th Radar Met. Conf., Am. Me. Soc., Boston, 35-38

Michaud, J.D. and Sorosshian, S. (1994). "Effect of rainfall-sampling errors on simulations of desert flash floods", *Water Resour. Res.*, 30, No. 10, 2765-2775.

Moore, R.J. (1987). "Towards a more effective use of radar data for flood forecasting", Chapter 15 in *Weather Radar and Flood Forecasting*, editor V.K. Collinge and C. Kirby, publ. John Wiley & Sons Ltd., 223-238.

Moore, R.J., Watson, B.C., Jones, D.A. and Black, K.B. (1991). "Local recalibration of weather radar", Paper 6 in *Hydrological Applications of Weather Radar*, editors I.D. Cluckie and C.G. Collier, publ. Ellis Horwood, Chichester, 65-73.

Obled, Ch., Wendling, J. and Bevan, K. (1994). "The sensitivity of hydrological models to spatial rainfall patterns: an evaluation using observed data", *J. Hydrology*, 159, 305-333.

Ogden, F.L. and Julien, P.Y. (1994) "Runoff model sensitivity to radar rainfall resolution", *J Hydrology*, 158, 1-18.

Roberts, G.K. (1987). "The use of radar rainfall data in urban drainage models in Manchester", *Public Health Eng.*, 14, No. 6, 61-64.

Rosenfeld, D., D. Atlas and Short, D.A. (1990). "The estimation of convective rainfall by area integrals, Part II: The height-Area Threshold (HART) Method", *J. Geophysics Res.*, 95, D3, 2161-2176.

Rosenfeld, D., Atlas, D., Wolf, D.B. and Amitai, E. (1992). "Beamwidth effects on Z-R relations and area-integrated rainfall", *J. App. Met.*, 31, 454-464.

Rosenfeld, D., Amitai, E. and Wolf, D.A.. (1995a) "Classification of rain regimes by the three-dimensional properties of reflectivity fields", *J. App. Met.* 34, 198-211.

(1995b) "Improved accuracy of radar WPMM estimated rainfall upon application of objective classification criteria", *J. App. Met.*, 34, 212-223.

Sachidananda, M and Zrnic, D.S. (1987). "Differential propagation phase shift and rainfall rate estimation", *Radio Science*, 21, 235-247.

Sauvagesot, H., (1994) "The probability density function of rain rate and the estimation of rainfall by area integrals", *J. App. Met.*, 33, 1255-1262.

Schultz, G.A. (1987). "Flood forecasting using rainfall radar measurements and stochastic rainfall forecasting in the Federal Republic of Germany", Chapter 13 in *Weather Radar and Flood Forecasting*, editors V.K. Collinge and C. Kirby, publ. J Wiley & Sons Ltd., 191-207.

Sekhon, R.S. and Srivastava, R.C. (1970). "Snow size spectra and radar reflectivity", *J. Atm. Sci.*, 27, 228-367.

Seliga, T.A. and Bringi, V.N. (1976) "Potential use of radar differential reflectivity measurements at orthogonal polarization for measuring precipitation", *J. App. Met.*, 15, 69-75.

Smith, P.L. (1984). "Equivalent radar reflectivity factors for snow and ice particles", *J. Climate App. Met.*, 23, No. 8, 1258-1260.

Zawadzki, I.I. (1984). "Factors affecting the precision of radar measurements of rain", Preprint Volume, 22nd Conf. on Radar Met., Zurich Switzerland, Am. Met. Soc., Boston, 251-256.

CHAPTER 9
APPLICATION OF REMOTE SENSING FOR HYDROLOGICAL MODELLING

F.P. DE TROCH, P.A. TROCH, Z. SU and D.S. LIN
Laboratory of Hydrology and Water Management
University of Ghent, Coupure links 653, B-9000 Ghent, Belgium

1. Introduction

It has long been recognised that the results obtained by hydrological modelling of a river basin depend heavily on the quality of the input data used. The main problem in many hydrological studies is that there are not enough adequate data to describe quantitatively hydrological processes with sufficient accuracy. Studies on hydrological effects of land use and climate changes in large river basins are possible only if detailed information about topography, geology, soil, vegetation, and climate are available. With the advances of remote sensing techniques hydrological relevant information about large river basins can be derived from different sensors. A major problem facing the user of these data is how to effectively incorporate remotely sensed data into hydrological studies and models (Peck et al., 1981; Rango, 1987; Schultz, 1988; Engman and Gurney, 1991).

In contrast to conventional methods of data collection the main advantages of remote sensing techniques may be summarised as: (1) no interference between data acquisition devices and the process being measured; (2) areally distributed measurements instead of point measurements; (3) rather high resolution in space and/or time; (4) data available in digital form; (5) information possible about remote inaccessible areas. The main disadvantage is that remote sensors do not directly provide data in a form needed in hydrological modelling. The information acquired by such sensors usually consists of measurements of electromagnetic signals that have to be converted into hydrological relevant data. Techniques for operationally interpreting remote sensing data into hydrological information are still under development. Difficulties still exist in choosing the most suitable spectral data for studying hydrological processes as well as in interpreting these data with appropriate methods.

One of the most straightforward applications of remote sensing methods related to hydrology and water resources management is land use classification. Based on visible and near-infrared observations, land surface characteristics important to hydrological modelling can be derived for extended areas (e.g. mesoscale river basins). However, lack of understanding the problem of scaling (from the hydrodynamic to the basin and the regional scale) of land-atmosphere exchange processes impedes further progress in applying remote sensing data products. Since surface soil moisture controls, and is affected by, water and energy exchange processes between the atmosphere and the land, a key instrument to study these scaling issues in hydrological modelling may be microwave remote sensing of soil moisture. Microwave remote sensing allows the observation of surface soil moisture since soil water content affects the dielectric properties of the surface soil layer. For two decades now, the use of both active and passive microwave instruments for surface soil moisture observation is under investigation. Airborne campaigns using microwave instruments (e.g. MACHYDRO'90, EMAC'94) have permitted the testing and calibration of different techniques for soil moisture observation and retrieval. Spaceborne active microwave sensing is particularly interesting because of good spatial and temporal resolution of the observations. Currently, several satellites produce Synthetic Aperture Radar (SAR) observations from space for large areas in the world (e.g. ERS-1, ERS-2, JERS-1, RADARSAT).

This chapter will first present an overview of the current state-of-the-art of remote sensing applications in hydrology and water resources. Then, important new developments and research results in soil moisture observations from microwave remote sensing will be reviewed. Finally, the importance of information about spatial and temporal soil moisture variations in distributed hydrological modelling is demonstrated by a few examples.

2. Remote Sensing for Hydrological Studies of River Basins: State-of-the-art

2.1. PHYSICAL CONSIDERATIONS OF REMOTE SENSING

Remote sensing is the technique of obtaining information about an object without physical contact, as opposed to in-situ sensing in which the measuring device is in touch with the object. The quantity most frequently measured in current remote sensing systems is the electromagnetic energy emanating from an object. Remote sensing systems may be classified in two categories: passive and active. In passive operation, devices merely detect and record the natural energy that arrives from the target. An active system both transmits the electromagnetic signal and receives the backscattered or reflected signal.

Electromagnetic radiation is the means by which electromagnetic energy is propagated in the form of waves. This is of great significance for remote sensing, because radiation is a form of energy that can travel either through a medium (e.g. water and atmosphere) in certain wavelengths, or through a vacuum (e.g. space). Radiation is emitted from all bodies that have absolute temperatures above zero and is characterised by a signal whose configuration is determined by the physical properties of its source. The signal is usually described in terms of wavelength (or frequency) and phase.

Properties of electromagnetic radiation relating to wavelength are: (1) the smaller the wavelength, the greater the energy; (2) the higher the (absolute) temperature of an object, the greater the total energy and the shorter the peak wavelength emitted; (3) the interaction of energy with matter is wavelength dependent. When radiation strikes an object, four types of interaction are possible, viz. transmission, absorption, reflection and/or scattering. Knowledge about the combined effects of these four ways of interaction will be of help in selecting the most appropriate wavebands for analysis of properties of objects by means of remote sensing techniques.

2.2. SOURCES OF REMOTE SENSING DATA FOR HYDROLOGICAL STUDIES

2.2.1. *Sensors and Platforms*

Remote sensors in most common use for hydrological studies may be divided into six groups:

(1) Photographic cameras, which are the most simple form of remote sensors. They exploit the visible (VIS) or near-infrared (NIR) regions of the electromagnetic spectrum and are primarily used for mapping purposes;

(2) Vidicon cameras, in which optical images are focused and retained temporarily on photo-conductive surfaces that are scanned electronically for recording and/or transmission in the form of continuous, variable electric signals;

(3) Scanning radiometers, which use rotation or oscillation of part of the instrument, or its platform, or adjustment of the phases of the received signals, to scan the target area and to build up strips of data as the platform advances along its path. Scanning radiometers have been designed to exploit VIS, infrared (IR) and microwave radiation wavebands. Many satellites carry sensors of this type, including the Landsat Multispectral Scanner (MSS), the Advanced Very High Resolution Radiometer (AVHRR) of the US National Oceanic and Atmospheric Administration (NOAA), and the Spectral Sensor Microwave Imager (SSM/I) of the US Defence Meteorological Satellite Program (DMSP);

(4) Pushbrooms, developed in order to reduce the geometric and mechanical complications that may arise when extended arrays are required. As the name suggests, an extended array of solid state sensors is mounted on a head that scans a surface area progressively along the sub-satellite track. This concept of pushing a broom-head of sensors has been adopted in the SPOT (Système Probatoire d'Observation de la Terre) satellites, allowing also to acquire stereoscopic images;

(5) Spectrometers, in which incoming radiation is selected and dispersed by means of prisms, mirrors, gratings or filters to provide multispectral data for detailed analysis of the spectral signature of the target;

(6) Microwave radars, which are, unlike the others, active radiation systems and measure the reflected echoes of radiation emitted from the devices themselves. Radars are especially useful in cloudy areas because some wavelengths of microwave radiation are not significantly attenuated by water in the atmosphere. Examples are the SAR instruments aboard the European Remote Sensing Satellites (ERS-1 and ERS-2), the Japanese JERS-1 and the Canadian RADARSAT.

For hydrological studies the following platforms are of interest: (1) ground-based observation platforms (e.g. weather radar systems); (2) airborne platforms: balloons, up to altitudes of about 30 km, aircraft: used for topographic surveys, hazard monitoring (e.g. flooding), disaster assessment and also hydrological studies. Piloted aircraft operates up to altitudes of about 15 km; (3) spaceborne platforms: shuttle spacecraft (200--300 km), satellites, either low altitude polar orbiting satellites (800--1500 km) or high altitude equatorial orbiting satellites at geostationary altitudes (35,500 km).

2.2.2. Remote Sensing Satellite Systems

For most practical purposes in hydrologic studies, it is convenient to differentiate between two broad classes of satellites: earth resources satellites and environmental satellites. The former, observing the same area relatively infrequently with a repeat cycle in the order of several days but with relatively high spatial resolutions (e.g. Landsat), have contributed to the mapping and general monitoring of surface features and conditions. The latter, observing frequently (in the order of hours) but at relatively low spatial resolutions, have contributed to hydrological studies by providing information on weather conditions (Meteosat, GOES) and large scale surface phenomena (NOAA/AVHRR).

Earth Resources Systems:

(1) Landsat MSS, TM: Landsat satellites, operated by the US National Aeronautics and Space Administration (NASA), have been providing since 1972 (Landsat 1, 2, 3) operational information for vegetation, crop and land cover inventories in four spectral bands by means of a multispectral scanner (MSS) with a resolution of 80 m and in seven spectral bands using the Thematic Mapper (TM) with a resolution of 30 m (except band six, having a resolution of 120 m) since 1984 (Landsat 4, 5). Landsat 6 launched in August 1993 and lost shortly after launch, carried aboard an Enhanced Thematic Mapper consisting of the seven Landsat 5 spectral bands and an additional panchromatic band with a 13 m x 15 m spatial resolution (Dornier, 1993).

(2) SPOT: The first SPOT satellite was launched in 1986 by the Centre National d'Etudes Spatiales (CNES), France. It is an important data source for geographic information and offers unique features in the field of spaceborne remote sensing with ground resolution of 10 m in panchromatic band and 20 m in multispectral band, extremely flexible acquisition possibilities for almost any point on the Earth's surface, possibility of stereoscopic viewing and excellent geometric accuracy. Currently SPOT-2 and SPOT-3 are in operation. SPOT satellites are equipped with two imaging instruments, the High Resolution Visible imagers HRV1 and HRV2, able to function

independently of each other. The HRVs are designed to operate in two modes of sensing: a 10 m resolution panchromatic mode and a 20 m resolution multispectral mode with 3 spectral bands.

(3) ERS-1/SAR: The European Remote Sensing Satellite 1 launched in 1991 by the European Space Agency (ESA) provides global and repetitive observations using advanced microwave techniques which enable all-weather observations of the Earth. The ERS-1 carries, among other instruments, an Active Microwave Instrument (AMI) which combines the functions of a synthetic aperture radar and a wind scatterometer. The SAR is a C-band (5.3 GHz) radar with vertical transmission and vertical receiving (VV) polarisation, operating in image mode for the acquisition of wide-swath, all-weather images over oceans, polar regions, coastal zones and land surfaces. In April 1995, ERS-2, carrying aboard the same instruments as ERS-1, was successfully launched. Similar to the ERS/SAR are the Japanese JERS-1 and the Canadian RADARSAT satellites. The JERS-1 carries a L-band (1.275 GHz) SAR and was launched in 1992. Very recently (November 1995) RADARSAT with a C-band SAR and switchable incidence angle has been launched.

Environmental Satellite Systems:

(1) NOAA/AVHRR: The NOAA/AVHRR has been providing since 1978 information for hydrologic, oceanographic, and meteorological studies in 5 spectral bands with a 1.1 km resolution.

(2) Geostationary Meteorological Satellites: These include ESA's Meteosat, US Geostationary Operational Environmental Satellites (GOES) West and East, Japan's Geostationary Meteorological Satellites (GMS) and India's INSAT.

(3) DMSP: Since 1978 the US Air Force Defence Meteorological Satellite Program (DMSP) F8 and F10 satellites carry a sensor of great importance for hydrological studies. The SSM/I is designed to measure ocean surface wind speed, ice coverage and age, cloud water content, precipitation and soil moisture in 4 bands (7 channels formed by dual polarisation in 3 bands (19.35, 37.0 and 85.5 GHz) and vertical polarisation in one band (22.23 GHz)).

2.3. HYDROLOGICAL APPLICATIONS OF REMOTELY SENSED DATA

2.3.1. *Precipitation*

Remote sensing based rainfall estimation techniques include ground based radar methods, satellite based cloud indexing methods using visible/infrared observations as well as passive microwave observations (Browning and Collier, 1989). Applications of ground based radar in rainfall monitoring have been reported, among many others, by Klatt and Schultz (1983) and by Collinge and Kirby (1987), mainly in the field of flood forecasting. Techniques using the visible and/or infrared regions of the electromagnetic spectrum rely on cloud top radiation. Passive microwave data based techniques provide a more direct measurement of rainfall characteristics. In the microwave region below about 20 GHz, evidence of rain is provided by absorption/emission processes, and above 60 GHz, evidence of rainfall comes predominantly from scattering processes. Passive microwave techniques are much superior to visible and infrared data based techniques (Barrett, 1993).

2.3.2. Snow and Ice

In many mountainous areas snow is the main source of streamflow during spring and summer. Since snow is often located in remote, inaccessible regions where extensive field measurements are very difficult and expensive to perform, remote sensing techniques are obviously advantageous. These techniques, developed in the last twenty years, include applications such as mapping of areal snow cover by visible and SAR sensors, measuring snow accumulation, snow water equivalent and snow albedo by microwave sensors, and snowmelt runoff forecasting combining the snow cover depletion curve derived from remotely sensed data with hydrological models (Chang et al., 1991; Martinec and Rango, 1991; Rango, 1993).

2.3.3. Evapotranspiration

Evapotranspiration is of great importance in water balance modelling of a river basin, but cannot be measured directly by remote sensing techniques. However, some parameters and variables needed for calculating evapotranspiration from the energy budget equation (such as incoming solar radiation, surface albedo, surface temperature, land cover, vegetation density and soil moisture) may be estimated using remote sensing data. Despite many studies (Menenti, 1983; Nieuwenhuis, 1986; Seguin et al., 1990; Feddes et al., 1993) there is still no real operational method to determine evapotranspiration based on remote sensing techniques.

2.3.4. Soil Moisture

Soil moisture is an important variable in many hydrological, agricultural, meteorological, and climatic studies. Conventional methods for soil moisture measurements are both time and labour consuming and are very difficult, if not impossible, to deploy over a large river basin. Remote sensing techniques offer the possibility of collecting spatially distributed near surface soil moisture estimates.

Remote sensing measurements of near surface soil moisture may be based on: (1) measuring bare soil reflectance in the visible and near infrared regions of the spectrum. This provides only a poor indication of soil moisture since soil reflectance is heavily influenced by soil texture and colour; (2) measuring the surface temperature in the thermal infrared region. Limitations to this type of measurements are due to effects of cloud cover, vegetation and meteorological factors; (3) measuring the brightness temperature in the microwave region. This passive technique utilises the distinctive difference of the dielectric constant of water and of dry soil for determining soil water content in the top soil layer of about 5 cm (Schmugge, 1985; Jackson, 1993; Hollenbeck et al., 1996). Because attenuation of the microwave radiation increases with increasing vegetation density this method is limited to sparsely vegetated areas; (4)

measuring the backscattering coefficient with active microwave sensors. Many research efforts are now being conducted in developing techniques for measuring near surface soil moisture (e.g. MACHYDRO'90, MACEUROPE'91, EMAC'94). Progress has been made with regard to the choice of microwave frequency and quantifying the influence of soil roughness and vegetation cover (Ulaby et al., 1978; Ulaby et al., 1984). Recently, procedures for modelling profile soil moisture from intermittent remotely sensed near surface soil moisture are being developed (Ragab, 1995). We will address some of the important issues in soil moisture mapping from remote sensors in Section 3 of this chapter.

2.3.5. Surface Water and Runoff

Remote sensing data can generally enhance conventional methods used in surface water inventory, including mapping changes of surface water coverage, flood plain and flood damage determination, and improved management of inland waters. Although runoff cannot be directly measured, remotely sensed data do play a very important role in providing input data to distributed hydrological models, in measuring state variables such as soil moisture, and in estimating model parameters, so that runoff can be simulated more accurately.

2.3.6. Catchment Characteristics

In combination with geographic information systems, satellite based remote sensing offers a possibility for mapping catchment characteristics (Su et al., 1992; Su and Schultz, 1993). When applying remotely sensed catchment characteristics in hydrological modelling, the model structure as well as the spatial and temporal resolution must be carefully considered. Since remotely sensed data are spatially distributed by nature, the use of distributed models would be most appropriate. However, since remote sensing data are usually describing surface features, model parameters related to subsurface processes often have to be calibrated for a subarea, based on at least some hydrological observations (e.g. measured streamflow). In this case semi-distributed models are often preferred. In order to utilise the large amount of data efficiently and to reduce model complexity, groups of pixels may be aggregated together into "hydrologically similar units", "representative elementary areas" (Wood et al., 1988) or "grouped response units" (Kouwen et al., 1993).

3. Microwave Remote Sensing of Soil Moisture

3.1. INTRODUCTION

Soil is the thin layer of porous material at the interface between the atmosphere and the geosphere. Retention of soil moisture and runoff formation resulting from rainfall and/or snowmelt, or from irrigation, are fundamental processes upon which civilisation depends for food and energy production, for water supply, and for many industrial and transport purposes. In addition, soil moisture is often a dominant factor in determining the ecosystem's response to the physical environment. Near surface soil moisture heavily controls the partitioning of available energy at the ground surface into sensible and latent heat exchange with the atmosphere, thus linking the water and energy balances at the land surface and the moisture and thermal states of the soil. Adequate knowledge about the soil moisture status, as well as evapotranspiration, is essential to the understanding and prediction of the reciprocal influences between land surface processes and weather and climate. Despite its importance, global measurements and analyses of soil moisture still remain deficient.

Recent studies have demonstrated that remote sensing techniques can be applied to measure soil moisture states at the ground surface under a variety of topographic and land cover conditions. To a certain extend, remote sensing of soil moisture may be accomplished in all regions in all regions of the electromagnetic spectrum. However, only the microwave region offers the potential of truly quantitative measurements from airborne or spaceborne instruments. The microwave sensors are attractive because of the strong dependency of the soil's dielectric properties on its moisture content and of their relative immunity against atmospheric interference. Microwave techniques for measuring soil moisture include both passive and active approaches each having distinct advantages.

3.1.1. *Passive systems*

All matter at temperatures above absolute zero emits electromagnetic radiation due to the motion of the charged particles of its atoms and molecules. Passive microwave systems use radiometric instruments to measure this radiation at frequency bands in the microwave region. The intensity of the naturally emitted radiation is commonly expressed as the target's brightness temperature, which is defined as the product of the target's physical temperature and its emissivity. A number of studies using microwave radiometers have verified the brightness temperature - soil moisture relationship for various targets and different sensor parameters (Newton et al., 1982; Njoku and O'Neill, 1982; Wang et al., 1983; Schmugge et al., 1992; Jackson, 1993). It has also been shown that two of the surface characteristics, roughness and vegetation, tend to reduce the sensors' sensitivity to soil moisture variations. The spatial resolution of a passive microwave system is a function of the distance to the target and of the antenna's dimension. Unless the antenna is very large, it is impossible to achieve meter-scale spatial resolution from a space platform.

3.1.2. *Active Systems*

In contrast to passive sensors, active microwave systems or radars emit pulses of energy and measure the signals reflected from the surface. The reflected, or backscattered, energy from the illuminated area is usually characterised as the backscattering coefficient, σ^0, which is defined as the average scattering cross-section per unit area.

The relationship between soil moisture and radar echoes has been studied by many investigators (Ulaby et al., 1978; Ulaby et al., 1982; Pultz et al., 1990; Wood et al., 1993). These studies indicate that active systems are even more sensitive to surface roughness and vegetation than passive systems (Ulaby et al., 1979). However, the spatial resolution of active systems is considerably better than that of passive systems. When the synthetic aperture antenna (SAR) technique is used, the system's resolution is basically independent of the altitude of the platform (Colwell, 1983).

3.2. SOIL MOISTURE RETRIEVAL ALGORITHMS

3.2.1. *The Inverse Problem*

Since microwave remote sensors do not measure soil moisture directly, a retrieval algorithm is needed to extract this information from the measured signals which are often contaminated with noise. From a mathematical point of view, this is equivalent to solving an inverse problem closely related to the forward modelling procedure. Forward modelling develops a set of mathematical relationships to simulate the instrument's response for a given set of model parameters. In the context of soil moisture remote sensing, these parameters generally include soil properties and the geometry and phenology of the overlying vegetation canopy. To solve the inverse problem, it is crucial to start from a forward modelling procedure that is capable to adequately describe the observations. It is also important to know the number of model parameters used to describe the objects being measured, and to know which parameters most sensitively influence the returned signal.

In the following sections we concentrate on some inversion algorithms for soil moisture retrieval from active microwave sensors. Then, we describe a sensitivity analysis in the case of soil moisture retrieval from bare soil, using ERS-1/SAR data.

3.2.2. *Review of Microwave Scattering Models*

Consider the problem of microwaves emitted by a radar's transmitter and impinging upon a layer of vegetation canopy overlying a rough ground surface. The waves penetrate the layer and interact with various parts of the inhomogeneous vegetation canopy and with the (top) soil matrix, resulting in a series of absorption and scattering reactions. A portion of the scattered waves is returned in the direction of the radar's receiver and carries within it information regarding the illuminated vegetation-soil medium.

In essence, this backscattering process can be subdivided into three components: (1) a component representing the scattering contribution of the vegetation canopy; (2) a component representing the surface-volume interaction contribution; (3) a component representing the ground backscattering contribution, including the two-way attenuation caused by the vegetation. The relative importance of every component depends on frequency, polarisation and incidence angle of the radar waves, on vegetation and soil water contents, on vegetation density and orientation, on soil surface roughness and soil texture, and on other land surface parameters.

The simplest models consisting of empirical relationships between radar measurements and some land surface characteristics are usually developed from fitting to experimental data. Examples of such models are abundant in literature, such as Ulaby et al. (1978), Ulaby et al. (1979) and Pultz et al. (1990), among others. More recently, Wood et al. (1993) have developed an empirical model relating NASA's airborne SAR (AIRSAR) backscattering signals to surface soil moisture for three different kinds of vegetation canopies. These models are simple in structure and easy to use. However, they suffer from a number of drawbacks: first, they use regression parameters or empirical coefficients which are not physical variables that can be measured in-situ; second, they are site specific and usually have a rather limited range of validity. In addition, since sampling from different platforms results in different responses, these empirical relationships are also instrument specific.

The problem of wave scattering from a randomly rough surface has been studied theoretically using both low- and high-frequency approximations. Among the high-frequency scattering models, the Kirchhoff formulation (KF) is the most commonly used (Beckmann and Spizzichino, 1963; Sancer, 1969). The basic assumption of this method is that the total scattered field at any point on the surface as if the incident wave is impinging upon an infinite plane tangent to the point. Analytic solutions have been developed for surfaces with a large standard deviation (s) of the surface heights, using the stationary phase approximation in conjunction with the Kirchhoff formulation (Wu and Fung, 1972), and for surfaces with small slopes and small s, using a scalar approximation (Ulaby et al., 1986).

For a ground surface whose s and correlation length are much smaller than the wavelength, the small perturbation method (SPM) (Valenzuela, 1967), which is a low-frequency solution, can be used to estimate the backscattering contribution. The region of validity of the SPM has been extended to higher values of s by Wineberner and Ishimaru (1985), using a perturbation expansion of the phase of the surface field.

Attempts have also been made to unite the KF and the SPM in order to extend the range of validity. This led to the development of two-scale models such as described by Wright (1968), Leader (1978), Brown (1978), Bahar (1985) and Fung and Pan (1987). More recently, Fung et al. (1992) have developed a surface scattering model based on the surface field integral equations, called the Integral Equation Model (IEM). The IEM reduces to the SPM when the surface is smooth, and to the standard Kirchhoff model when s is larger than the incident wavelength. However, since the IEM deals with surface scattering, care should be taken to its application in situations where volume scattering occurs on the ground surface (e.g. for low moisture content and/or at low frequencies (Le Toan et al., 1994)).

Microwave scattering models for a vegetation canopy can be categorised into two classes: empirical (or phenomenological) models, and physical (or theoretical) models. The empirical models are based on intuitive understanding of the relative importance of various vegetation parameters, then summing up the contributions from each component believed to be important (Ulaby et al., 1979; Engheta and Elachi, 1982; Mo et al., 1984; Richards et al., 1987). The physical models are based upon the modelling of the interactions between microwaves and the various scattering elements of a vegetation canopy. The major difficulties in modelling these interactions are the determination of the canopy geometry and the multiple-scattering pattern. It is common practice to model the vegetation canopy either as a continuous medium with specific dielectric properties, or as a mixture of discrete scatters randomly distributed in an inhomogeneous layer (Attema and Ulaby, 1978).

3.2.3. Retrieving Soil Moisture Over Bare Soil From ERS-1/SAR: A Sensitivity Analysis
In a recent study (Altese et al., 1995) the Integral Equation Model of Fung et al. (1992) has been used to analyse the sensitivity of radar echoes, in terms of the backscattering coefficient, to the surface parameters of random rough bare soil fields, under the ERS-1/SAR sensor configuration (5.3 GHz frequency, VV polarisation and 23° incidence angle). In the IEM, the backscattering coefficient is expressed as a function of the radar configuration (frequency, polarisation and incidence angle), the soil dielectric constant, and the soil roughness parameters (surface root mean square (rms) height, s, correlation function, $\rho(\xi,\zeta)$, and correlation length, L). In the study some more simplifying assumptions are made: only the real part of the relative dielectric constant, ε, is taken into account and the surface correlation function is assumed isotropic and represented by either the Gaussian, or the exponential distribution.

In Figure 1, the dependency of the backscattering coefficient, σ^0, on the rms height is shown, using both the Gaussian (a) and the exponential (b) correlation function. The curves indicate that the sensitivity of σ^0 to the surface roughness is very strong at low rms height (s<1 cm), and that this sensitivity decreases as roughness increases. It is also noted that the behaviour of the model is highly dependent on the choice of the correlation function and that the sensitivity is stronger with the Gaussian correlation function than with the exponential function.

Figure 2 shows the IEM behaviour when the correlation length L is varied from 3 to 15 cm, while s is kept constant at 0.6 cm. In this case the curves show a fairly different behaviour in the two cases: using the Gaussian correlation function (a), the sensitivity of σ^0 to surface roughness is strong at all correlation lengths, but particularly at high values of L (variations of more than 10 dB for an increment of 3 cm in L); using the exponential correlation function (b), the sensitivity is much lower (variations of 1 dB for each increment of 3 cm in L).

From both Figures 1 and 2, it can also be seen that the sensitivity of σ^0 to the dielectric constant ε is not very strong. For a variation in ε from 5 to 25 (which compares to values for dry and wet soils), a variation in σ^0 of about 5 dB is observed, almost independent of the roughness parameter values. Further analysis with the IEM has also shown that the radar configuration of ERS-1/SAR (C-band) is nearly optimal from a point of view of soil moisture retrieval possibilities (Altese et al., 1995).

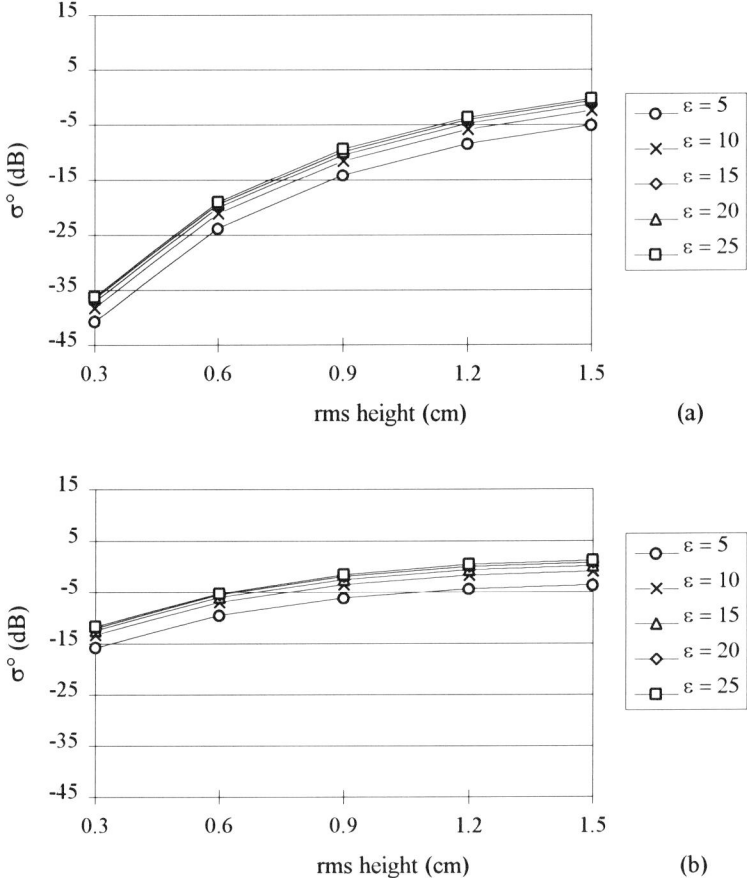

Figure 1. Backscattering coefficient σ^0 versus rms height as a function of dielectric constant ε for both Gaussian (a) and exponential (b) correlation function and for fixed correlation length (L=10 cm); radar configuration is of ERS-1 SAR (5.3 GHz, VV polarisation, incidence angle 23°).

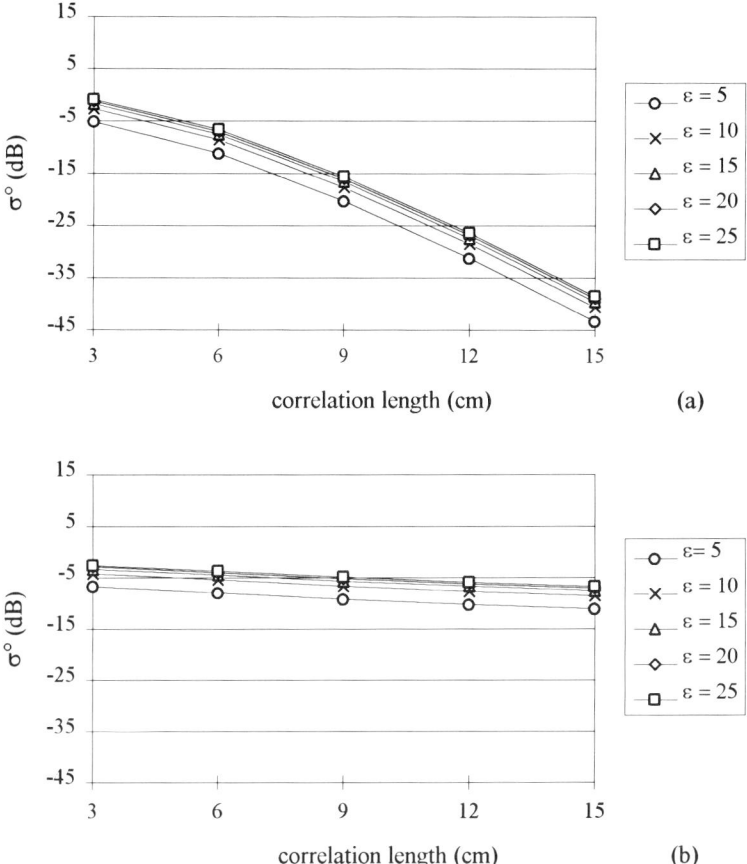

Figure 2. Backscattering coefficient σ^0 versus correlation length as a function of dielectric constant ε for both Gaussian (a) and exponential (b) correlation function and for fixed rms height (s=0.6 cm); radar configuration is of ERS-1 SAR (5.3 GHz, VV polarisation, incidence angle 23°).

The results presented above clearly indicate that, given measurements of surface roughness for a specific field, it is difficult to obtain reliable estimates of soil moisture from single frequency, single polarisation microwave observations whenever the surface is smooth. Figure 3 shows the range of variation of the backscattering coefficient σ^0, assuming an error of 0.2 cm in the measurement of the rms surface height values. It is noted that $\Delta\sigma^0$ decreases significantly as s increases. For values of s usually encountered in agricultural bare soil fields (1 to 3 cm), the range of variation of σ^0 is about 1 to 3 dB.

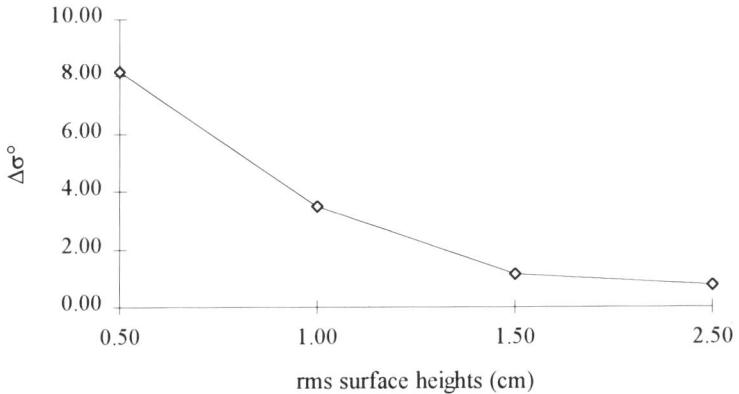

Figure 3. Range of variation ($\Delta\sigma^0$) of backscattering coefficients calculated assuming an error of 0.2 cm in the measurement of the surface rms height for ERS-1 SAR configuration.

4. Remotely Sensed Soil Moisture and Distributed Hydrological Modelling

4.1. INTRODUCTION

Hydrologic models are an indispensable tool in the development and testing of theories and hypotheses, in the analysis of data and the determination of what data should be collected, but they do not compensate for a lack of understanding of the natural processes. Observations still play a vital role at the heart of all basic problems concerning hydrologic simulations. One of the critical state variables in hydrological modelling is the spatial distribution of soil moisture which exerts a major control on the land surface water and energy balance. Despite its importance, direct usage of this information in hydrologic models has not gained widespread application. This is in part due to the fact that most hydrologic models do not treat soil moisture as a measurable state variable. Also, it is difficult to measure soil moisture in a consistent and spatially comprehensive way using conventional methods. It will be essential for the progress of hydrology as a science that the hydrologic science community becomes aware of the utility of microwave remote sensing technology for deriving surface soil moisture distributions at the basin scale and that efforts are made to modify existing hydrologic models, or to devise new ones, to allow the incorporation of spatially referenced data in addition to conventional point data (Engman, 1990). In the following we present some recent attempts to compare and/or incorporate remotely sensed soil moisture information in hydrological simulation studies.

4.2. MULTISCALE MODELLING OF WATER AND ENERGY BALANCE

An interesting modelling strategy to investigate spatial and temporal characteristics of soil moisture at the field, catchment and regional scale is described in detail by Famiglietti and Wood (1994). They developed water and energy balance models at these scales, by aggregating simple soil-vegetation-atmosphere transfer schemes (SVATS) across scales in a topographic framework.

The local model partitions the land surface into bare soil and vegetated components. Both evaporation and transpiration are computed for the wet and dry canopy, whereas evaporation is computed for the bare soil component of the land surface. Runoff generation in the model occurs by both the infiltration excess and the saturation excess mechanisms. The subsurface soil column is partitioned into two layers: an upper, more active, root zone and a lower, less active, transmission zone. The model is driven by standard meteorological data at a time resolution high enough to represent adequately the diurnal dynamics of land-atmosphere interaction.

The spatially distributed model formulation uses a digital elevation model (DEM) to represent catchment topography. The catchment is discretised into grid elements based on the resolution of the DEM and the local SVATS is applied to each grid element. Spatially distributed fields of the model parameters and the inputs are co-registered with the DEM, so that spatial variability in the model outputs can also be represented explicitly. Since the SVATS need the local water table depth as a lower boundary condition, the spatially distributed model framework requires the spatial pattern of water table depths to couple grid elements together at the catchment scale. The topographic-soil index of Beven (1986) is utilised to parameterise spatial variability in topographic and soil properties, and accordingly water table depth, between the grid elements.

To aggregate to the macroscale, Famiglietti and Wood (1994) assumed that subgrid-scale variations in topography and soil properties dominate the process of spatial redistribution of soil water over large land areas. A second assumption at the macroscale is that a threshold modelling scale has been exceeded, so that the exact pattern of topographic and soil heterogeneity needs not to be represented explicitly within the macroscale model structure: at this scale a statistical representation of the variability will suffice. Therefore, a statistical distribution of the topographic-soil index is employed as the framework of the macroscale modelling. The distribution of the index is dicretised into a number of intervals and the local SVATS is applied at each interval.

4.3. RECENT REMOTE SENSING EXPERIMENTS IN HYDROLOGY

4.3.1. MACHYDRO '90
MACHYDRO'90 was a multi-sensor airborne campaign that took place in July 1990 over a portion of the Mahantango Creek catchment, a 7.4 km^2 research watershed operated by the Northeast Watershed Research Center of the US Department of Agriculture. The study area included a subwatershed (WD38) of about 60 ha in the eastern part of Mahantango Creek catchment. WD38 contains a mixture of land uses (maize, wheat, oat, pasture and hay fields), bounded in the south by forest. During a 12-day period (July, 9-20, 1990), detailed hydrologic and meteorological data have been collected in the catchment. In addition, field surveys were organised to collect ground truth data (soil moisture, surface roughness, vegetation characteristics). Microwave sensors flown during the experiment were the passive pushbroom microwave radiometer (PBMR) and the active AIRSAR. The PBMR operated at L-band with cross-track resolution of approximately 90 m. AIRSAR is a full-polarisation radar with frequencies of 0.44, 1.25 and 5.33 GHz. The azimuth and slant range resolutions of the processed AIRSAR images are 12.1 m and 6.6 m, respectively.

Figure 4 shows the computed (based on the catchment scale model presented in Section 4.2.) and the observed watershed average surface soil moisture for the WD38 subwatershed. Rainfall records are plotted in Figure 4 for comparison. Two types of model initialisation were used: first, the initial wetness of the catchment was computed by means of the technique developed by Troch et al. (1993a), which is based on initial water depths computed from baseflow analysis; second, the catchment's initial wetness was estimated from remote sensing observations on the first day of the experiment. The model results for the watershed average volumetric soil moisture are shown for the two types of initialisation by the dashed line and the solid line, respectively. It can be seen that, while both simulations reflect the variation resulting from the weather conditions quite correctly, the model predictions based on the baseflow derived initial conditions appear to be too wet. Application of the remotely sensed initial data improves the simulation accuracy for surface soil moisture significantly. This result underscores the potential of remote sensing techniques in hydrologic simulations (Troch et al., 1993b).

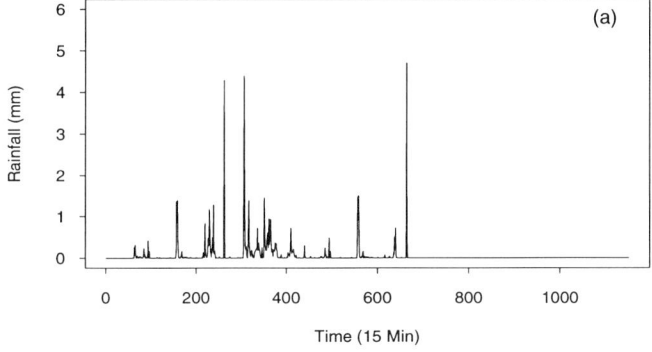

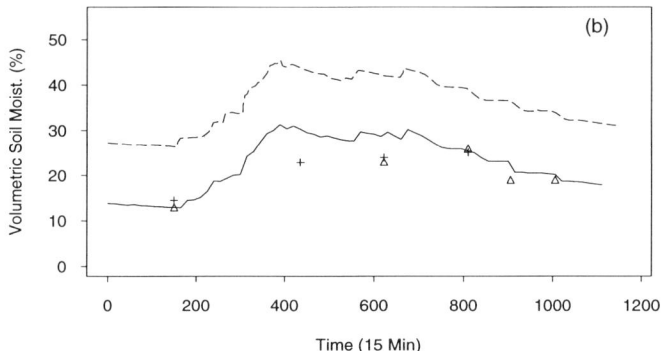

Figure 4. Time series of rainfall (a) and watershed averaged surface soil moisture. The dashed line is obtained using the baseflow-derived initial condition; the solid line represents the remote sensing based simulation. PBMR and SAR based soil moisture estimates are given as triangles and crosses, respectively.

4.3.2. *MACEUROPE'91*

A similar experiment (MACEUROPE'91) was organised in two European catchments in summer 1991. The first catchment, Slapton Wood, is located in the county of Devon, Southwest England. The drainage area is about 94 ha and the watershed contains a wide variety of land cover: 14% forest, 25% arable crops and the remainder permanent and ley pasture. Soils within the catchment are mainly silty loam and loam. The second catchment, Virginiolo, is located in Tuscany, Italy. It is a first order watershed of the Arno river, with a drainage area of 4.5 km^2 and it shows the typical hilly landscape of the central Italian Apennines. The main land cover is olive trees, vineyards and bare soil. Soils within this catchment are mainly sandy clay.

Only the AIRSAR instrument was flown during this experiment. Measurements were made in three wavelength bands (C, L and P) and for three polarisations (HH, VV and HV). For Slapton Wood, two flights were executed on June 29 and July 5, 1991. For each flight, images were taken at multiple azimuth (330° and 45°) and mid-swath incidence angles (25° and 40°). Over Virginiolo, three flights were executed: June 22, June 29 and July 14, 1991. The flight route was fixed at 275° west direction, such that the radar illumination was almost aligned with the valley direction. The incidence angle at mid-swath was 25°.

For the grass covered areas of the Slapton Wood catchment, Lin et al. (1993) found that an estimate of soil moisture could be obtained from L-band images through simple regression algorithms. Mancini et al. (1993) obtained soil moisture estimates for the Virginiolo catchment using an empirical inversion model developed by Oh et al. (1992). Giacomelli et al. (1995) compared the SAR derived soil moisture distribution with hydrologic simulations, using the distributed model of Famiglietti and Wood (1994). They report acceptable results when comparing sampled, modelled and SAR derived soil moisture at the field scale, but also that the spatial patterns of soil moisture predicted by the distributed model and by the SAR images are quite different. They claim that these differences are due to the model structure and its initial conditions. Further studies still need to be undertaken, to enhance the detail of the model by taking into consideration the variation of important land surface parameters, such as vegetation and soil hydraulic properties.

4.3.3. *EMAC'94*

EMAC'94 (European Multi-sensor Airborne Campaign) is a collaborative programme of the European Space Agency and the Joint Research Centre (JRC), EC. The objectives of EMAC'94 are multi-disciplinary, with interests including agriculture, forestry, hydrology, snow and ice, coastal and ocean studies. The airborne instruments used in the programme include synthetic aperture radar (ESAR), imaging spectrometer (ROSIS) and microwave radiometer instruments.

The ESAR is an "experimental" multifrequency, dual polarisation SAR, developed by the German Aerospace Research Establishment (DLR), Institute for Radio Frequency Technology, and designed for medium size turbo-prop aircraft. For EMAC'94, ESAR was flown in four frequencies (X, C, L and P band) with vertical and horizontal co-polarisation. Frequency and polarisation are switchable during flight. Cross polarisation is not available. For the EMAC'94 campaign wide swath mode was used (standard scene size: 6 km x 6 km, geometric resolution: 4.5 m x 4.5 m, mid-swath incidence angle: 52°). The Reflective Optics System Imaging Spectrometer (ROSIS) is a compact airborne spectrometer that can operate in different modes. For data collection during EMAC'94 the imaging mode was used.

Within the "Vegetation and Soils" thematic group of EMAC'94, a test site in Belgium has been selected: Zwalm Creek, a tributary of the Scheldt river. The Zwalm catchment is located in Flanders, about 20 km south of Ghent, and has a total drainage area of 114 km² (Figure 5). ESAR flights were executed on April 9, June 30 and August 19, 1994 and one ROSIS flight was flown on July 12, 1994 (cloud free conditions).

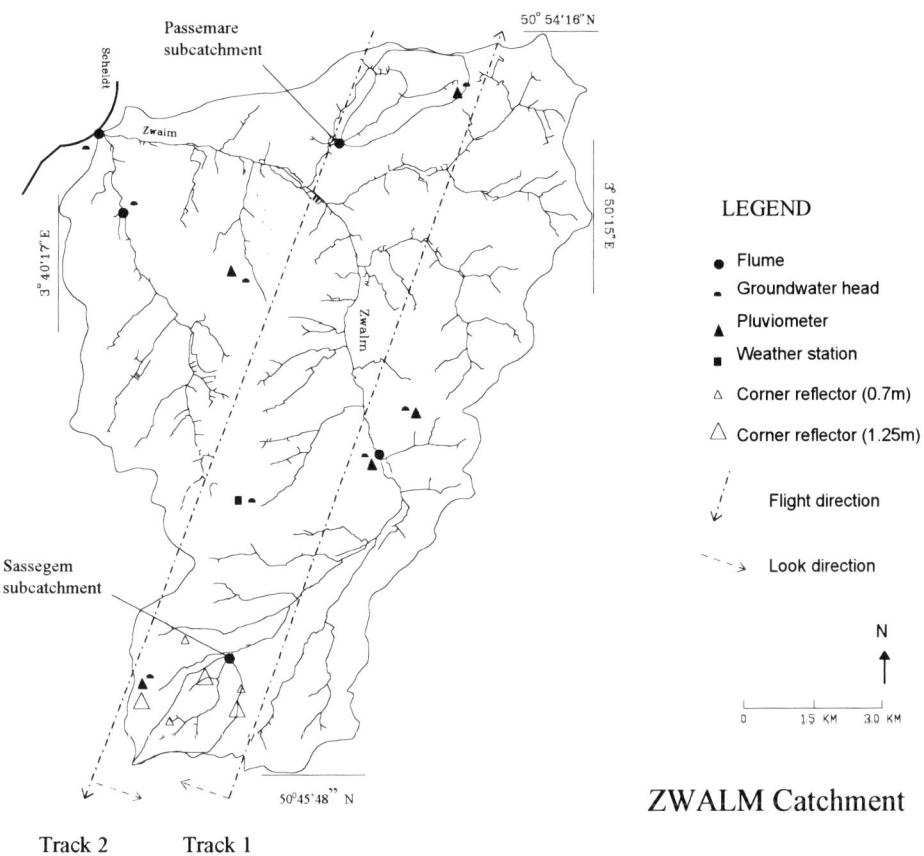

Figure 5. Location and data acquisition network of Zwalm catchment.

During the remote sensing campaign dates ground truth data collection took place in two subcatchments, viz. the Passemarebeek and the Sassegembeek. The Passemarebeek is located in the north of the catchment, with a drainage area of 2.5 km² and an average slope of about 5%. Land use is mainly agricultural, but the east of the watershed is urbanised. The Sassegembeek is located in the south of the catchment. It has a drainage area of 2.7 km² and is partly forested (40%) and partly agricultural. The average slope is about 8%. Data from the ROSIS instrument were used to develop detailed land use maps of the subcatchments.

Preliminary results on soil moisture retrieval from ESAR data, based on the approximate version of the IEM as described by Altese et al. (1995), have been obtained (Su et al., 1996). As indicated in Figure 6, the IEM, using the measured surface soil moisture data (as dielectric constants) as an input, predicts the averaged ESAR measured backscattering coefficients in the different fields reasonable well, although in several fields they tend to be underestimated, both for C and L-band.

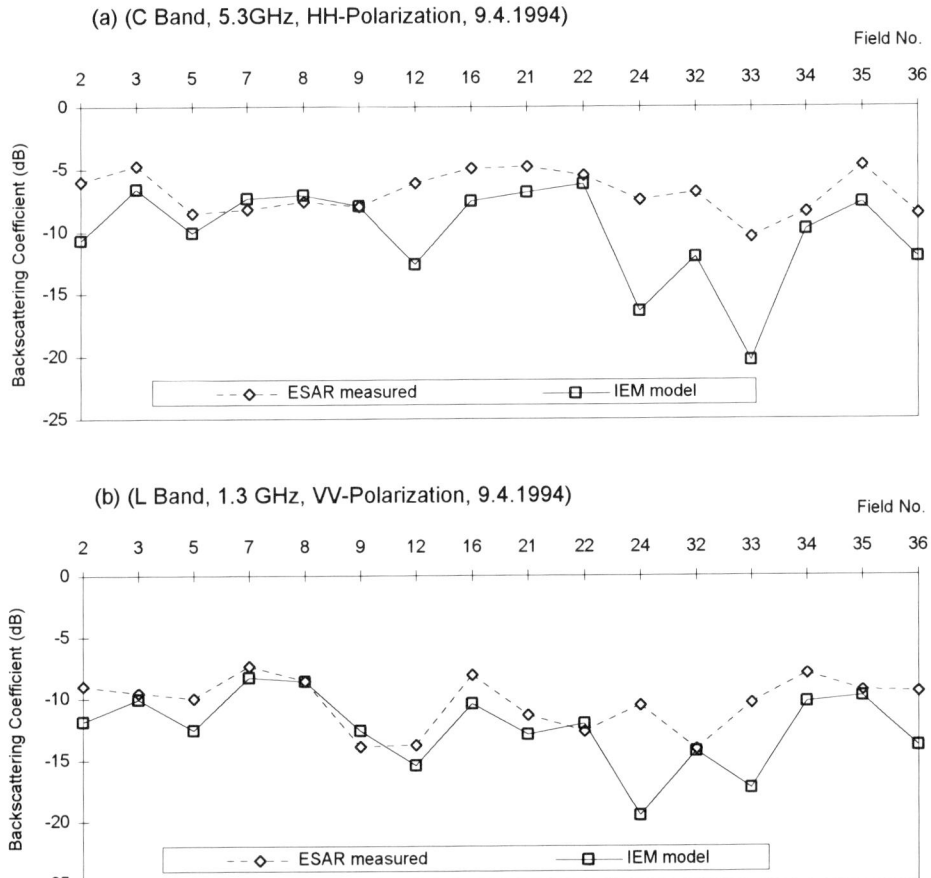

Figure 6. Comparison of observed and retrieved backscattering coefficients using non-calibrated and calibrated IEM for different fields in Passemarebeek on April 9, 1994: (a) C-band; (b) L-band.

Considering the sensitivity of the IEM to roughness parameters, it is assumed that this underestimation is mainly due to the conventional and rather inaccurate measurements of field roughness, using a metal board of 1 m length with 5 cm x 5 cm grids.

However, the purpose of an inversion algorithm is to retrieve soil moisture from SAR measured backscattering coefficients. Hence, it was decided to use one ESAR data set (C-band in this case) as an input in the IEM to invert for the roughness parameters, and to use another ESAR data set (L-band) together with the "calibrated" roughness parameters - instead of the metal board measured ones - as an input for soil moisture retrieval. Both data sets were taken from the same day (April 9, 1994).

The improved results in terms of backscattering coefficients, as shown in Figure 7, confirm the validity of this approach.

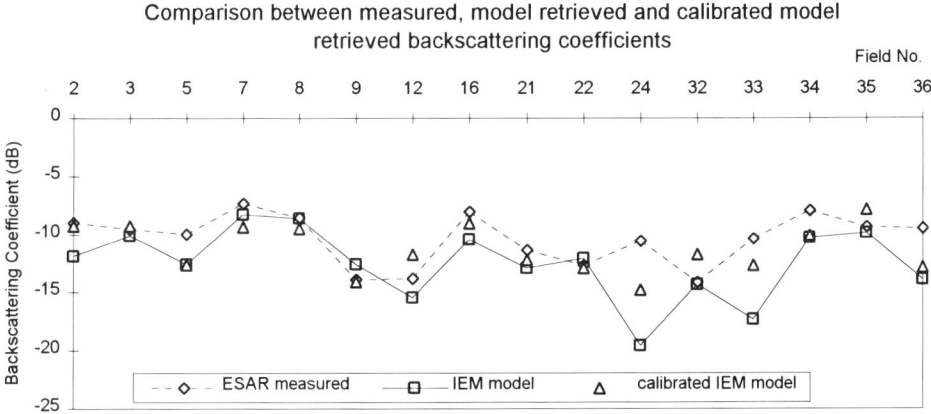

Figure 7. Comparison of observed and retrieved backscattering coefficients using non-calibrated and calibrated roughness parameters for different fields in Passemarebeek on April 9, 1994.

Figure 8 compares the retrieved soil moisture values with the in-situ surface soil moisture measurements, showing an acceptable performance of the proposed method. Moreover, by reversing the data sets in the sequence of the analysis (L-band to invert for the roughness parameters, C-band for moisture retrieval) very similar results were obtained, indicating the robustness of the method. Some fields (no. 2, 24 and 33 in Figure 8) produced results where inverted roughness parameters were out of the range of validity of the IEM.

In order to extend this "calibrated roughness parameter" method to spaceborne SAR data such as acquired by ERS-1 and/or ERS-2 SAR, additional consideration should be taken into account due to the fact that these SAR instruments are single frequency and single polarisation. In this case multitemporal data sets must be used, requiring that soil moisture is the only time varying parameter during the time span of the multitemporal data sets. The current tandem operation of ERS-1 and ERS-2 may prove valuable in this respect. The proposed method also needs to be examined for extension towards vegetated areas. In this case an appropriate model should take the place of the IEM.

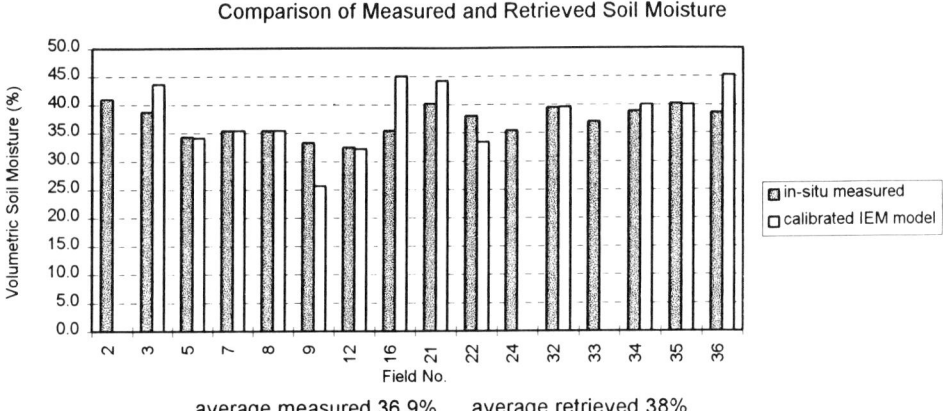

Figure 8. Comparison of in-situ measured and calibrated IEM retrieved volumetric soil moisture content (%) for different fields in Passemarebeek on April 9, 1994.

4.3.4. *SIR-C/X-SAR*

SIR-C is part of the Shuttle Imaging Radar-C / X-band Synthetic Aperture Radar (SIR-C/X-SAR) joint project of NASA, the German Space Agency (DARA) and the Italian Space Agency (ASI). It is a next step in a series of spaceborne imaging radars, beginning with Seasat in 1978 and continuing with SIR-A (1981), Germany's Microwave Remote Sensing Experiment (1983) and SIR-B (1984). It is the precursor of the Earth Observing System (EOS) SAR and the Global Topography Mission (GTM), planned for later in this decade. The SIR-C antenna includes a two-frequency radar in L-band (1.3 GHz) and C-band (5.2 GHz), with four polarisations and X-SAR is a one-frequency radar in X-band, with one polarisation. Two missions aboard NASA's space shuttle "Endeavour" were flown in April 1994 (STS-59) and September-October 1994 (STS-68).

SIR-C/X-SAR data acquired over the Zwalm catchment are currently under investigation at the Laboratory of Hydrology and Water Management, University of

Ghent. It is hoped that the methodology described in the previous section can be verified and extended to the larger scale of the complete Zwalm catchment.

5. Conclusions

In this chapter we have reviewed the current state-of-the-art of remote sensing applications, including sensors, platforms and remote sensing systems for hydrological studies, as well as applications of remotely sensed data in studies of precipitation, snow and ice, evapotranspiration, soil moisture, surface water and runoff, and catchments characteristics.

Much attention has been paid to microwave, especially active microwave, remote sensing of soil moisture. After reviewing some representative electromagnetic backscattering models a case study on the sensitivity of ERS-1/SAR data for soil moisture retrieval from bare soil fields was presented. Using the Integral Equation Model, it was shown that it is difficult to obtain accurate soil moisture estimates for smooth bare soil fields using single frequency, single polarisation measurements. Further it was also shown that the sensitivity of radar measurements to surface roughness quickly reduces as roughness approaches values observed in common agricultural fields. This indicates that retrieval of soil moisture for normal agricultural fields is feasible given that the roughness parameters are known a priori and with sufficient accuracy.

We have also outlined some recent remote sensing experiments in hydrology, viz. MACHYDRO'90, MACEUROPE'91, EMAC'94 and SIR-C/X-SAR, and presented some results obtained using data acquired from these experiments. It has been shown that applying remote sensing data to initialise a distributed hydrological model significantly improved the accuracy of simulated surface soil moisture. Based on ESAR multifrequency data sets, a methodology was proposed to retrieve surface soil moisture using "calibrated soil roughness" characteristics. This method provides an alternative to overcome the difficulties encountered in in-situ measurement of surface roughness parameters for input into theoretical backscattering models and hence provides opportunities for operational application of remotely sensed soil moisture in hydrological modelling.

Summarising, it may be concluded that remote sensing data can be utilised in different ways in distributed modelling: (a) as parametric input data, including land cover data, such as land use classes and soil properties, mainly acquired from passive remote sensing instruments, and precipitation data, mainly obtained using ground based weather radar systems; (b) as data on initial conditions, such as initial catchment wetness, preferably determined by active microwave sensors; and (c) as data on hydrological state variables, such as soil moisture, vegetation status relating to evapotranspiration, and snow cover extend. Perspectives are that methods to retrieve these data from remote sensing systems still have to be developed further.

Whereas application of the data of types (a) and (b) in distributed models is, in principle, quite straightforward, application of type (c) data requires new approaches for distributed models, such as data assimilation (Ottlé and Vidal-Madjar, 1994), where uncertainties in both remote sensing inferred data and model estimates ultimately have to be incorporated.

It is hoped that this chapter has provided the reader with a comprehensive - though not exhaustive - view of current possibilities of remote sensing applications in hydrology. It is especially hoped that results from recent and future remote sensing campaigns may contribute towards improving the understanding of the basic hydrological processes.

6. Acknowledgement

Part of the results presented here has been obtained with support of ESA and the Belgian "Federale Diensten voor Wetenschappelijke, Technische en Culturele Aangelegenheden (DWTC), through grant n° T3/02/21 (for EMAC'94) and of the EC Environment and Climate Research Programme, under contract n° EV5V-CT94-0446, Climatology and Natural Hazards (for ERS-1/SAR).

7. References

Altese, E., Bolognani, O., Mancini, M. and Troch, P.A. (1995) Retrieving soil moisture over bare soil from ERS-1 SAR data, a sensitivity analysis based on a theoretical surface scattering model and field data, *Water Resources Research*, in press.

Attema, E.P.W. and Ulaby, F.T. (1978) Vegetation modeled as a water cloud, *Radio Science* 13(2), 357-364.

Bahar, E. (1985) Scattering by anisotropic models of composite rough surface, full wave solution, *IEEE Trans. Antenna Propagation* 33, 106-112.

Barrett, E.C. (1993) Precipitation measurements by satellites: Towards community algorithms, *Adv. Space Res.* 13, 5119-5136.

Beckman, P. and Spizzichino, A. (1963) *The Scattering of Electromagnetic Waves From Rough Surfaces*, Macmillan Inc., New York.

Beven, K. (1986) Runoff production and flood frequency in catchments of order n: An alternative approach, in V.K. Gupta (ed.), *Scale Problems in Hydrology*, D. Reidel, Norwell, Mass., pp. 107-131.

Brown, G.S. (1978) Backscattering from a Gaussian distributed perfectly conducting rough surface, *IEEE Trans. Antenna Propagation* 26, 472-482.

Browning, K.A. and Collier, C.G. (1989) Nowcasting of precipitation series, *Rev. Geophysics* 27(3), 345-370.

Chang, A.T.C., Foster, J.L., Rango, A. and Joseberger, E.G. (1991) The use of microwave radiometry for characterizing snow storage in large river basins, *IAHS Publ.* 205, 73-80.

Collinge, V. and Kirby, C. (1987) *Weather Radar and Flood Forecasting*, John Wiley and Sons, Chichester.

Colwell, R.N. (ed.) (1983) *Manual of Remote Sensing*, 2nd ed., American Society of Photogrammetry, Fall Church.

Dornier (1993) *Erderkundungs-Daten-Service: Landsat 6 Informationen*, Dornier GmbH.
Engheta, N. and Elachi, C. (1982) Radar scattering from a diffuse vegetation layer over a smooth surface, *IEEE Trans. Geosci. Remote Sens.* 20, 212-216.
Engman, E.T. (1990) Progress in microwave remote sensing of soil moisture, *Can. Journ. of Remote Sens.* 16(3), 6-13.
Engman, E.T. and Gurney, R.J. (1991) *Remote Sensing in Hydrology*, Chapman and Hill, London.
Famiglietti, J.S. and Wood, E.F. (1994) Multiscale modeling of spatially water and energy balance components, *Water Resources Research* 30(11), 3061-3078.
Feddes, R.A., Menenti, M., Kabat, P. and Bastiaanssen, W.G.M. (1993) Is large scale inverse modelling of unsaturated flow with areal average evaporation and surface soil moisture as estimated from remote sensing feasible? *J. Hydrol.* 143, 125-152.
Fung, A.K., Li, Z. and Chen, K.S. (1992) Backscattering from a randomly rough dielectric surface, *IEEE Trans. Geosci. Remote Sens.* 30, 356-369.
Fung, A.K. and Pan, G.W. (1987) A scattering model for perfectly conducting random surfaces, I, Model development, *Int. J. Remote Sensing* 8, 1579-1593.
Giacomelli, A., Bacchiega, U., Troch, P.A. and Mancini M. (1995) Evaluation of surface soil moisture distribution by means of SAR remote sensing techniques and conceptual hydrological modelling, *J. Hydrol* 166, 445-459.
Hollenbeck, K.J., Schmugge, T.J., Hornberger, G.M. and Wang, J.R. (1996) Identifying soil hydraulic heterogeneity by detection of relative change in passive microwave remote sensing observations, *Water Resources Research* 32(1), 139-148.
Jackson, T.J. (1993) Measuring surface soil moisture using passive microwave remote sensing, *Hydrological Processes* 7(2), 139-152.
Klatt, P. and Schultz, G.A. (1983) Flood forecasting on the basis of radar rainfall measurements and rainfall forecasting, in *Hydrological Applications of Remote Sensing and Remote Data Transmission*, *IAHS Publ. no. 145*, 307-315.
Kouwen, N., Soulis, E.D., Pietroniro, A., Donald, J. and Harrington, R.A. (1993) Grouped response units for distributed hydrologic modeling, *J. Water Res. Planning and Management* 119 (3), 289-305.
Leader, J.C. (1978) Incoherent backscatter from rough surfaces, The two scale model re-examined, *Radio Science* 13, 441-457.
Le Toan, T., Smacchia, P., Souyris, J.C., Beaudoin, A., Merdas, M., Wooding, M. and Lichteneger, J. (1994) On the retrieval of soil moisture from ERS-1 SAR data, *Proc. Second ERS-1 Symposium: Space at the Service of our Environment*, ESA SP-361, 883-888.
Lin, D.S., Wood, E.F., Saatchi, S. and Beven, K. (1993) Soil moisture estimation during MAC-EUROPE'91 using AIRSAR, *Proc. 25th Intern. Symp. Remote Sensing and Global Environ. Change*, I-172, Graz.
Mancini, M., Rosso, R., Lin, D.S., Wood, E.F. and Troch, P.A. (1993) AIRSAR capability in soil moisture content for different climate scenarios, *Proc. 25th Intern. Symp. Remote Sensing and Global Environ. Change*, I-185, Graz.
Martinec, J. and Rango, A. (1991) Indirect evaluation of snow reserves in mountain basins, *IAHS Publ. no. 205*, 111-119.
Menenti, M. (1983) A new geophysical approach using remote sensing techniques to study groundwater table depths and regional evaporation from aquifers in deserts, *ICW report 9*, Wageningen.
Mo, T., Schmugge, T.J. and Jackson, T.J. (1984) Calculations of radar backscattering coefficient of vegetation covered soils, *Remote Sens. Env.* 15, 119-133.
Newton, R.W., Black, Q.R., Makanvand, S., Blanchard, A.J. and Jean, B.R. (1982) Soil moisture information and thermal microwave emission, *IEEE Trans. Geosci. Remote Sens.* 20, 275-281.
Nieuwenhuis, G.J.A. (1986) Integration of remote sensing with a water balance simulation model (SWATRE), *ICW Techn. Bulletin 59*, Wageningen.
Njoku, E.G. and O'Neill, P.E. (1982) Multifrequency microwave radiometer measurements of soil moisture, *IEEE Trans. Geosci. Remote Sens.* 20, 468-475.

Oh, Y., Sarabandi, K. and Ulaby, F.T. (1992) An empirical model and an inversion technique for radar scattering from bare soil surfaces, *IEEE Trans. Geosci. and Remote Sensing* 30(2), 370-381.

Ottlé, C. and Vidal-Madjar, D. (1994) Assimilation of soil moisture inferred from infrared remote sensing in a hydrological model over the HAPEX-MOBILHY region, *J. Hydrol.* 158, 241-264.

Peck, E.L., Keefer, T.N. and Johnson, E.R. (1981) Strategies for using remotely sensed data in hydrologic models, *NASA-CR-66729*.

Pultz, T.J., Leconte, R., Brown, R.J. and Brisco, B. (1990) Quantitative soil moisture extraction from airborne SAR data, *Canad. J. Remote Sens.* 16, 56-62.

Ragab, R. (1995) Towards a continuous operational system to estimate the root-zone soil moisture from intermittent remotely sensed surface moisture, *J. Hydrol.* 173, 1-25.

Rango, A. (1987) New technology for hydrological data acquisition and applications, *IAHS Publ. No. 164*, 511-517.

Rango, A. (1993) Snow hydrology processes and remote sensing, *Hydrological Processes* 7, 121-138.

Richards, J.A., Sun, G.Q. and Simonett, D.S. (1987) L-band radar backscatter modelling of forest stands, *IEEE Trans. Geosci. Remote Sens.* 23, 487-498.

Sancer, M.I. (1969) Shadow-corrected electromagnetic scattering from a randomly rough surface, *IEEE Trans. Antenna Propagation* 17, 577-589.

Schmugge, T.J. (1985) Remote sensing of soil moisture, in M.G. Anderson and T.P. Burt (eds.), *Hydrological Forecasting*, John Wiley and Sons, Chichester, pp. 101-124.

Schmugge, T.J., Jackson, T.J., Kustas, W.P. and Wang, J.R. (1992) Passive microwave remote sensing of soil moisture: Results from HAPEX, FIFE and MONSOON'90, *J. Photogrammetry Remote Sens.*

Schultz, G.A. (1988) Remote sensing in hydrology, *J. Hydrol.* 100, 239-265.

Seguin, B., Savane, M. and Guillot, B. (1990) Estimation of large area evaporation from thermal infrared meteorological satellite data: A case study with Meteosat and NOAA for France, *Proc. Int. Symp. Remote Sensing and Water Resources*, Enschede, 215-228.

Su, Z., Neumann, P., Fett, W., Schumann, A.S. and Schultz, G.A. (1992) Application of remote sensing and geographic information systems in hydrological modelling, *EARSeL Adv. Remote Sensing* 1(3), 180-185.

Su, Z. and Schultz, G.A. (1993) A distributed runoff prediction model developed on the basis of remotely sensed information, *Proc. EARSeL Specialist Meeting*, Dundee, 50-64.

Su, Z., Troch, P.A. and De Troch, F.P. (1996) Remote sensing of soil moisture using EMAC/ESAR data, *Int. J. Remote Sensing*, submitted.

Troch, P.A., De Troch, F.P. and Brutsaert, W. (1993a) Effective water table depth to describe initial conditions prior to storm rainfall in humid regions, *Water Resources Research* 29(2), 427-434.

Troch, P.A., Mancini, M., Paniconi, C. and Wood, E.F. (1993b) Evaluation of a distributed catchment scale water balance model, *Water Resources Research* 29(6), 1805-1818.

Ulaby, F.T., Batlivala, P.P. and Dobson, M.C. (1978) Microwave backscatter dependence on surface roughness, soil moisture, and soil texture: Part I, Bare soil, *IEEE Trans. Geosci. Remote Sens.* 16, 286-295.

Ulaby, F.T., Bradley, G.A. and Dobson, M.C. (1979) Microwave backscatter dependence on surface roughness, soil moisture, and soil texture: Part II, Vegetation covered soil, *IEEE Trans. Geosci. Remote Sens.* 17, 33-40.

Ulaby, F.T., Aslam, A. and Dobson, M.C. (1982) Effect of vegetation cover on radar sensitivity to soil moisture, *IEEE Trans. Geosci. Remote Sens.* 20, 476-481.

Ulaby, F.T., Allen, C.T. and Eger, G. (1984) Relating the microwave backscattering coefficient to leaf area index, *Remote Sensing Environ.* 14, 113-133.

Ulaby, F.T., Moore, R.K. and Fung, A.K. (1986) *Microwave remote sensing: Active and passive, vol. III, From theory to applications*, Arctech House, Inc., Dedham, MA.

Valenzuela, G.R. (1967) Depolarization of EM waves by slightly rough surfaces, *IEEE Trans. Antenna Propagation* 15, 552-557.

Wang, J.R., O'Neill, P.E., Jackson, T.J. and Engman, E.T. (1983) Multifrequency measurements of the effect of soil moisture, soil texture and surface roughness, *IEEE Trans. Geosci. Remote Sens.* 21, 44-51.

Wineberner, D. and Ishimaru, A. (1985) Investigation of a surface field phase perturbation technique for scattering from rough surfaces, *Radio Sci.* 20, 161-170.

Wood, E.F., Lin, D.S., Mancini, M., Thongs, D., Troch, P., Famiglietti, J. and Jackson, T.J. (1993) Intercomparison between passive and active microwave remote sensing and hydrological modelling for soil moisture, *Adv. Space Res.* 13(5) 167-176.

Wood, E.F., Sivapalan, M., Beven, K. and Band, L. (1988) Effects of spatial variability and scale with implications to hydrological modeling, *J. Hydrol.* 102, 29-47.

Wright, J.W. (1968) A new model for sea clutter, *IEEE Trans. Antenna Propagation* 16, 217-223.

Wu, S.T. and Fung, A.K. (1972) A noncoherent model for microwave emission backscattering from the sea surface, *J. Geophys. Res.* 77, 5917-5929.

CHAPTER 10
GEOLOGICAL MODELLING

M. HANSEN AND P. GRAVESEN
Geological Survey of Denmark & Greenland

1. Introduction

This chapter describes the collection and processing of geological data and the subsequent establishment of three-dimensional geological models to be used as input to three-dimensional hydrogeological modelling.

A geological model is a idealized simplification set up to aid the understanding of complex natural phenomena and processes. The visualization of the geological model help us to see the relationship of environments to each other and to picture the processes and resulting products that should be expected to find in these invironments. This chapter describes three different approaches ranging from manually correlated and constructed geological cross sections and fence diagrams over layer-based models generated by interpolation of values extracted from a geological database to statistic models generated using stochastic modelling. Also, the combination of "hard" geological data from borehole samples with "soft" data, such as geological knowledge and indirect geophysical data, into a three-dimensional geological model is described.

When making three-dimensional flow- and transport modelling the normal situation is that the by far most uncertain factor is the geological model. Therefore it is crucial to extract as much information as possible about the spatial geological setting from the hard data and to make efficient use of soft data when correlating between the points of observation.

While creating geological models it is important to remember that the nature is usually more complex than can be seen from borehole data. Figure 1 shows a typical Danish cliff section from the island of Ærø, southern Denmark. This is a good figure to keep in mind when trying to construct geological models from borehole data.

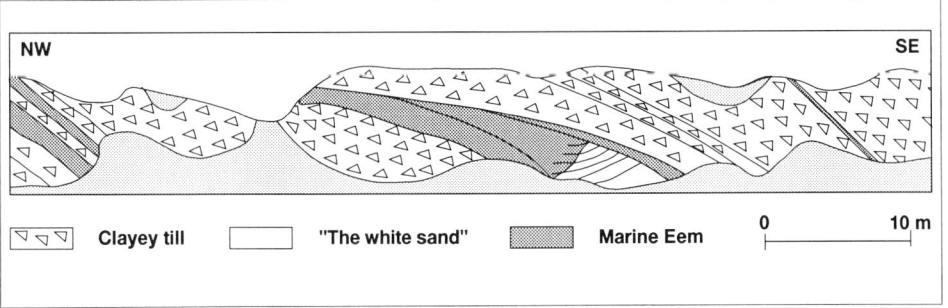

Figure 1: A typical Danish coastal cliff section from the island of Ærø, southern Denmark (from Hansen, 1987).

The examples in this chapter are from the Grundfør - Hinnerup area in eastern Jutland and the island of Amager south west of Copenhagen and the hard geological input is mainly preexisting data from the geological database ZEUS at the Geological Survey of Denmark. The presence of this geological database with its large amount of data has made geological modelling at the Geological Survey of Denmark focus on use of existing data stored in a highly structured geological database.

2. Geological data

2.1. TYPES OF DATA

Geological models can be generated using both hard and so-called soft data. In this chapter hard data are considered to be direct field information such as drill hole samples and outcrop data. Soft data are indirect information such as geological knowledge and surface-based geoelectrical measurements of the subsurface properties.

The examples described in this chapter are all from Denmark and the hard geological data used in the models are mainly preexisting data from the geological database ZEUS (Gravesen & Fredericia, 1984). The borehole information in the ZEUS database is from the Well Data Archive at the Geological Survey of Denmark. This archive includes information about 250.000 boreholes. Of these, data from about 175.000 boreholes are stored in the ZEUS database. Most of the borings in the Zeus database were drilled for the purposes shown in table 1:

Table 1: Distribution of wells in the Zeus database sorted after drilling purpose.

Groundwater supply	61%
Geotechnical investigation	20%
Raw material investigation	9%
Geophysical shot holes	7%
Monitoring wells	3%

These different types of boreholes contain much information relevant to manual and computerized geological modelling. This geological information is primarily based on analyses of borehole samples and data collected by the driller. The samples are investigated with respect to rock/sediment type, lithification, grain-size, colour, mineralogy, fossils, depositional environment or formation and age. In the Grundfør - Hinnerup area information interpreted from gamma logs and resistivity logs are also important inputs to the models. In this area the geological borehole data are "extended" by different types of geoelectrical and geomagnetical measurements.

Boreholes drilled at different times or for different purposes have different resolutions both for the sampling rate and regarding information contained in each sample. Because the ZEUS database includes borehole information of different origin and age (from the last hundred years), the modelling procedure must be able to handle this inhomogeneity in data.

Hard data are often easy to incorporate in geological models at the point of sampling however, the spatial distribution and variability of the hard data are difficult to depict. Volumetric geoelectrical measurements (Christensen & Sørensen, 1994) and geological knowledge (e.g. the location and orientation of pre-Quaternary valley systems) are examples of soft data that can easily be incorporated into manually developed models. Such soft data are difficult to include in computer generated geological models.

2.2. STORAGE AND RETRIEVAL OF GEOLOGICAL DATA

New geological data are expensive so it is essential to utilize as many preexisting data as possible and to analyse where new borehole data will give as much information as possible. It is also crucial to store the geological data in a way that ensure an accurate retrieval of data and allow searching on both areal and geological information. If geological models are generated using data extracted from a geological database, it is important that data and modifications are stored so that the model can be readily regenerated using the cross sections, data files and programme scripts.

3. Choice of modelling approach

Although the scale and the processes being studied should be the main criteria when selecting a modelling approach, it is often the amount and quality of the geological data that determine which model approach is selected.

Constructing deterministic geological models with very few hard geological data is possible. For instance, it is sufficient to know that about 5 metres of till covers a small sand layer on top of the chalk but of course the model will become more detailed with many data. If the amount of data becomes very large, it is not possible to utilize all the information and some kind of selective sorting of the data becomes necessary.

With the contoured model approach, it is necessary to have more information about each layer and there is no problem in using large amounts of data.

For statistical models, hard data are crucial. To be able to make a statistical description of the geological units and their variability it is necessary to have data sampled smaller distances than the correlation length of the units being modelled. Because it is not typical to have such detailed data sets, statistical models often must rely on data collected in analogue areas or outcrops. Table 2 show in a schematic form need for hard data for the different modelling approaches.

Table 2: The different types of models need for hard data.

	Need for hard data	
	Few	Many
Deterministic approach	---	
Contoured model		--
Statistic model		--------------------------------

4. Traditional deterministic geological models

The traditional approach to description of the aquifers is the establishment of a manually constructed geological model which is produced by the interpretation of the data by a skilled and experienced geologist. Basic geological, hydrogeological and geophysical data are input to the geological model and groundwater chemistry data can be included in the evaluation and verification of the model. The most important data are the geological logs including analyses of sediment or rock borehole samples with detailed description of the texture, colour, petrography and fossils. Also, the interpretations of geophysical borehole logs contribute to the geological model. Furthermore, the interpretation or determination of depositional environment and age of the deposits are critical since these data are needed for the correlation of layers from borehole to borehole. The correlated units correspond to the reservoir sediments or rocks and the confining layers above, below and adjacent to the reservoirs.

The construction of two- and three-dimensional geological models normally involve a large element of interpretation. For example the distribution of isolated aquifer bodies between the boreholes are based on interpretation of sedimentary facies models which only can give indications of the position of layers. The heterogeneity of the aquifers is very difficult to describe and explain in detail but abundant borehole data and geophysical surveys give fair criteria for defining the lithofacies parameters. Knowledge and experiences from the established conceptual sedimentary facies models can contribute to the aquifer description and in, e.g. the area of glacial fluvial and alluvial fan environment detailed models are presented (Miall, 1977, Eyles et al., 1983). The concept of three-dimensional sedimentary architecture (Miall, 1985, Miall & Tyler, 1991) are especially developed for these environments. Also, other clastic and carbonate depositional environments can be modelled in the subsurface based on the concept of facies models (Tillman & Weber, 1987). The sedimentary facies models can be used and extended into construction of hydrogeological facies models as in the case of glacial and glaciofluvial sediments (Anderson, 1989), and they can describe large scale heterogeneities but not heterogeneities at the small scale within individual facies or layers. However, the variability of grain-size distributions and structures within the models often determined in outcrops can be used in the subsurface as important guidelines in the interpretation.

Also, the considerations on the type and scale of the overall heterogeneity based on depositional models can support the model interpretation (Gravesen, 1994). However, the prediction of the heterogeneity in the aquifers from the interpretation of conceptual facies models is often difficult because the ideal depositional sequences are unlikely to be found at any given location (Anderson, 1990). Anderson (1990) also stresses, it is possible and necessary to use conceptual models to predict regional successions of aquifer deposits. In smaller areas, which are investigated in detail with abundant borehole data together with detailed outcrop surveys, the models can give data input with respect to the aquifer heterogeneities, especially for mathematical model simulations. Also, the use of present information on interconnections of geological sand units in the evaluation of heterogeneity can be included. Examples of modelling of fluvial and alluvial-fan sediments can be found in Poeter & Gaylord (1990), Ritzi et al. (1994) and Neton et al., (1994). Furthermore, new geophysical methods contribute to the description of the structural heterogeneity and these data support the information on geological description from outcrops and borehole samples (Auken et al., 1994).

4.1. TWO AND THREE-DIMENSIONAL MODELS

The construction of two-dimensional geological models is most often shown on cross sections where all the relevant borehole data are incorporated and the interpretation is supported by outcrop studies and geophysical surveys. As a part of the model contoured surface maps from selected levels and of different themes are typically produced. The following working process is described based on recent investigations at Grundfør, Denmark (figure 2 a-e):

1 The first step in the process is using the borehole information including the borehole sample descriptions and the interpretation of logs as gamma logs and resistivity logs combined with outcrop studies (figure 2 a). The task is to connect layers of the same genesis and age which can be very difficult because many deposits are without diagnostic age elements as fossils. Often the layers are correlated and connected based on lithology parameters and therefore it can be very valuable to study outcrops, if present in the area for information of lithology types and the style and scale of sedimentary and tectonic structures. The outcrop study also gives important information on the heterogeneity and interconnectiveness. Based on the knowledge of the possible depositional environment and the distribution of the layers and their variability in this model, the section is constructed being aware that the areas between the boreholes can only be correlated based on qualified estimates and experience.
2 The second step is to incorporate geophysical data into the model. In Denmark new geoelectrical and electromagnetical surface methods are important because they can map the horizontal and vertical resistivity of the sediments down to at least 200 m below surface (Christensen & Sørensen, 1994) but also seismic, gravimetric and georadar surveys can produce results to the model. These surveys contribute to the description of the mesoscopic scale structural heterogeneities in the aquifers and the possible connections between aquifer units. Interpretating the resistivity data into rock and

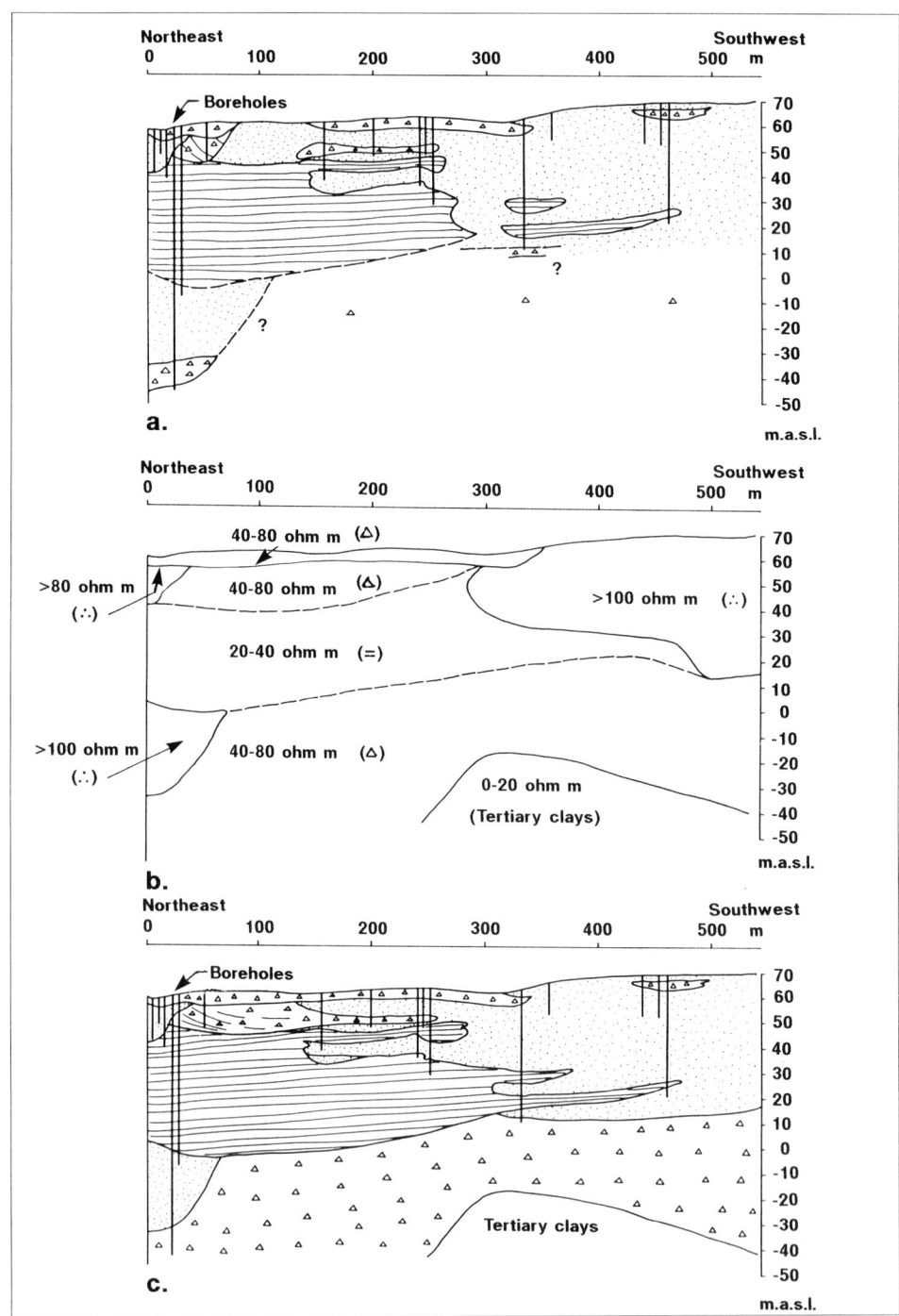

Figure 2a-2e: Figure captions on next page.

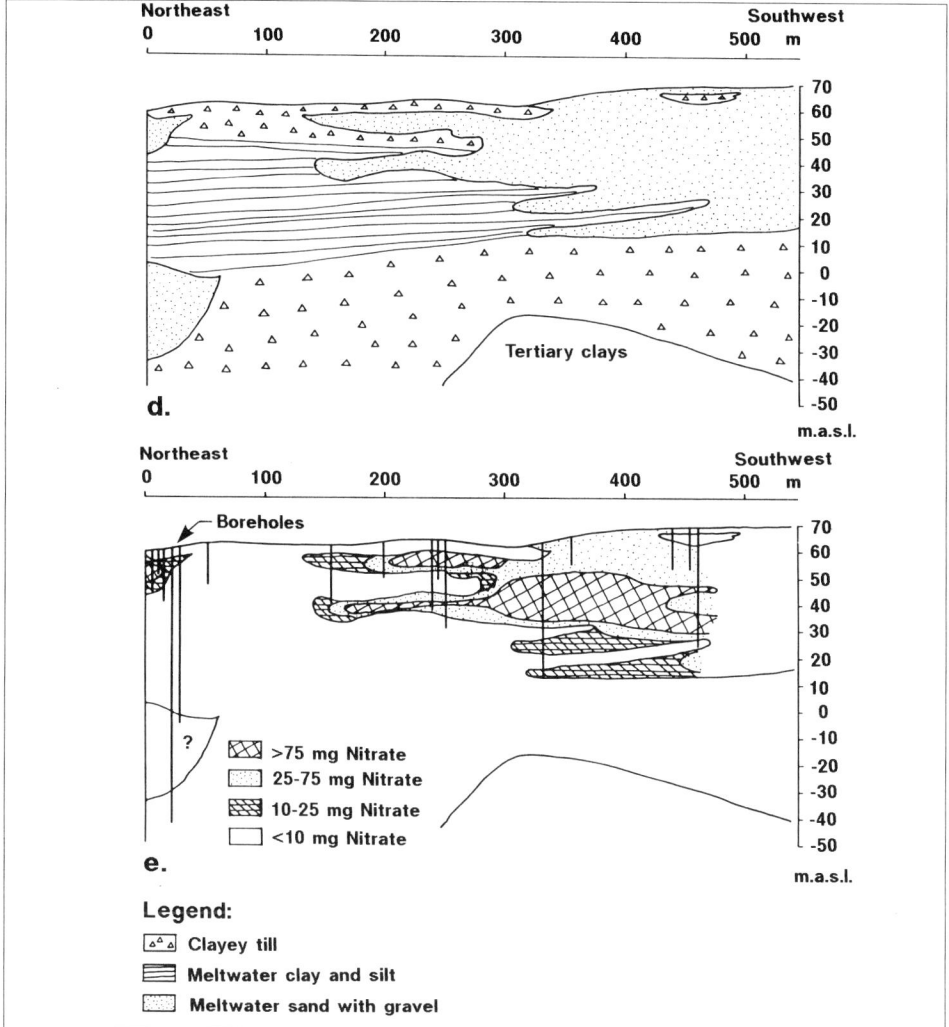

Figure 2a-2e: The geological model.
a   Cross section interpreted from borehole data: Analysed samples and geophysical logs together with information from outcrops. The section line is situated just east of the village Grundfør, eastern Jutland.
b   Interpretation of surface geophysical surveys in the same section line as a. The resistivity values are measured in geoelectrical and transient electromagnetic surveys. The values are interpreted in term of lithological units which are delineated from each other.
c   The information from the two section on a and b are put together in one section and a total interpretation of the distribution of the layers and their boundaries are performed.
d   A simplified section of c can be prepared by e.g. merging the clays layers into fewer bodies and the important sand-gravel layers then comprise the major aquifer.
e   The geological model in the Grundfør area can be verified by the distribution of the chemical parameters in the groundwater. The nitrate distribution is in the section drawn on the geological model and it fit very well to the overall geological conditions.

sediment types supports the description and distribution of the major lithological units. Geophysical data are shown on maps in different levels down through the aquifer, and also as vertical cross sections. Geophysical sections in same position as the geological sections are then produced from a combination of the two geophysical methods where the geoelectrical measurements are used down to 30 m below ground surface and the electromagnetic measurements are used down to the bottom of the sections. The resistivity values are interpreted in terms of sediment types according to known values from the Danish area and the main lithological units are then interpreted and delineated in the geophysical section (figure 2 b).

3 In the third step the two types of sections are combined and the information is used to interpret the distribution of the lithological units for constructing the final detailed geological section. The positions of lithological units not documented in the boreholes are now integrated with the geophysical analyses (figure 2 c).

4 Step four involves the simplification of the detailed information obtained in sections of the geological model. The extent of this simplification depends on the type of investigation. The method involves merging together relevant data from clay-silt bodies stressing the interconnections of preferential flow of the groundwater in the sand, gravel or limestone layers (figure 2 d). Sometimes the distribution of the chemical parameters in the groundwater can help verify the model as in the Grundfør area where the nitrate concentrations in the aquifer occur in a pattern matching the sand -clay distribution (figure 2 e).

The spatial representation of the aquifers is then constructed in the three-dimensional model which is constructed by joining cross sections to a fence-diagram or block-diagram. An important part of this work involves fitting the points of the crossing sections with respect to the lithological composition and the vertical and lateral distribution of the layers (figure 3).

The spatial model can then be input into the mathematical hydrogeological model where the geological structures are the framework for groundwater flow. The lithological types and characteristics will also be a starting point for evaluation of the hydraulic parameters such as hydraulic conductivity. In the more advanced models it is possible to include the digitized geological sections.

4.2. FINAL REMARKS

The construction of the deterministic geological model is a straightforward process based on the experience of the geologist. The degree of detail depends on the amount of existing data. New point data (boreholes) are often expensive to obtain but very critical for understanding of the spatial framework with respect to the correlation of units. However, combining borehole data with geophysical data, which cover the areas between the boreholes in horizontal and vertical directions, seems to be a very promising area of work.

The weakness of the deterministic model is that the work process gives only one suggestion to the final model and this model is difficult to change in order to incorporate

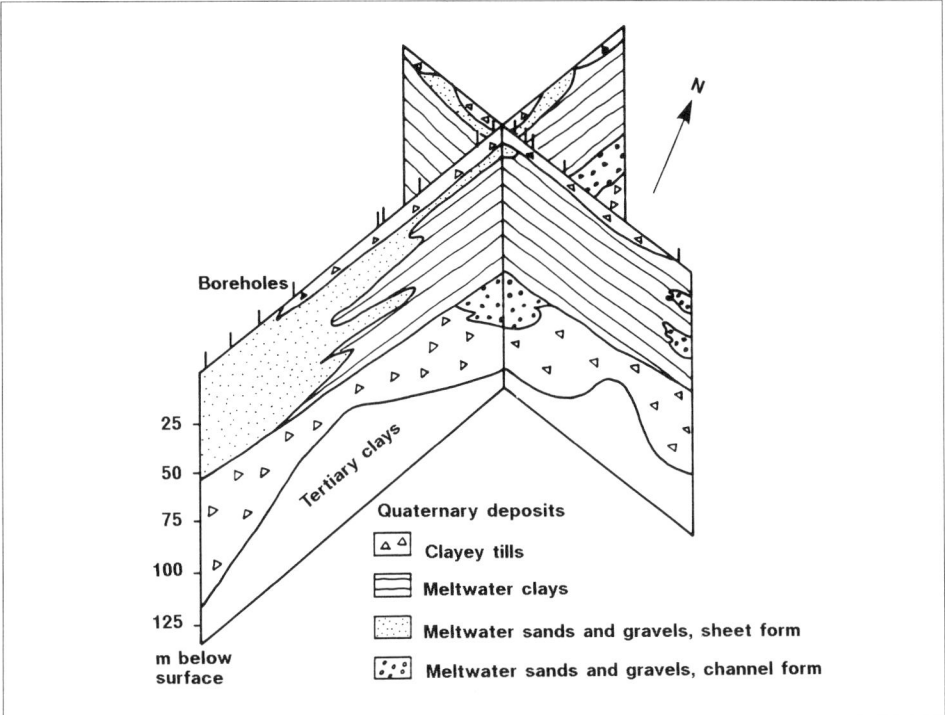

Figure 3: Fence-diagram from the Grundfør area showing the three-dimensional conditions east of the village. Two major aquifers are found. A lower restricted channel sand-gravel aquifer situated in clayey tills and capped by thick meltwater clays occur in a buried valley in the Tertiary clays. The upper sheet sandur sand-gravel aquifer intercalates with the meltwater clays but occur also above the clay. The sand-gravel is covered by clayey tills but geological windows or holes in this cover occur in several places (partly after Gravesen et al., 1995).

new data into the model. Before the geological models can be applied to mathematical hydrogeological modelling the data have to be transferred to the modelling programme, either in the form of digitized sections or as geological or hydraulic parameter values defined within boxes or cells.

Table 3: Strengths and weaknesses of the traditional deterministic modelling approch.

Strengths	Straight forward.
	Easy to incorporate geological knowledge.
	No need for advanced software.
Weaknesses	Difficult to incorporate new data.
	Difficult to model three-dimensions.
	Difficult to transfer to a hydrogeological modelling programme.
	Only one model produced, being highly dependent on the experience of the geologist who made it.

5. Surface models / contoured models

This method is often considered as "quick and dirty". While the traditional deterministic approach forces the geologist to consider each piece of geological information this method allows large amounts of geological data to pass without scrutiny from the geological database to the geological model.

Surface modelling is that type of geological modelling that most easily handles many data. The geological model can be represented by surfaces, isopach maps or a combination of these. In this chapter the example presented is from the island of Amager south west of Copenhagen, Denmark. Amager is one of the most intensively drilled areas of Denmark with more than 3,500 boreholes within an area of 95 km^2. In 1992 the Geological Survey of Denmark developed a regional three-dimensional geological model of this island (DGU & GI, 1992) to be used as input for three-dimensional hydrological modelling using MIKE SHE (DHI, 1993).

Because of a clear division in geological / hydrogeological units and the large amounts of preexisting geological data, it was decided to create the geological model using contoured surfaces of values extracted from the geological database ZEUS. The contouring of the different layers was done using MIKE SHE's contouring routines.

5.1. CONCEPTUAL MODEL

At Amager the geology can be divided into the following four main groups:

* Pre-Quaternary chalk / limestone.
* Meltwater sand lying in direct contact with the chalk.
* Alternating layers of meltwater sand and clayey till.
* A surface layer of clayey tills.

As the geological data vere stored in a geological database the conceptual model had to sum up knowledge about both very detailed boreholes and boreholes with very sparse information, and the search criteria had to be able to distinguish between the model units. In this area the sediments that are distinguishable from both old and new borehole description forms are:

* Chalk.
* Meltwater sand and gravel.
* Clayey till.

It was decided to make a geological model with six layers (see figure 4) dividing the geology into layers reflecting both the lithology and the hydraulics of the sediments (layers from bottom and up):

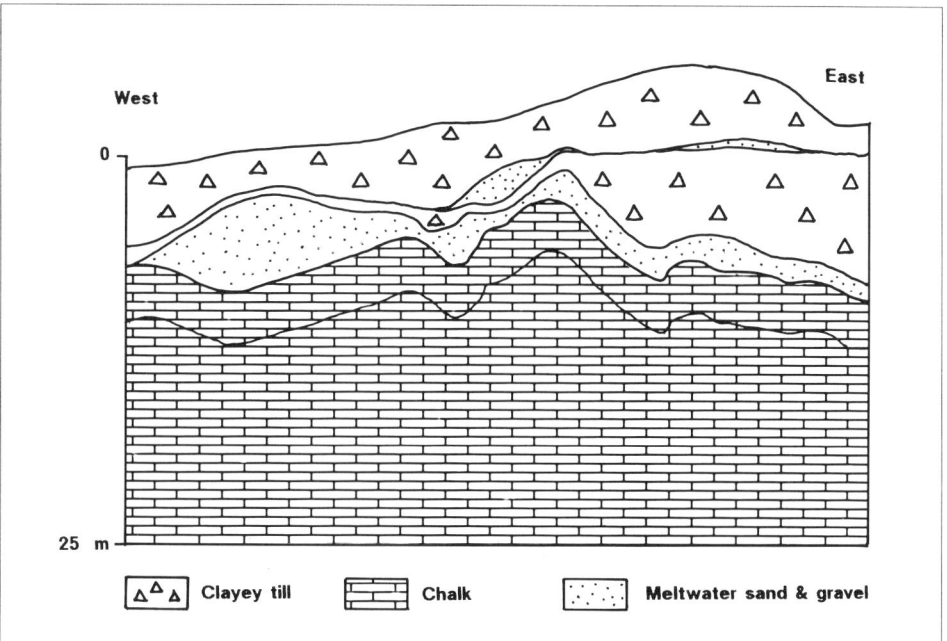

Figure 4: Cross section of the geological model of Amager.

1 An unfractured chalk.
2 An uppermost 3 metres thick chalk fractured by glacial processes. The thickness of this zone was taken as an average thickness. There was an unsuccessful attempt to find the thickness of this zone by using the length of the casing in the chalk with the assumption that casing was only used in the chalk if the chalk was weak (i.e., fractured).
3 Meltwater sand lying directly on the chalk or at most separated from the chalk by 0.5m of till / clay. This sand is considered in hydraulic contact with the chalk and then part of the main reservoir.
4 A secondary layer of till with a thickness equal the sum of all till layers except the surface till. The bottom of this layer was defined as the bottom of the lowermost layer of till used to calculate the thickness.
5 A till at the surface, layers of sand thinner than 0.5m ignored.
6 Ground surface. With the amount of data in this area the easiest way to get a digital surface map was to contour the terrain level measurements from the drill cites.

5.2. DATA EXTRACTION ROUTINES

When geological models are generated directly from a geological database the search routines have to include the geological knowledge and have to handle "insignificant" layers / samples. This is often difficult and the routines made for one area most often will not work for another area.

When using data from a geological database to extract the model layers it is important to know if some type of boreholes have to be excluded from some of the calculations. If a borehole was drilled in a shallow excavated well then there is no information about the uppermost sediments and the borehole information must be excluded from the calculation of the thickness of the upper till layer. The inclusive and exclusive criteria were used for the Amager model are summarized in table 4.

Table 4: Criteria to include or exclude borehole data from data extraction routines.

Layer	Include when:
	Exclude when:
1 Chalk	Boreholes reaching the chalk.
	Boreholes with only unknown lithology from surface to first chalk sample.
2 Fractured chalk	Boreholes reaching the chalk.
	Boreholes with only unknown lithology from surface to first chalk sample.
3 Sand on chalk	Boreholes reaching the chalk.
	Boreholes starting in an excavated well and boreholes with unknown lithology in the Quaternary sequence.
4 Secondary till	Boreholes reaching the chalk.
	Boreholes starting in an excavated well and boreholes with unknown lithology in the Quaternary sequence.
5 Surface till	Boreholes penetrating non till layers thicker than 0.5 metres
	Boreholes starting in an excavated well and boreholes with unknown lithology in the Quaternary sequence.
6 Ground surface	All Boreholes.
	None.

5.3 VERIFICATION OF CONTOUR MAPS

After generating the contour maps they must be tested against the geological knowledge, inconsistent data must be removed, and new data points (ghost data) must be added to guide the contouring programme in areas of few data. Actually contoured maps are an excellent way to find inconsistent data. Errors in contour maps are not always due to errors in the geological data, but sometimes due to different data quality. Thus it is often better to change the data files than the data in the database.

As contouring programmes have problems with valleys and steep gradients between data pairs it is nearly always necessary to add "ghost" data to force the contouring programme to create valleys instead of holes and to prevent steep gradients between data pairs to create artificial holes and hills.

GEOLOGICAL MODELLING

If the data files from the database and the files with the corrections are kept in a well organized way, it is easy change the model and the geological model can be regenerated using a simple script or programme when new data become available. Table 5 to 7 shows an efficient way to handle data extracted from a database, correction values, and ghost data.

Table 5: Sample file of data for top of pre-Quaternary surface extracted from ZEUS.

Borehole ID	Coordinates		Data values
	X	Y	
208.2001	725809	6170764	-45
208.2007	726472	5170863	-30

Table 6: Sample file containing correction values to pre-Quaternary surface

Borehole ID	Data values
208.2001	-30
208.2007	-15

Table 7: Sample file of ghost data to be added to the pre-Quaternary surface before contouring.

Borehole ID	Coordinates		Data value
	X	Y	
gh-0001	725300	6170550	-23
gh-0002	725350	6170600	-18

Because the surfaces of the model are contoured separately, a common problem is to have layers crossing each other or to have negative thickness of layers. Therefore it is necessary to have procedures that ensure the integrity of the model. In the Amager case all cell values smaller than 0.5 m were changed to 0.5 m in the thickness files and when layers crossed each other the boundary between the layers was set halfway between the layers. Figure 5 shows a flowchart describing the processes involved in generating contoured surfaces models.

5.4. FINAL REMARKS

If the data needed for a geological model are stored in a geological database, surface models are the easiest way to develop three-dimensional geological models on a format that can easily be transferred to hydrogeological modelling tools. If the data (including changes) data and procedures needed to ensure the integrity of the model are kept in a structured way it is easy to update the models when new data appear. The drawback of this method is that the geology must contain a limited number of layers or lenses which often results in rather simplified geological models.

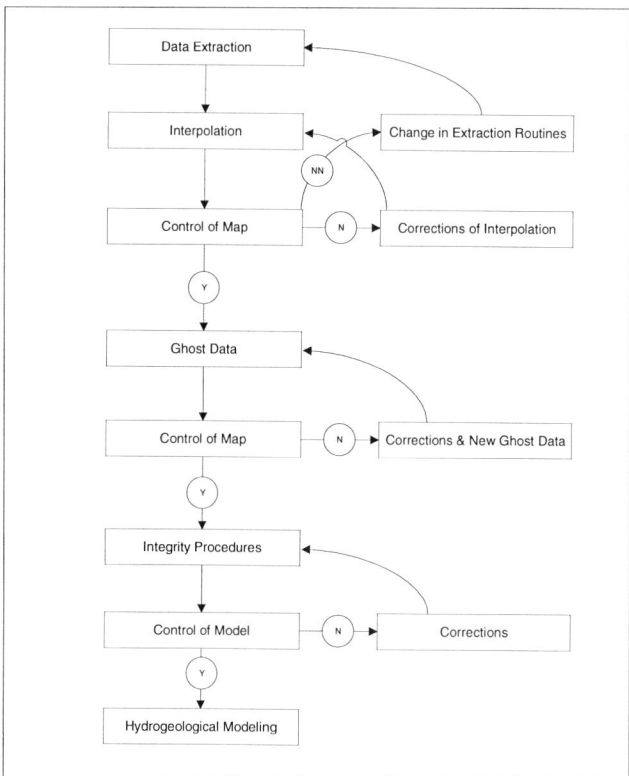

Figure 5: Flow chart describing the processes in developing contoured surfaces models.

Table 8: Strengths and weaknesses of the surface / contoured modelling approch

Strengths	Quick.
	Easy to transfer model to hydrogeological modelling programmes.
	Easy to update with new data.
	Handles huge amounts of data.
Weakness	Often very simplified geological model.
	Geological "units" must to be layers or lenses.
	Geological knowledge must be transferred to search statements.
	Contouring need heavy guiding by the geologist.

6. Geostastistic models

Geostatistic methods have been used for decades in Mining Engineering (David, 1977, Journel, 1974 and Journel & Huijbregts, 1978) and for some years in petroleum reservoir modelling (Begg et al., 1994). With the increased use of geostatistic methods in petroleum reservoir modelling, the type of programmes used for the modelling have developed from smaller single-routine programs such as ISIM3D (Gómez-Hernandez & Srivastava 1990) and programme libraries such as Geostat Toolbox (Froidevaux, 1988) and GSLIB (Clayton & Journel, 1992) to integrated commercial programmes such as Heresim (Galli et al., 1990), Storm (Bratvold & Waagbø, 1994) and $(RC)^2$ ($(RC)^2$, 1994) with more user friendly interfaces, visualisation tools, statistical analysis tools and export routines to generate input files for reservoir simulation programs. In the recent years, these techniques have also been used in examination of groundwater reservoirs. Geostatistic methods in hydrogeology have mainly been used as a tool to examine the variability of the lithology (de Marsily, 1986; Ritzi et al., 1994; Ribeiro, 1993 & Schafmeister & Pekdeger, 1993; Webb, 1994), but the method has potential to be used as a tool to generate three-dimensional geological models that can be directly as input to three-dimensional hydrogeological modelling tools.

In petroleum reservoir engineering geostatistical models have been used with success as input to petroleum reservoir flow simulation (e.g. Tyler et al., 1994). Although petroleum engineers are mainly concerned about bulk flow values they are also concerned about first arrival times and breakthrough of water or gas. Therefore some of their methods can be used also in groundwater modelling. With geostatistic modelling it is possible to generate geological models that includes the geological variability and that take into account the numerous small layers that control flow-path and arrival times.

6.1. WHY GEOSTASTISTIC MODELS ?

Stochastic modelling is the only way to include the geological variability in geological models. This variability may not be that important when doing regional hydrogeological modelling but if the modelling includes transport processes then small layers may greatly affect the arrival time of pollutant to a water supply well. Also, statistical modelling is capable of honouring the statistical distribution of the data. This means that the resulting geological models have the same overall distribution of the different types of sediments as the data upon which the models are based. In some of the geostatistical packages it is possible to include the geological knowledge as trends (e.g. vertical probability curves in ISIM3D) or maps in two and three dimensions (e.g. GSLIB) that describe the variation in frequency of the different sediment types with the location or by maps showing the areal distribution of porosity based on a conceptual geological. This concept of using soft data has been illustrated by Almeida & Frykman (1995) in simulation of porosity in the Dan field (Danish North Sea) where a trend map was used to guide a simulation of porosity values assuming a given correlation between the two data sets. In hydrogeology a similar method could be used to guide a simulation of sediment types with surface based geoelectrical or geomagnetical measurements of the subsurface.

6.2. TYPES OF GEOSTATISTICAL MODELS.

In geostatistical modelling three main modelling schemes are Boolean (object) modelling, sequential Gaussian modelling and sequential indicator modelling.

Boolean modelling is mainly used to model distribution of bodies with a known size and orientation distribution such as fractures, sand filled channels or clay layers. If the amounts of data are large, it is usually hard to make this type of model honour the hard data.

Sequential Gaussian modelling assigns a real number to each model cell and is often used in petroleum reservoir modelling to model the distribution of porosity and permeability values in reservoirs.

Sequential indicator modelling assigns an integer value to each model cell and is most often used to model the distribution of different types of sediments and facies. As with Gaussian modelling, sequential indicator modelling is used to model a limited number of porosity or permeability classes. Gaussian and Indicator modelling both honour the hard data but the Gaussian method does not fully represent the extreme values. Therefore the Gaussian method is less applicable in cases where extreme values (highs or lows) significantly contribute to the problem under evaluation (e.g. first arrival of pollution to a water supply well).

6.3. STOCHASTIC MODELLING EXAMPLE

In the Grundfør - Hinnerup area a series of two and three dimensional stochastic models have been made using ISIM3D (Gómez-Hernandez & Srivastava, 1990; Hansen, 1994). The programme is fairly CPU intensive when doing three-dimensional simulations. With the size of three-dimensional matrices used in this paper, each of the 50 simulations takes about three hours on a 66-mhz 486 PC. The variograms (see later) are fitted using PREVAR2D, VARIO2DP & MODEL (Pannatier, 1993). For the three-dimensional simulation, the variogram was calculated only in two-dimensions due to the lack of data. The third dimension was added as a factor of anisotropy assuming that the horizontal correlation was 30 times the vertical. For visualizing the matrices the programmes SPYGLASS (Spyglass, 1994) and IMAGEX (Hansen, 1995) was used.

The data needed for the simulations are:

For two-dimensional simulations:
x- and y-coordinates and a Boolean value (true or false) indicating whether the uppermost 2, 4 or 6 metres of sediment are clay. A variogram (Isaaks & Srivastava, 1989) that describes the correlation between the observations.

For three-dimensional simulations:
x-, y- and z-coordinates and a Boolean value for the geology (sand / clay) for each cell with a geological observation. A variogram and vertical probability curve (VPC-curve) showing the actual possibility for finding clayey sediments at each level.

This data extraction and reformatting were done using dBase procedures from the PC ZEUS database (Hansen, 1992). The procedures disregard the soil layer and layers of Postglacial peat. Layers of clayey till, sandy till, meltwater clay and meltwater silt are classified as clay, and layers of sand and gravel as sand.

The two-dimensional models for the probability maps cover an area of 7 km * 6 km and

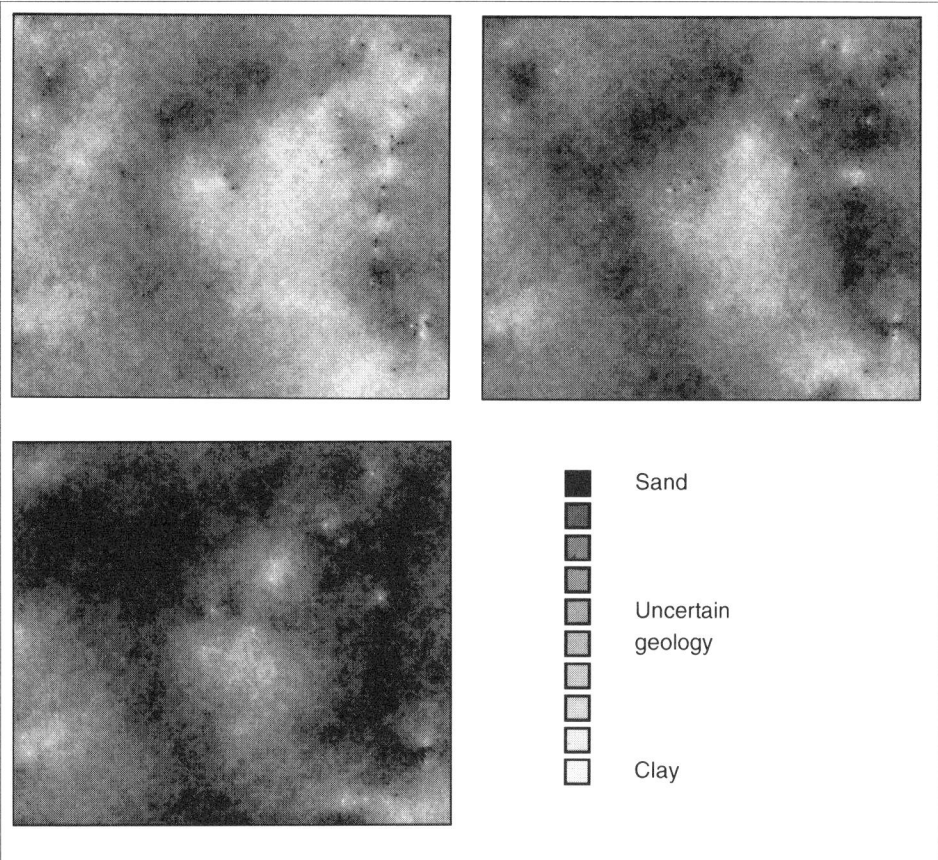

Figure 6: Probability maps showing the areal probability for having an upper clay layer with minimum thickness at 2 (top left), 4 (top right), and 6 metres (bottom).

the size of the regular grid are 280 * 240 cells (total of 67,200 cells) with the dimensions of 25 m * 25 m. For each of the three maps 50 simulations were made, the probability maps were then calculated at each cell unit by counting the number of cells with clay and the number without clay in all 50 matrices. If cell [1,1] has clay in 20 out of the 50 simulations, the probability of clay in this cell is 40% (100*20/50). Figure 6 shows the three probability maps.

The three-dimensional simulations for the three-dimensional geological model represent a part of the same area as the two-dimensional simulations (5 km * 4 km) and the interval from 80 metres above mean sea level to mean sea level. The number of cells used for the simulations are 200 * 160 * 40 (total of 1,312,000 cells) with dimensions 25 m * 25 m * 2 m. A three-dimensional probability matrix is calculated in the same way as the probability map described above.

The three-dimensional simulations fill the entire matrix with sand and clay, including the volume above ground level and the one below the pre-Quaternary surface. These volumes are removed by replacing all the cells above (or below) a contoured surface with blanks. Figure 7 shows an east-west profile from one of the 50 simulations and figure 8 shows the corresponding profile from the three-dimensional probability calculation.

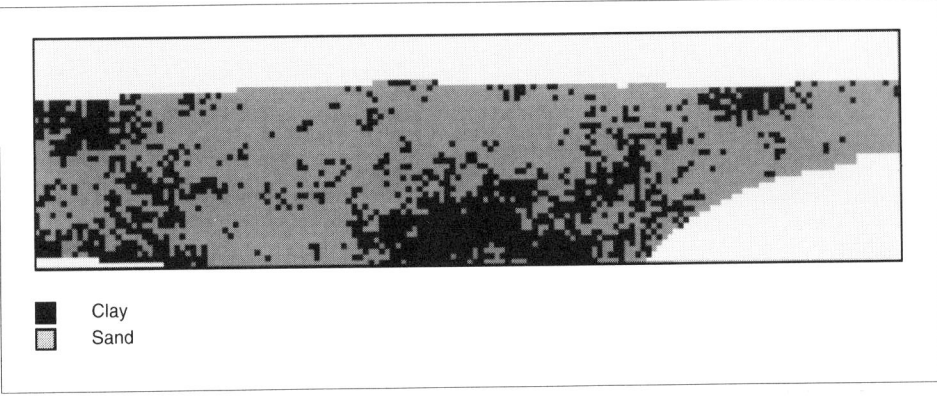

Figure 7: Vertical east - vest cross section of one three-dimensional realization.

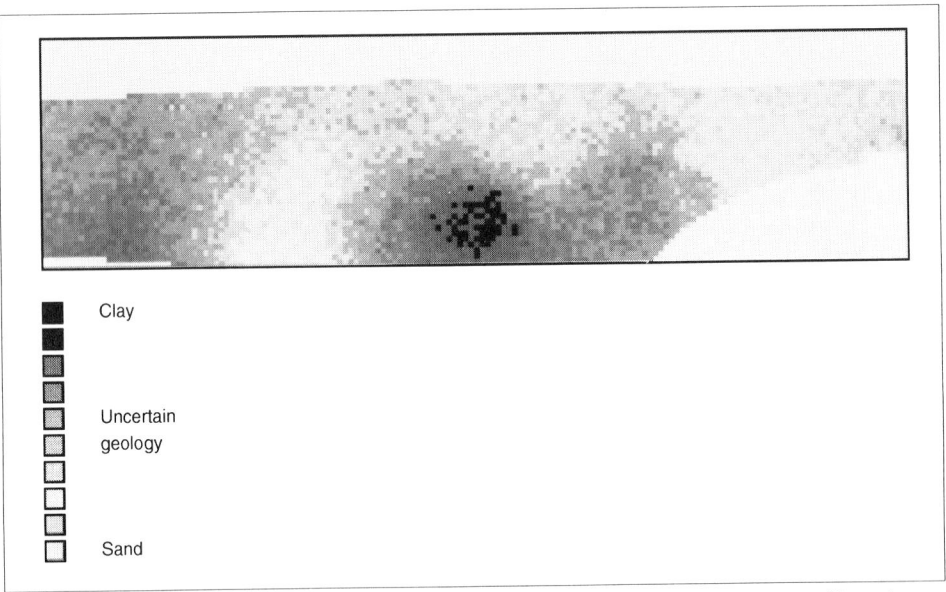

Figure 8: Vertical east - vest cross section of three-dimensional probability map (same profile as in figure 7).

Doing a series of 50 simulations with different seed number for initiating the modelling routine gives 50 different realizations of the geological model. These different models all honour the hard well data and all show the same overall statistical distribution of the clay and sand layers. By combining a series of 50 simulations it is possible to calculate the areal or spatial probability for the presence of clay layers. These probability maps (in two and three dimensions) show where clay or sand is most likely. They also show where the description of the geology is most uncertain and in which area further collection of data will benefit most to the geological knowledge.

The three-dimensional types of stochastic models give a better picture of the variability of the geological setting compared to traditional two-dimensional geological models showing possible distribution of restricted and thin clay layers, which can have a great influence on the local flow processes.

One of the major problems with this type of modelling is the lack of statistical data on the restricted Quaternary layers. Due to the small size of these layers, it is unrealistic to collect the necessary data from well information. Field measurements on outcrops of the size, orientation and spatial distribution of these layers are needed in order to incorporate this information from analogue outcrops in the models.

6.4. RANKING

As stochastic modelling does not generate one "true" model but a series of equi-probably models, it is necessary to analyse the effect of the different models. A few years ago the ranking was done visually by looking at the connectivity of low or high permeability layers. This could be done with coarse two-dimensional models. With the increase in computer power, the models have become three-dimensional and more complex and the ranking has to be modified. In petroleum reservoir modelling this has been done by ranking the realizations according to break through curves modelled with a simple two-dimensional flow simulator from an injection well to an extraction well (Tyler, 1994). This method is site-specific since the ranking is according to flow from one well to another. If the location of the wells change, the ranking sequence will most likely change. This method is often too simple to be used in hydrogeological modelling where the problem is not flow from one well to another but includes several pumping wells and many observation wells.

Because the ranking has to be based on the controlling parameters and are process specific, there exists no universal ranking method. The ranking method must depend on the use of the model. If the problem is an examination of groundwater recharge, the ranking could be according to the extension and coherence of a low permeability cover (regional ranking). If the model has to be used in a risk analysis of transport of pollutants from a deposit with unknown location to water supply wells, the connectivity and direction of high permeability bodies could be the ranking criteria (regional ranking). If the location of a deposit is known, the ranking could be done using particle tracing from the deposit to a number of extraction wells (semi-local ranking).

6.5 FINAL REMARKS

Stochastic modelling is not yet a common tool for standard hydrogeological modelling. The following problems still need attention:

1. Change of support (upscaling). To describe the geological variability, it is necessary to model the geology with cell size much smaller than the size of the geological classes that are described. In this way the geological model becomes much too detailed for hydrogeological modelling. The discussion of upscaling routines has been going on for years in petroleum reservoir modelling. The problems are the same in hydrogeology.
2. Lack of data within the correlation length of the geological units / classes. If the purpose of the modelling is transport of pollutants, the transport controlling layers often have the size as the layers on figure 1. Even in very detailed investigations having more than one borehole in each clay layer would be rare. To use stochastitic modelling at this scale it is necessary to have knowledge about the size, shape and orientation of these layers.
3. Calibration of the hydrogeological model. Because the stochastic modelling assigns a value to each cell in the model it is difficult to change the model if the water table is too high in one part of the area.
4. Because stochastic modelling results in several equally probable models and because it is not practically possible to do the hydrogeological modelling on all these models it is necessary to rank the models according to significant hydraulic criteria. This could be the connectivity of flow controlling layers (high or low permeability layers).

Table 9: Strengths and weaknesses of geostatistic modelling approch

Strengths	Honour the hard geological data both in the data points and in a statistical sense.
	Displays the variability of the data and shows the possible variation of the geology.
	Is the only way to include many restricted and thin layers as detailed variability.
Weakness	Input (variograms) hard to relate to the geological setting.
	Results in several "equally probably" models.
	Geological knowledge must be represented as maps or curves.

7. Conclusions

The three geological modelling concepts outlined in this chapter are very different both in their demands for data and in the resulting models. Using the traditional deterministic approach, it is easy to incorporate geological knowledge. Contoured models are easy to transfer to hydrogeological modelling tools and geostatistical models can include the geological variability into the geological models. The major draw back for the methods are the static and "hard to change" nature of the traditional deterministic models, the simplicity of the contoured models and the large demand for data for statistical models.

Although the usage of the geological models should be the main criterium in the selection of modelling approach, it is often the amount of data (or lack of data) and the data quality that end up being the major argument.

The three modelling schemes outlined in this paper can all be used as input to hydrogeological modelling programmes. Because most three-dimensional hydrogeological modelling tools are base on geological layers, the contoured models are most easy to use as geological input. The models constructed using the traditional deterministic approach must be digitized and then transferred to the hydrogeological modelling tool. Using the stochastic approach, creating geological models where each cell in the stochastic model represent a calculation cell in the hydrogeological modelling programme is possible.

8. References

Almeida, A. S. & Frykman, P. (1995) *Geostatistical Modelling of Chalk Reservoir Properties in the Dan Field, Danish North Sea.* In Yarus, J. M. & Chambers, R. L. (Editors), *"Stochastic Modelling and Geostatistics; Principles, Methods and Case Studies"*, AAPG Special Publication.

Anderson, M. P. (1989) *Hydrogeologic facies models to delineate large-scale spatial trends in glacial and glaciofluvial sediments.* Bull. Geol. Soc. America, vol 101, pp. 501-511.

Anderson, M. P. (1990) *Aquifer Heterogeneity; A Geological Perspective.* Proceedings, Fifth Canadian American Conference on Hydrology, Calgary, Alberta Research Council and National Water Assoc. S. Bachu (ed.), pp. 3-22.

Auken, E., Christensen, N. B., Sørensen, K. I. & Effersø, F. (1994) *Large scale Hydrogeological Investigation in the Beder Area: A Case Study.* Proceedings of the symposium on the Application of Geophysics to Engineering and Environmental problems, Boston, USA, pp. 615-628.

Begg, S. H., Kay. A. Gustason, E. R. & Angert, P. F. (1994) *Characterization of Complex Fluvial-Deltaic Reservoir for Simulation.* Abstract in Stanford Center for Reservoir Forecasting, Report 7, May 1994.

Bratvold, R. B. & Waagbø. K. (1994) *STORM: Integrated 3D stochastic Reservoir Modelling Tool for Geologist and Reservoir Engineers.* Stanford Center for Reservoir Forecasting, Report 7 May 1994.

Christensen, N. B. & Sørensen, K. I. (1994) *Integrated use of electromagnetic methods for hydrogeological investigations.* Proceedings of the symposium on the Application of Geophysics to Engineering and Environmental Problems, Boston, USA, pp. 163-176.

Clayton, V. D. & Journel A. G. (1992) *GSLIB: Geostatistical Software Library and User's Guide.* Oxford University Press. ISBN 0-19-507392-4. 340 p..

Davis, M. (1977) *Geostatistical Ore Reserve Estimation.* Elsevier. Amsterdam.

DGU & GI (1992) *Øresund - landanlæg Amager; Hydrogeologiske vurderinger 1992.* Geological Survey of Denmark (DGU 073-023) and Danish Geotecnical Institute (GI 154-07056), 72 p. (In Danish).

DHI (1993) *MIKE SHE WM- A short description.* Danish Hydraulic Institute.

Eyles, N., Eyles, C. H. & Miall, A. D. (1983) *Lithofacies types and vertical profile models: an alternative approach to the description and environmental interpretation of glacial diamict and diamict sequences.* Sedimentology, vol 30, pp. 393-410.

Froidevaux, R. (1988) *Geostat Toolbox.* PC programme. Stanford University, 101 p.

Galli, A., Guérillot, D., Ravenne, C. & HERESIM Group (1990) *Combining Geology, geostatistics and Multiphase Fluid Flow for 3D Reservoir Studies.* In: 2nd European Conference on the Mathematics of Oil Recovery, Guérillot, D. & Guillon, O. (Editors) and Éditions Technip, Paris 1990, pp. 11 - 19.

Gómez-Hernandez, J. & Srivastava, R. M. (1990) *ISIM3D: An ANSI-C three dimensional multiple indicator conditional simulation program.* Computers & Geosciences, 16(4) pp. 395-440.

Gravesen, P. (1994) *Modelling of the aquifers at Grundfør and Hinnerup, Eastern Jutland, Denmark- Traditional Approach and Geological Variability.* Nordic Hydrological Conference, NHP-Report no. 34, pp. 305-314.

Gravesen, P. & Fredericia, J. (1984) *ZEUS-geodatabase system. Borearkivet.* Danm. Geol. Unders. Serie D no. 3 1984 (in Danish but with English code terms) 259 p..

Gravesen, P., Kelstrup, N. & Leth, J. (1995) *Geological models in engineering geology: Data and process.* Proceedings Eleventh European Conference on Soil Mechanics and Foundation Engineering, Copenhagen. The Interplay between Geotechnical Engineering and Engineering Geology, Bull. Danish

Geotech. Soc., 11, vol. 6, pp. 39-44.
Hansen, M. (1987) *En kvartærgeologisk beskrivelse af Ærø*. Unpublished master thesis. Geological Institute. University of Copenhagen (in Danish), 86 p..
Hansen, M. (1992) *PC Zeus; Vejledning i anvendelse af PC Zeus*. Danm. Geol. Unders. (in Danish), 90 p..
Hansen, M. (1994) *Stochastic Modelling of Quaternary Settings*. Nordic Hydrological Conference, NHP-Report no. 34, pp. 407-414.
Hansen, M. (1995) *ImageX; A PC programme for visualizing large 3-dimensional matrices (up to 10^7 cells)*. Submitted to Computers & Geosciences in July 1995.
Isaaks, H. E. & Srivastava, M. R. (1989) *An Introduction to Applied Geostastitics*. Oxford University Press, Inc. ISBN 0-12-208916-2. 440 p..
Journel, A. G. (1974) *Geostatistics for Conditional Simulation of Ore Bodies*. Economic Geology, 69(5), pp. 673-687.
Journel, A. G. & Huijbregts, C. J. (1987) *Mine Geostatistics*. Academic Press, New York.
de Marsily, G. (1986) *Quantitative Groundwater Hydrology for Engineers*. Academic Press, Inc. ISBN 0-12-208916-2 440 p..
Miall, A. D. (1977) *A review of the braided river depositional environment*. Earth-Science Reviews, vol 13, pp. 1-62.
Miall, A. D. (1985) *Architectural-element analysis. A new method of facies analysis applied to fluvial deposits*. Earth-Science Reviews, vol 22, pp. 261-308.
Miall, A. D. & Tyler, N. (1991) *The Three-Dimensional Facies Architecture of Terrigenous Clastic Sediments and Its Implications for Hydrocarbon Discovery and Recovery*. Soc. Sed. Geol., Concepts in Sedimentology and Palaeontology, vol 3, 309 p..
Neton, M. J., Dorsch, J., Olson, C. D. & Young, S. C. (1994) *Architecture and directional scales of heterogeneity in alluvial-fan aquifers*. J. Sed. Research, vol B64, no. 2, pp. 245-257.
Pannatier, Y. (1993) *PREVAR2D, VARIO2DP, & MODEL: Three PC programs for 2-dimensional variograms modelling*. Institute of Mineralogy, University of Lausanne, BFSH 2, 1015 Lausanne, Switzerland, E-Mail: Yvan.Pannatier@imp.unil.ch.
Poeter, E. & Gaylord, D. R. (1990) *Influence of Aquifer Heterogeneity on contaminant Transport at the Hanford Site*. Ground Water, vol 28, no.6, pp. 90-909.
$(RC)^2$ (1994) *$(RC)^2$ Reservoir Characterization Research and Consulting, Inc.* 3790 El Camino Real, Suite 316, Palo Alto, CA 94360.
Ribeiro, L. (1993) *Evaluating the Uncertainty of Leakage Rates of Tejo and Sado Groundwater System*. In Soares, A (Editor), *Geostatistics Tróia '92 Volume 2, Quantitative Geology and Geostatistics*, Kluvwer Avademic Publisher, ISBN 0-7923-2156-1, 1088 p..
Ritzi, R. W. Jr., Dale, J. F., Zahradnik, A. J. Jr., Field, A. A. & Fogg, G. E., (1994) *Geostatistical Modeling of Heterogeneity in Glaciofluvial, Buried-Valey Aquifers*. Ground Water, vol 32, no. 4, pp. 666-674.
Schafmeister, M.-Th. & Pekdeger, A. (1993) *Spatial structure of Hydraulic Conductivity in Various Porous Media - Problem and Experiance*. In Soares, A (Editor), *Geostatistics Tróia '92 Volume 2, Quantitative Geology and Geostatistics*, Kluvwer Avademic Publisher, ISBN 0-7923-2156-1, 1088 p..
Spyglass (1994) *Slicer; The Quickest Tool for Visualizing Volumetric Data*. Spyglass, Inc. P.O. Bax 6388, Champaign, IL 61862. E-Mail support@spyglass.com, 146 p..
Tillman, R.W. & Weber, K.J. (eds.), (1987) *Reservoir sedimentology*. Soc. Econ. Pal. Min. Sp. Pub. No. 40, 357 p..
Tyler, K. J. (1994) *Heterogenities: Challenges in Reservoir Management*. Doctor theses Høgskolen i Stavanger, Norway. Skrifter nr. 9/1994, 61 p..
Tyler, K. J., Svanes, T. & Henriquez, A. (1994) *Heterogenity Modelling Used for Production Simulation of a Fluvial Reservoir*. SPE Formation Evaluation, June 1994. pp. 85-92.
Webb, E. K. (1994) *Simulating the three-dimensional Distribution of Sediments Units in Braided-Stream Deposits*. J. Sed. Research, Vol. B64, No. 2, May, 1994, pp. 219-231.

CHAPTER 11

USE OF GIS AND DATABASE WITH DISTRIBUTED MODELLING

F. DECKERS[1] AND C.B.M. TE STROET[2]
[1] TNO Institute of Applied Geoscience
P.O.Box 6012, 2600 JA, Delft, The Netherlands
[2] Univ. of Techn. Delft, Dept. Civil Engineering
P.O.Box 5048, 2600 GA, Delft, The Netherlands

1. Introduction

Integrated water management has developed into a complex task as a result of the conflicting claims made by competing groups of users on a limited resource - water. To obtain sound, realistic and economically feasible solutions for the increasing problems, the hydrologist needs to co-operate intensively with experts in the field of ecology, agriculture, urban planning and economics (Engelen and Kloosterman, to be published). They have to evaluate the solutions he proposes for all their possible impacts. Due to conflicting interests, the feasibility of an increasingly larger set of scenarios has to be examined and even the minor effects of these scenario's have to be quantified. This can be achieved by simulating the hydrological measures implied by these scenarios. A trend can be observed to decrease the subjective elements in the simulation process and to give insight in the quality of the model in a stochastic context (Te Stroet, 1995). When many scenario's have to be evaluated against each other, this becomes essential because a quantification of minimum and maximum effects prevents the hydrologist from making the wrong choices based on an erroneous model. Nowadays, the pure numerical results of a simulation are no longer the final products delivered by the hydrologist. The results have to be translated systematically into hydrological effects and subsequently into socially relevant quantities. As an example changes in agricultural production could be related to decreases in crop yields as a result of water shortness in summer periods and water excesses caused by insufficient drainage in winter periods. Also, the hydrologist should be able to provide a clear insight about the quality of the model to the interested parties.

Due to the growing complexity of the hydrological issues, more data from different fields of expertise and different sources have to be taken into consideration. New data acquisition methods like remote sensing get into common use in hydrology (Schultz, 1993). This all implicates a considerable increase in the data volumes that are involved in the modelling process. During the past decennium, the model codes have evolved sufficiently to meet the new quantitative and qualitative requirements that result from the trend towards integrated water management. The problem of providing these models in an efficient way with the up-to-date and accurate information they need has remained or has even aggravated. Depending on the scale of the problem, the large data volumes make the conception of a hydrological model from the basic data quite a challenge for the hydrologist. The fast and accurate translation of the modelling results through combination with the abundancy of other thematical data is a second challenge he is facing.

It is obvious that the hydrologist can no longer depend on tabular representations of his data. The introduction of modern, graphical tools has become a necessity. The - combined - use of software tools like data bases and geographical information systems (GIS) could at least provide a partial solution to his problems. The successfull introduction of such instruments within his modelling environment depends on different factors like the willingness to get acquainted with their use, the scale of the hydrological problems faced and the way in which the modelling process has been structured.

2. Databases in hydrology

When large amounts of data are used on an operational basis by a person or an organization, the storage of these data in a database management system can have a number of advantages. The accessibility of the data increases, its reliability increases through the use of built-in controls and data redundancy can be reduced. At this moment relational database management systems (Date, 1983) are the de facto industry standard. Although the introduction of object-oriented databases in the late eighties and early nineties seemed to offer an attractive alternative, until now this new technology did not succeed in acquiring a substantial part of the database market. Commercial relational systems are also commonly used in hydrology. The awareness has grown that investments made in the development of dedicated database systems for hydrological data - justified by the alleged particular character of this data - do not pay off. Although the use of database systems in a modelling context may not be common practice at this moment, it is expected that the further development of the modelling tools into 5th generation modelling systems will place information management at a central position in the modelling process (Nachtnebel et al., 1993). This implicates that databases will become the central component in the architecture of these systems. A set of tools, including model codes, will access this central resource in order to extract information from it or to add new information to it. The database will be used to build and maintain a digital form of a hydrological model of a certain area (Deckers, 1994). The tools - model codes included - provide different views on this model and will allow to carry out experiments with it (Fürst et al., 1993; Fedra, 1993).

3. GIS in hydrology

Geographical information systems emerged as independent, dedicated software systems in the late seventies, early eighties from the mapping and CAD programs that existed earlier on (Burrough, 1986). The increase in popularity of these systems in the late eighties and early nineties can be related to the introduction of powerfull hardware architectures (e.g. RISC processors) which provided workstations and later on PC's with the required power to fully exploit the capabilities of this software. Although GIS are still often considered as graphic or cartographic tools, this qualification is far too restrictive. Their visualisation capacities have since long been extended with a powerfull set of spatial analysis functionality.

For reasons of performance early GIS systems made a clear distinction between spatial and non-spatial data. Where non-spatial data often resided in a relational database, spatial data was stored in dedicated, proprietary formats that allowed fast retrieval based on spatial criteria. The functional aspect of the GIS was considered as being far more important than the data maintenance aspect. In recent years the conscience has grown among GIS vendors that sufficient attention should be payed to the data maintenance aspect. This resulted in the emergence of a number of GIS products which can be described as geographical database management systems. Also traditional relational database vendors have shown an increasing interest in the GIS market (Nieuwenhuijs, 1995). The evolution from closed dedicated GIS applications to open geographical database management systems and GIS toolboxes which we notice in recent years will only be reinforced by these trends.

The use of GIS in hydrology evolved along with the growing popularity of GIS in general. Where GIS were mainly used as hydrological mapping tools in the early days, nowadays they play an increasingly important role in hydrological modelling studies (Devantier and Feldman, 1993; Ross and Tara, 1993). GIS has become one of the major tools used by the hydrological community. The emergence of events like HydroGIS, a conference solely dedicated to the use of GIS in hydrology, seems to prove this. Although in many organisations GIS is still used in an experimental way, often in relation to one specific modelling study (Batelaan *et al.*, 1993; Biesheuvel and Hemker, 1993; Hay *et al.*, 1993; Kern and Stednick, 1993; Klaghofer *et al.*, 1993; Stibnitz *et al.*, 1993), some more ambitious attempts are made to integrate ranges of model codes with central information systems built upon a GIS (Hoogendoorn *et al.*, 1993) in order to obtain modelling environments that further increase productivity through their integrated GIS capabilities.

In the process of modelling, GIS can play an important role both in the preparation of the model input and in the presentation of the results in so called decision variables. With the aid of a GIS different types of information can easily be combined in a visual appealing way. By presenting the surface water system in one overlay with contour lines of the surface level and polygon information about the land use, logical relations become immediately apparent. Woods are mainly occuring on the higher ground which also show a less dense water drainage system. Larger rivers draining larger catchments appear in the valleys, grass- and farmland occur in the areas with better drainage indicated by a denser pattern of ditches. The mere process of visual combination of data often provides a lot of insight in the different sub-systems and their related hydrological processes. GIS can also be used to transform the basic input data during the model building stage. Re-calculating transmissivities from the thickness of an aquifer and its hydraulic permeability may serve as an example. The thickness of the aquifer is on its turn derived from the top of the aquifer e.g. surface level and its bottom, defined by the occurrence of a clay seam. The calculated transmissivities are in this way spatially defined in the GIS. They are dependent on the spatial variability of the basic data and can be transformed into input data on the grid of a numerical groundwater flow model.

To illustrate the use of GIS in a modelling context, the next paragraphs are dedicated to a detailed description of a modelling study in which GIS has been used extensively (Hoogendoorn and Te Stroet, 1994). It concerns the optimization of the water management in the area of Wierden, The Netherlands.

4. Modelling study 'Wierdense veld'

4.1. MAIN LINE OF THE STUDY

In the eastern part of the Netherlands, in a triangle between the cities Nijverdal, Wierden and Hoge Hexel (figure 1), conflicting demands concerning the use of groundwater exist :

- the nature area 'Wierdense veld' is drying up;
- the drinking water supply of the cities Wierden and Hoge Hexel has to be secured;
- the groundwater and surface water levels have to be controlled with regard to the agricultural needs and the urban areas in the neighbourhood.

The goal of the study was to obtain a balanced package of measures satisfying the different - conflicting - interests. The scenarios that had to be evaluated were:

- change the locations of groundwater abstractions near the city of Wierden or adapt the configuration of the wells;
- modify the surface water system (e.g. fill in water courses or change water levels).

To compare the scenario's, a numerical groundwater model is used as a tool to simulate the different measures. The geohydrological model is based on the model code MODFLOW (Mc.Donald and Harbaugh, 1984), that uses a finite difference discretization, so parameter values and model results are representative for average values over rectangular blocks. In first instance a steady state model is built and calibrated, which is extended with dynamical parameters for calibration/validation of the transient model. Because of the complexity of the model and to minimize subjective elements, the calibration procedure is structured (based on: (Olsthoorn, 1989; Jakeman et al., 1990; LaVenue and Pickens, 1993) by:

1. deleting the non-sensitive unknown parameters with respect to the value of the objective function from the calibration procedure, retaining their prior uncertainty;

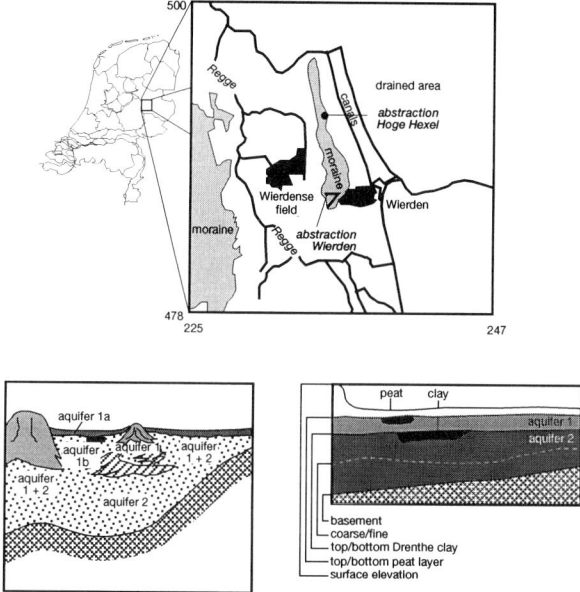

Figure 1. Location and characterisation of the area studied

2. decreasing the prior uncertainty ranges by generating random models within the free space determined by the prior ranges and selecting the models in a Monte Carlo analysis having an maximum likelihood objective function value (Carrera and Neuman, 1986a; Carrera and Neuman, 1986b; Carrera and Neuman, 1986c) less than a certain criterion (generally more than one model is accepted to be representative for the real system);
3. evaluating the parameter sets of the accepted models by an analysis of the posteriori probabilities (moments of the accepted parameter values) to gain insight in the identifiability of the different parameters and possible dependencies between parameters.

The posteriori statistics are used as new prior ranges and proceed again from step 1. The use of GIS for model-building and calibration is the subject of section 2.2.

On the basis of the insight gathered during the building and calibration of the model, four different scenario's are selected. They have to be investigated in more detail to weight the different management measures against each other. These scenario's are investigated using the transient model with time steps of 14 days, but for presentation and comparison purposes two important time moments are chosen, one at the end of the wet season and one at the end of the dry season respectively. A selection of results of this part of the study using the GIS as a presentation and analysis tool is presented in section 4.3.

For the translation of calculated groundwater heads to effects on agricultural production the BODEP program (Landinrichtingsdienst, 1992) is used. This program describes the relation between the moisture content in the soil and the productivity of certain crops. The groundwater flow patterns and residence times are calculated with PATH3D (Zheng, 1989) to identify the capture zones of the abstractions. On the basis of the analysis of capture zones the effects on the quality of the drinking water is predicted. The effects of changes in groundwater flow regime on the ecological potential values is investigated by the KIWA institute (Jansen, 1994). To compare the different effects, they are translated into economical gains or losses. These gains and losses are most of the time easily quantified, like for instance the costs of constructing or relocating production wells, the costs related to reconstructing the surface water system, or gains or losses in agricultural production. Sometimes however this quantification is not so easy and a more or less political choice has to be made. This will for instance be the case when the economical effects on nature conservation have to be evaluated. The role a GIS can play in the translation of model results into economical quantities is elaborated in more in detail in section 4.4.

All data preparation, manipulation and visualisation was realised within the GIS system ARC/INFO (ESRI, 1992).

4.2. BUILDING THE MODEL

The hydrogeological schematization into an upper and lower aquifer and the geometry of the aquifer boundaries is considered to be relatively accurate. Together with location of water courses, withdrawals, withdrawal rates and measured water levels they form the so called 'hard' information. The remaining 'soft' information like conductivities, vertical resistances, groundwater recharge are zoned or divided into several classes like bed resistances, non-measured water levels, etc. All areal information (geometry, locations of boreholes, water courses, wells, zoning information) is entered into GIS data structures, called coverages. The different parameters are stored as related attributes in the GIS database (borehole information, land use, summer/winter levels, bed-elevation, bed-resistances, water supply, rates etc.).

The advantages of gathering all this information into one system are:

— a good insight is obtained in the quantity and quality of available data and the complexity of the model under study;
— data can be checked on wrong values, consistency (e.g. basement should be lower than top of layer), and on logical relationships (e.g. dense surface water system in areas with lowest surface elevation, variable ice-pushed clay layers are below moraines);
— changing, manipulating and interpolating data (e.g. adapting it to other zonation choices, derivation of geological geometry from borehole descriptions etc.) becomes straightforward.

Four levels of data are distinguished :

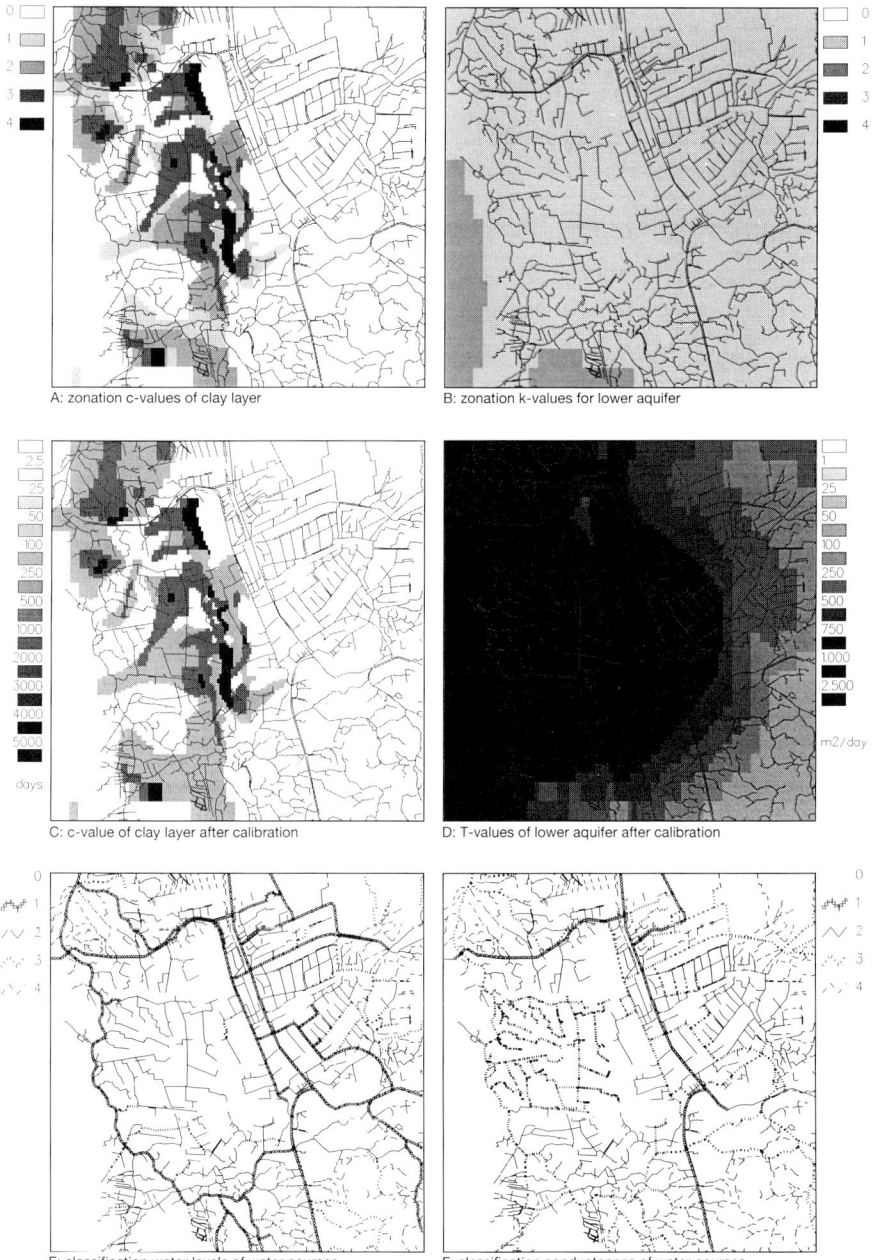

Figure 2. Zonation of soft parameters: transmissivity, vertical resistance lower aquifer (areal-distribution) and the surface water system (line-distribution)

1. the 'real world level' i.e. the available real world data like geometry of the surface water system, borehole locations and descriptions, measured surface elevation, land use, etc.;
2. 'derived real world data' i.e. interpolated surface elevation, distribution of clay lenses derived from borehole information, etc.;
3. 'combined derived real world data' i.e. data that is obtained through a combination of different data like transmissivities derived as conductivities*(top-bottom-aquifer), groundwater recharge as a function of land use, thickness of unsaturated zone, etc.;
4. 'model input data' i.e. data of levels 1 till 3 spotted on the grid of the model.

Data manipulation takes place at the first level only; the derivation to other levels of data is realised using GIS tools, by translating logical relations into scripts and by combining different GIS coverages.

A part of the derived parameter information on modelling level, before and after calibration, is summarized in figure 2. The model discretization is presented in figure 3, together with locations of measurements and withdrawals and some of the measured time series.

4.3. ANALYSIS OF THE GEOHYDROLOGICAL SYSTEM

The GIS helped to get insight in the relative influence of the 'soft' parameters on the model results and on the value of the objective function. An example of results from the sensitivity analysis is given in figure 4.

First, an analysis of the model uncertainty is presented. In addition to the mean groundwater head (figure 4a) of the 26 accepted models (selected out of more than 10.000 simulations), a range of minimum and maximum values can be determined. Although the average difference over 189 measurement locations of the modelled and measured head was approximately 21 cm for all accepted models, locally differences of more than 2 m existed between the individual models (this is a measure for the second moment of the posterior uncertainty; figure 4b). The GIS helped to get insight in the origins of this uncertainty and in its spatial distribution. With this insight, it became easy to decide what additional information should be gained to improve the model most effectively. The computed ranges of heads can be combined with other available information in the GIS to derive other variables. The uncertainty in the depth of the groundwater table is presented in figure 4c, the uncertainty in the lowering of the water table as a result of withdrawals in figure 5.

Furthermore, the GIS greatly improved the quality of the analysis of the flow system. It was very easy to get clear insight in the dynamical behaviour of the flow system by comparing the situations at the end of the wet and dry season (as an example the difference of the heads after wet and dry season is shown in figure 6). On the basis of this analysis areas could be defined with permanent infiltration, with seepage or with intermittant infiltration and seepage. This classification is an important element in the determination of the potential ecological value (see figure 7 together with the drainage by water courses as an example of the flux analysis of the

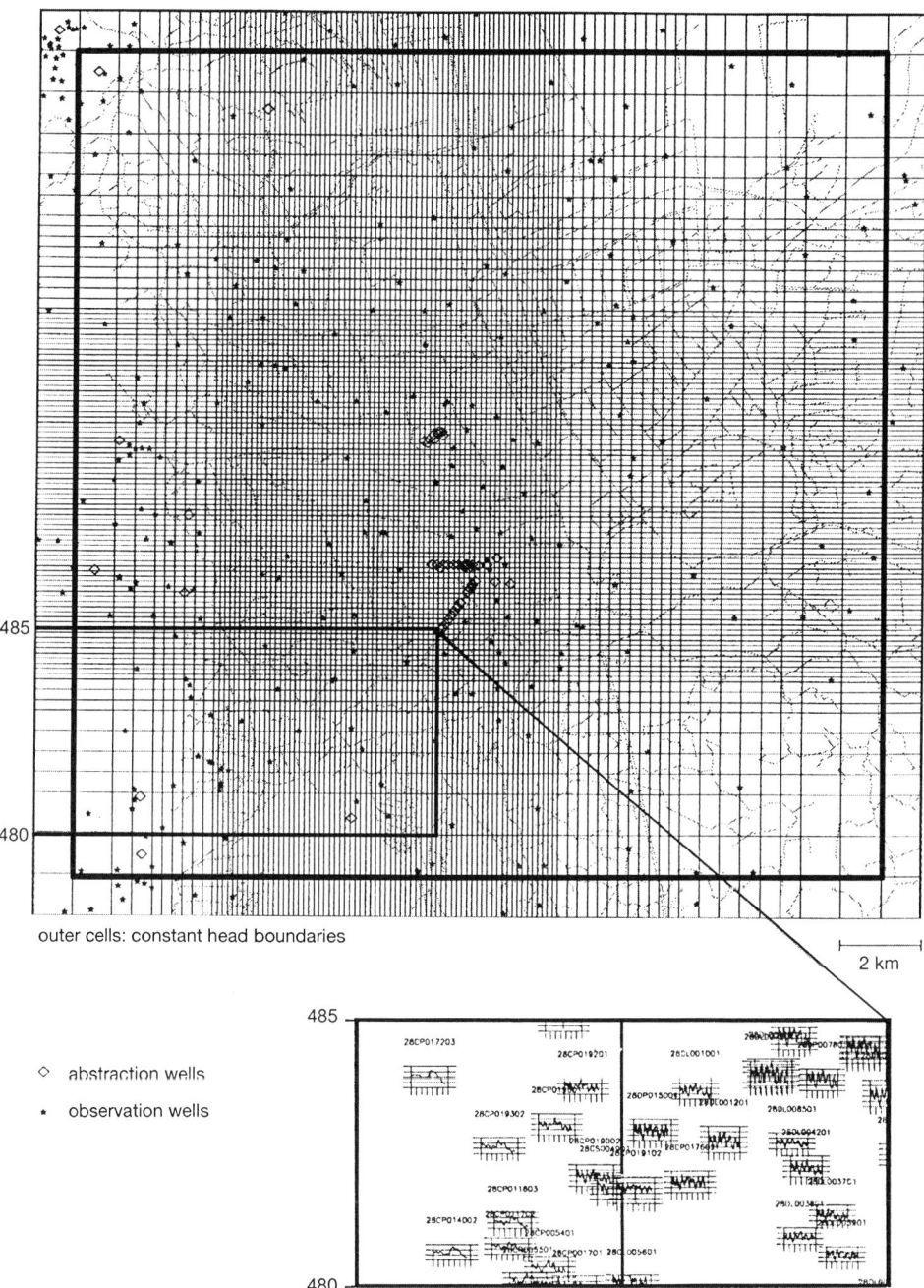

Figure 3. Grid spacings with locations of pumping wells and measurement locations

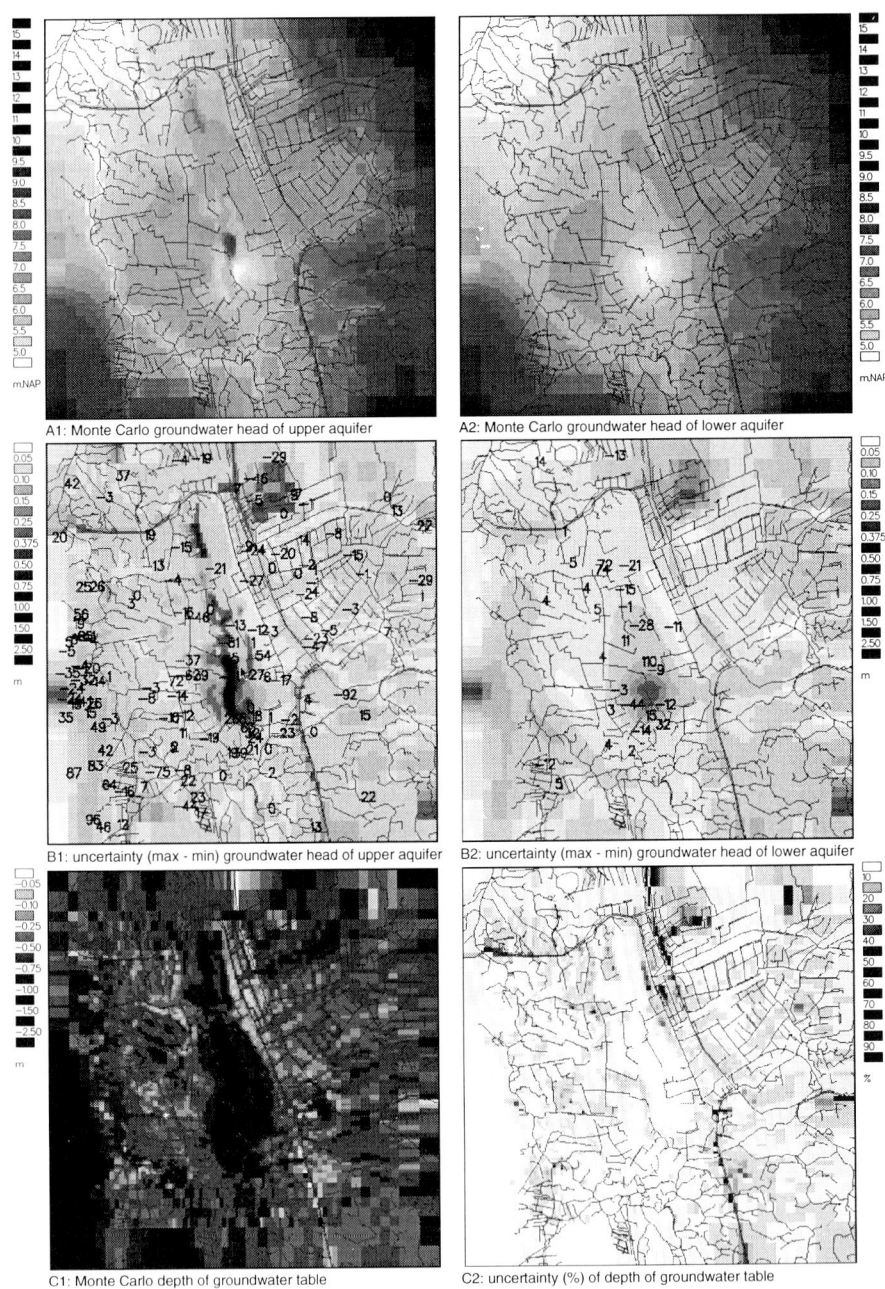

Figure 4. Uncertainty of modelled groundwater heads ("Monte Carlo" means average over 29 accepted models)

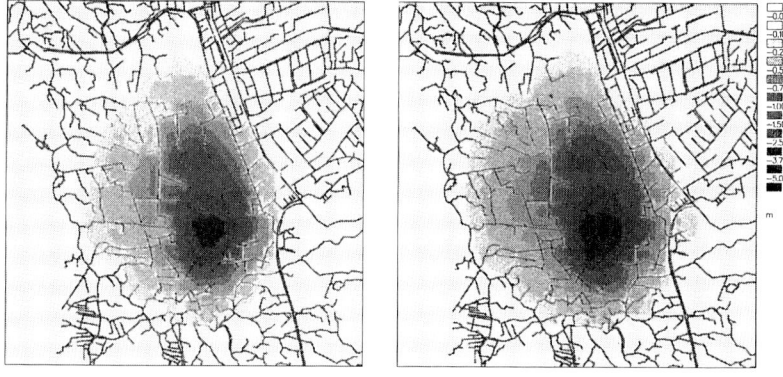

Minimum/maximum lowering of the water table

Figure 5. Uncertainy of lowering groundwater table because of withdrawals

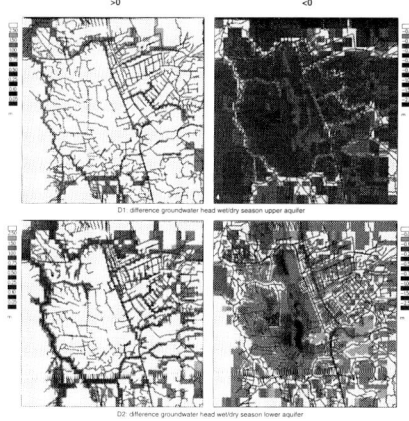

Figure 6. Analysis of the geohydrological system: heads

system). To determine the effectiveness of possible measures that can be part of a scenario, the effects of human interventions on the behaviour of the geohydrological system are investigated. In so called 'historical scenarios' the influences of artificial water courses and withdrawals were calculated and visualised with the GIS (figure 8).

4.4. ANALYSIS OF THE MANAGEMENT SCENARIO'S

The general trend of feasible water management measures was clearly outlined by the analysis of the geohydrological system : locally change the water levels at Wierdense veld and relocate the Wierden pumping station. Four different scenarios were evalu-

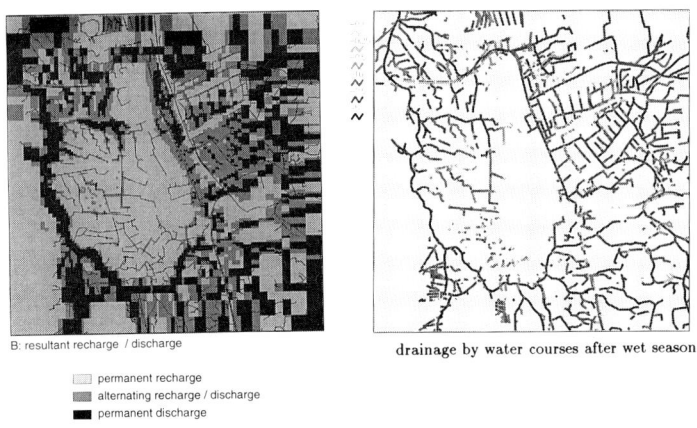

B: resultant recharge / discharge   drainage by water courses after wet season

- permanent recharge
- alternating recharge / discharge
- permanent discharge

Figure 7. Analysis of the geohydrological system: fluxes

ated in detail and using the GIS the effects on the groundwater head, the infiltration and seepage areas, the agricultural production, the capture zones of withdrawals and the ecological potential of the nature reserve Wierdense veld were evaluated. The best scenario located the pumping station near the city of Vriezenveen, increased the water levels by 20 cm at Wierdense veld and filled in one nearby main water course. A summary of the selected scenario is given in figure 9.

4.5. EVALUATION

The study around the city Wierden in the Netherlands, shows that many aspects are important in finding an answer for a relatively simple question like 'how can we prevent a nature reserve from drying up'. It also shows that use of a GIS in the phase of model building is clearly an advantage. The manipulation of data can be restricted to the level of the real world data, from which model input can be derived automatically. This increases the level of control the hydrologist has over the model building process and therefore decreases the risk of introducing errors. Furthermore, the fact that the model builder is forced to give insight into the increase or decrease in detail made by the translation from real-world data to model input, exposes the representability of the model. The internal structure of the model becomes more transparant, which is important for the principal party who wants to get a good impression about the quality of the product he is paying for.

Although the Wierden case clearly indicates the advantages of the use of GIS within hydrological modelling studies, one should also be warned against the thoughtless application of GIS (Brilly *et al.*, 1993; Lam and Swayne, 1993; Grayson *et al.*, 1993; Kemp, 1993). Therefore keep in mind that:

- the presentation of modelling results in appealing pictures often results in a misleading feeling of correctness of the results.

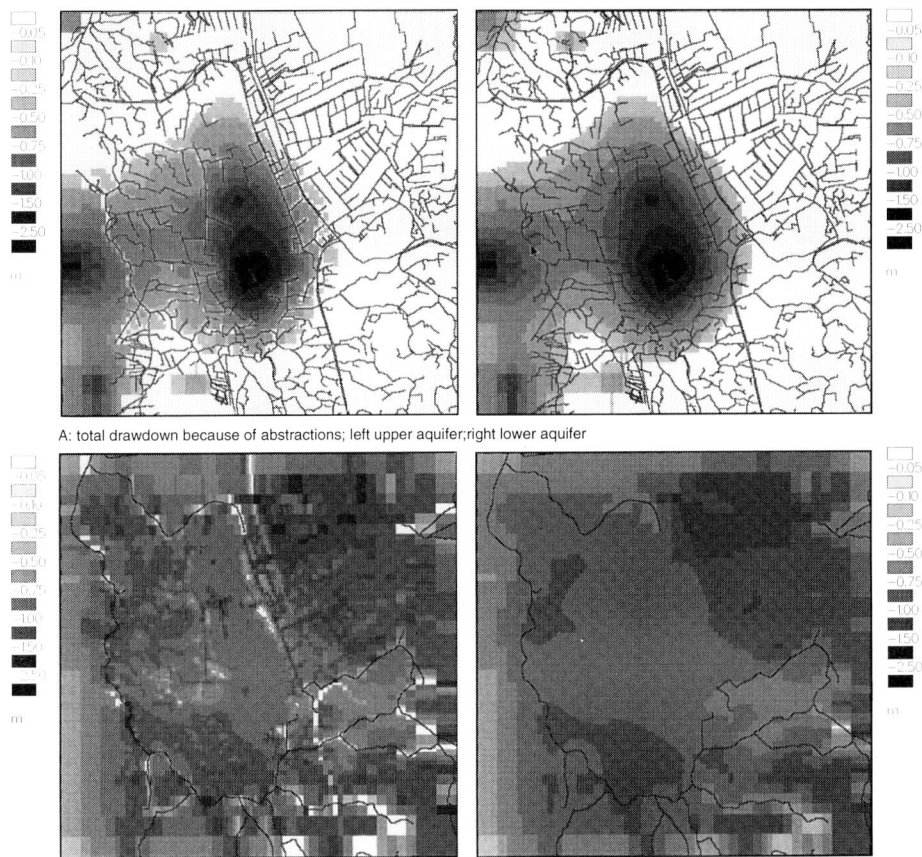

A: total drawdown because of abstractions; left upper aquifer;right lower aquifer

B: total drawdown because of surface water courses; left upper aquifer;right lower aquifer

Figure 8. Effects of withdrawals vs. effects of surface water system on groundwater head

- results are often derived by interpolation or extrapolation and the methods that are available for this purpose in a GIS are often inferior to proven geostatistical methods.
- data derived by the use of a GIS should never be considered to be real data.
- the traditional ways to represent spatial variability in a GIS are always based on discretized representations of spatial continuous data.

5. Incorporation of GIS and databases in a modeling environment

The quality of a model and therefore of the hydrological management which is based upon it is primarily depending on the quality of the data and on the way they are used in the modelling process. The problems encountered with data preparation within modelling studies tend to change with the scale of the studies and the fre-

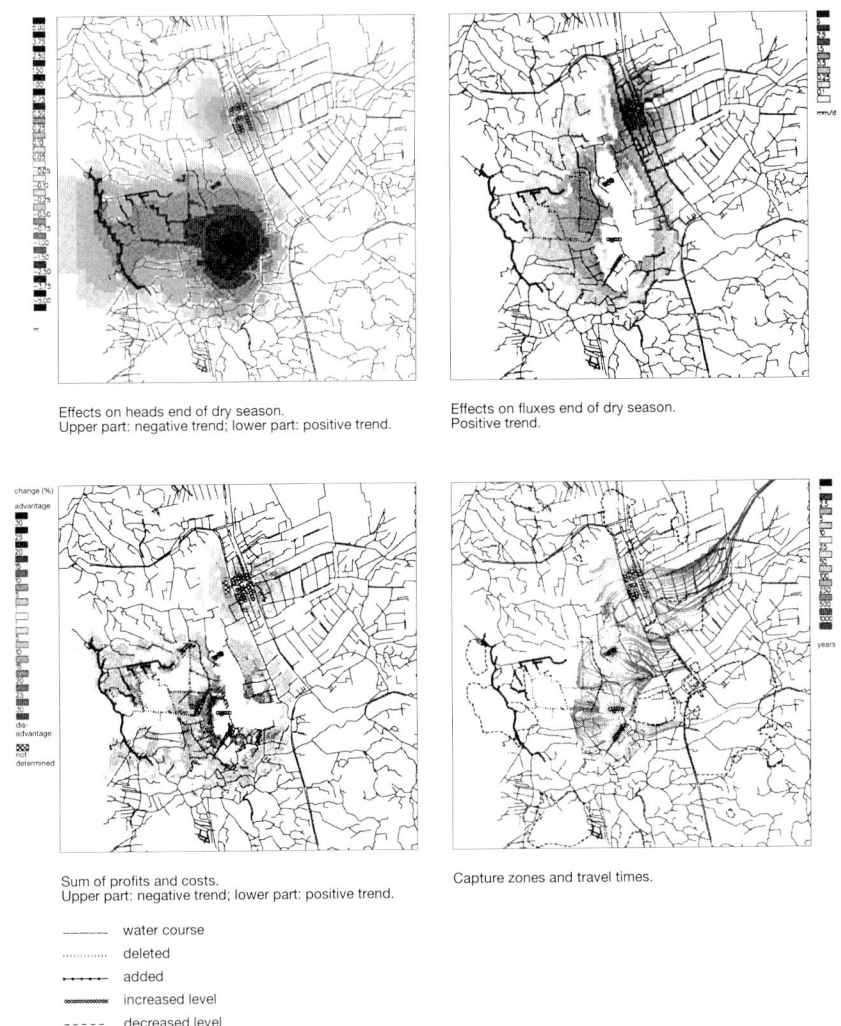

Figure 9. Effect of the selected scenario at the end of the wet season: on the groundwater head (a), the infiltration/seepage (b), the agricultural production (c) and the capture zone (d)

quency with which they are conducted. When modelling studies are carried out in different geographical contexts and on an incidental basis, the introduction of a - eventually partially - automated form of data preparation like proposed in the Wierden case may not be considered as a necessity. When modelling studies are related to the same geographical area and are conducted with higher frequency, establishing a continously available and up-to-date set of data that can be used in the different modelling studies becomes more lucrative. One may even consider building a permanent digital hydrological model of the area which is continuously updated with newly acquired data and adapted to new hydrological insights in a separate, independent process. This requires the introduction of a (geo)hydrological management system based upon a combination of database system and GIS, into the organization (Deckers, 1993).

From these observations, it seems that two extreme alternatives of modelling software environments can be defined (figure 10). In the first environment the model code occupies a central position. When used at all, the GIS and the data base management system are considered as separate tools that play a more or less important role in the pre- and post processing of the input data and the modelling results. A (geo)hydrological information system plays the central role in the second alternative. It is used to maintain a digital representation of the hydrological model. One or more model codes are integrated along with other tools in the systems' application environment. Functionality to set up scenario's and to convert hydrological features into model code objects is provided by the information system. The model codes are driven from within the information system through a driver mechanism.

Different alternatives can be derived from these two extreme cases. For the second alternative a much looser coupling between information system and model codes can be imagined. The information system will only be used to provide the model codes with basic data and control of the model codes is kept completely outside of the information system. An example of such a system developed for the modelling of the Danubian Lowland in the Slovak Republic is described in (Sorensen et al., 1995) One could also imagine an alternative in which the hydrological information system is replaced by a permanent set of individual data files and a set of procedures that describe how data and modelling results should be manipulated. Subsequently small software elements (GIS macro's, conversion programs) with limited functionality can be introduced that implement - parts of - these procedures. In this way conformation to the procedures is encouraged through the availablity of these software elements. Ultimately this will give way to a partially automated modelling environment which can be further formalized. Such a semi-formalized modelling environment has been developed for instance by TNO using the experience acquired in the Wierden case (Hoogendoorn and Te Stroet, 1992).

6. Conclusions

GIS and to a lesser extent data base management systems are certainly valuable assets in the processes of data preparation and result evaluation that are part of a modelling study. The current tendency towards integrated modelling studies will only

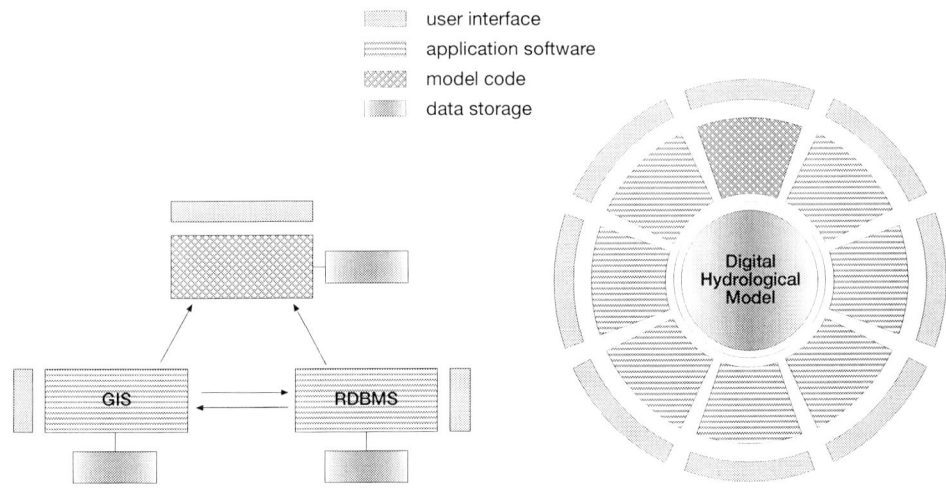

Figure 10. Two extreme alternatives for a modelling software environment

augment the added value of instruments like GIS. A clear development in the direction of integration of databases, GIS and model codes into the operational modelling environment of organizations can be observed. This can be achieved in a number of ways and how this will be done depends on the frequency, scale and geographical context of the modelling studies conducted by the organization. Two extreme cases of integration between model codes, data bases and GIS have been considered, one based on strong integration by the introduction of a (geo)hydrological information system and another one in which the integration is virtually non-existent. Between these two extremes a whole range of different alternatives can be implemented.

References

Batelaan, O. , De Smedt, F. , Otero Valle, M.N. and Huybrechts, W. . Development and application of a groundwater model integrated in the GIS GRASS. In Kovar, K. and Nachtnebel, H.P. , editors, *Application of Geographic Information Systems in Hydrology and Water Resources Management*, pages 581–589. IAHS Pub. No. 211, April 1993.

Biesheuvel, A. and Hemker, C.J. . Groundwater modelling and GIS: integrating MICRO-FEM and ILWIS. In Kovar, K. and Nachtnebel, H.P. , editors, *Application of Geographic Information Systems in Hydrology and Water Resources Management*, pages 289–296. IAHS Pub. No. 211, April 1993.

Brilly, M. , Smith, M. and Vidmar, A. . Spatially oriented surface water hydrological modelling and GIS. In Kovar, K. and Nachtnebel, H.P. , editors, *Application of Geographic Information Systems in Hydrology and Water Resources Management*, pages 547–558. IAHS Pub. No. 211, April 1993.

Burrough, P.A. . *Principles of Geographical Information Systems for Land Resources Assessment*. Clarendon Press - Oxford, 1986.

Carrera, J. and Neuman, S.P. . Estimation of aquifer parameters under transient and steady state conditions, 1, maximum likelihood method incorporating prior information. *Water Resour. Res.*, 22:199–210, 1986.

Carrera, J. and Neuman, S.P. . Estimation of aquifer parameters under transient and steady state conditions, 2, uniqueness, stability, and solution algorithms. *Water Resour. Res.*, 22:211–227, 1986.

Carrera, J. and Neuman, S.P. . Estimation of aquifer parameters under transient and steady state conditions, 3, application to synthetic and field data. *Water Resour. Res.*, 22:228–242, 1986.

Date, C.J. . *An Introduction to Database Systems*. Addison Wesley Publishing Company Inc., 1983.

Deckers, F. . EGIS, a geohydrological information system. In Kovar, K. and Nachtnebel, H.P. , editors, *Application of Geographic Information Systems in Hydrology and Water Resources Management*, pages 611–619. IAHS Pub. No. 211, April 1993.

Deckers, F. . A geohydrological information system based on object-oriented technology. In Verwey, A. , Minns, A.W. and Babovic, A.W. , editors, *Hydroinformatics '94*, pages 247–252, September 1994.

Devantier, B.A. and Feldman, A.D. . Review of GIS application in hydrologic modelling. *Journal of Water Res. Planning and Management*, 119(2), 1993.

ESRI, . ARC/INFO: programmapakket voor opslag en bewerking van ruimtelijke gegevens. Technical report, Environmental System Research Institute Inc., 1992.

Fedra, K. . Models, GIS, and expert systems; integrated water resources models. In Kovar, K. and Nachtnebel, H.P. , editors, *Application of Geographic Information Systems in Hydrology and Water Resources Management*, pages 297–308. IAHS Pub. No. 211, April 1993.

Fürst, J. , Girstmair, G. and Nachtnebel, H.P. . Application of GIS in Decision Support Sysytems for groundwater. In Kovar, K. and Nachtnebel, H.P. , editors, *Application of Geographic Information Systems in Hydrology and Water Resources Management*, pages 13–21. IAHS Pub. No. 211, April 1993.

Grayson, R.B. , Bloschl, G. , Barling, R.D. and Moore, I.D. . Process, scale and constraints to hydrological modelling in GIS. In Kovar, K. and Nachtnebel, H.P. , editors, *Application of Geographic Information Systems in Hydrology and Water Resources Management*, pages 83–92. IAHS Pub. No. 211, April 1993.

Hay, L.E. , Battaglin, W.A. , Branson, M.D. and Leavesley, G.H. . Application of GIS in modelling winter orographic precipitation. In Kovar, K. and Nachtnebel, H.P. , editors, *Application of Geographic Information Systems in Hydrology and Water Resources Management*, pages 491–499. IAHS Pub. No. 211, April 1993.

Hoogendoorn, J.H. and Te Stroet, C.B.M. . Groundwatermodellering met behulp van MODFLOW en ARC/INFO, met incorporatie van MONTE CARLO techniek. PN 92-05, TNO Institute of Applied Geoscience, 1992.

Hoogendoorn, J.H. and Te Stroet, C.B.M. . Optimalisatie Waterbeheer Wierden/Wierdense Veld. OS 94-14-B, TNO-GG, 1994.

Hoogendoorn, J.H. , Linden, W. van der and Te Stroet, C.B.M. . The importance of GIS in regional geohydrological studies. In Kovar, K. and Nachtnebel, H.P. , editors, *Application of Geographic Information Systems in Hydrology and Water Resources Management*, pages 375–383. IAHS Pub. No. 211, April 1993.

Jakeman, A.J. , Ghassemi, F. and Dietrich, C.R. . Calibration and reliability of an aquifer system model using generalized sensitivity analysis. In Kovar, K. , editor, *ModelCare 90: Calibration and Reliability in Groundwater Modelling*, pages 43–51, The Hague, 5-9 September 1990. IAHS-publ.195.

Jansen, A.J.M. . Bepaling vna de effecten van vier waterhuishoudkundige varianten op de vegeatie van het Wierdense Veld (Ov.). SWO 93.214, KIWA, 1994.

Kemp, K. . Environmental modelling and GIS: dealing with spatial continuity. In Kovar, K. and Nachtnebel, H.P. , editors, *Application of Geographic Information Systems in Hydrology and Water Resources Management*, pages 107–115. IAHS Pub. No. 211, April 1993.

Kern, T.J. and Stednick, J.D. . Identification of heavy metal concentrations in surface waters through coupling of GIS and hydrochemical models. In Kovar, K. and Nachtnebel, H.P. , editors, *Application of Geographic Information Systems in Hydrology and Water Resources Management*, pages 559–567. IAHS Pub. No. 211, April 1993.

Klaghofer, E. , Birnbaum, W. and Summer, W. . Linking sediment and nutrient export models with a geographic information system. In Kovar, K. and Nachtnebel, H.P. , editors, *Application of Geographic Information Systems in Hydrology and Water Resources Management*, pages 501–506. IAHS Pub. No. 211, April 1993.

Lam, D.C.L. and Swayne, D.A. . An expert system approach of integrating hydrological database, models and GIS: application of the RAISON System. In Kovar, K. and Nachtnebel, H.P. , editors, *Application of Geographic Information Systems in Hydrology and Water Resources Management*, pages 23–33. IAHS Pub. No. 211, April 1993.

Landinrichtingsdienst, . BODEP; Berekening Opbrengst DEPressies; een programma pakket voor berekening van de invloed vna de waterhuishouding op de landbouwkundige productie. Technical report, Landinrichtingsdienst, Utrecht, 1992.

LaVenue, M.A. and Pickens, J.F. . Application of a coupled Adjoint-Sensitivity and Kriging Approach to Calibrate a Ground-Water Flow Model. *Water Resour. Res.*, -(-):–, 1993.

Mc.Donald, M.G. and Harbaugh, A.W. . A Modular Three-Dimensional Finite-Difference Ground Water Flow Model. manual 83-875, U.S. Geological Survey, 1984.

Nachtnebel, H.P. , Furst, J. and Holzmann, H. . Application of geographical information systems to support groundwater modelling. In Kovar, K. and Nachtnebel, H.P. , editors, *Application of Geographic Information Systems in Hydrology and Water Resources Management*. IAHS Pub. No. 211, April 1993.

Nieuwenhuijs, S. . Oracle MultiDimension : new frontiers in spatial data management. *GIS Europe*, (June):40–42, 1995.

Olsthoorn, T.N. . Grondwaterstandsverlagingen ten gevolge van de Duitse bruinkoolwinning in de Roerdalslenk. Kwantitatieve analyse van de verschillen tussen de Duitse en de Nederlandse modelstudie. Technical Report 728610001, RIVM, 1989.

Ross, M.A. and Tara, P.D. . Integrated Hydrologic Modelling with Geographic Information Systems. *Journal of Water Res. Planning and Management*, 119(2), 1993.

Schultz, G.A. . Application of GIS and remote sensing in hydrology. In Kovar, K. and Nachtnebel, H.P. , editors, *Application of Geographic Information Systems in Hydrology and Water Resources Management*, pages 127–140. IAHS Pub. No. 211, April 1993.

Sorensen, H.R. , Storm, B. , Mucha, I. , Deckers, F. and Waardenburg, F.D.E. . Hydrological Information Management for the Danubian Lowlands, Integration of ARC/INFO and MIKE SHE. In *1995 ESRI User Conference*, pages 653–664. Environmental Systems Research Institute Inc., May 1995.

Stibnitz, M. , Patzelt, Z. and Wolfbauer, J. . Database development and GIS application in support of groundwater management: case study at the Austrian-Bohemian Border. In Kovar, K. and Nachtnebel, H.P. , editors, *Application of Geographic Information Systems in Hydrology and Water Resources Management*, pages 263–270. IAHS Pub. No. 211, April 1993.

Te Stroet, C.B.M. . *Calibration of stochastic groundwater flow models. Estimation of system noise statistics and model parameters*. PhD thesis, Univ. of Techn. Delft, Delft, 1995.

Zheng, C. . PATH3D a Ground-Water Path and Travel-Time Simulator, Version 2.0. Technical report, S.S. Papadopulos and Associates Inc., 1989.

CHAPTER 12
AN ENGINEERING CASE STUDY - MODELLING THE INFLUENCES OF GABCIKOVO HYDROPOWER PLANT ON THE HYDROLOGY AND ECOLOGY IN THE SLOVAKIAN PART OF THE RIVER BRANCH SYSTEM OF ZITNY OSTROV

H.R. SØRENSEN[1], J.KLUCOVSKA[2], J. TOPOLSKA[2], T. CLAUSEN[1], AND J.C. REFSGAARD[1]
[1] *Danish Hydraulic Institute, Hørsholm, Denmark*
[2] *Ground Water Consulting, Ltd, Bratislava, Slovakia*

1. Introduction

1.1. THE DANUBIAN LOWLAND AND THE GABCIKOVO HYDROPOWER SCHEME

The Danubian Lowland (Fig. 1) between Bratislava and Komárno is an inland delta formed in the past by river sediments from the Danube. The entire area forms an alluvial aquifer, which throughout the year receives in the order of 30 m^3/s infiltration water from the Danube in the upper parts of the area and returns it into the Danube and the drainage channels in the downstream part. The aquifer is an important water resource for municipal and agricultural water supply.

Human influence has gradually changed the hydrological regime in the area. Construction of dams upstream of Bratislava together with exploitation of river sediments has significantly deepened the river bed and lowered the water level in the river. These changes have had a significant influence on the conditions of the ground water regime as well as the sensitive riverside forests downstream of Bratislava. In spite of this basically negative trend the floodplain area with its alluvial forests and the associated ecosystems still represents a very unique landscape of outstanding importance.

The Gabcikovo hydropower scheme was put into operation in 1992. A large number of hydraulic structures has been established as part of the hydropower scheme. The key elements are a system of weirs across the Danube at Cunovo 15 km downstream of Bratislava, a reservoir created by the damming at Cunovo, a new lined canal running parallel to the Old Danube over a stretch of approximately 30 km for navigation and with intake to the hydropower plant, a hydropower plant and two shiplocks at Gabcikovo, and an intake structure at Dobrohost diverting water from the new canal to the river branch system. The entire scheme has significantly affected the hydrological regime and the ecosystem of the region. The scheme was originally planned and the major parts of the construction were carried out as a joint effort between Czecho-

Slovakia and Hungary. However, today Gabcikovo is a matter of controversy between Slovakia and Hungary, who have referred some disputed questions to the International Court of Justice in Haag.

Comprehensive monitoring and assessments of environmental impacts have been made, see Mucha (1995) for an overview.

1.2. THE PHARE PROJECT 'DANUBIAN LOWLAND - GROUND WATER MODEL'

To understand and analyze the complex relationships between physical, chemical and biological changes in the surface- and subsurface water regimes in the Danubian Lowland requires multidisciplinary expertise in combination with field data and advanced mathematical modelling techniques. For this purpose the project "Danubian Lowland - Ground Water Model" was defined within the PHARE programme agreed upon between the European Commission and the Government of the Slovak Republic.

The overall project objective was to establish a comprehensive modelling and information system suitable as a decision support tool for water resources management in the area. The aim of the developed integrated modelling system was to provide a reliable tool for analyzing the environmental impact of alternative management strategies and hence support in the formulation of optimal management strategies leading to a protection of the water resources and to a sound ecological development for the area.

The PHARE project was executed by the Slovak Ministry of the Environment. Specialists from the following Slovakian organisations were involved in various aspects of the project implementation: Comenius University, Faculty of Natural Science (PRIF UK); Water Research Institute (VUVH); Irrigation Research Institute (VUZH); and Ground Water Consulting, Ltd (GWC).

A Danish-Dutch consortium of six organisations, headed by DHI, was selected as consultant for the project. The project was initiated in the beginning of 1992 and was completed by the end of 1995.

1.3. THE PRESENT CHAPTER

The aims of the present chapter are to illustrate the overall structure and functioning of the developed integrated modelling and information system, to outline the modelling approach for such complex case and to show some selected applications from the flood plain and river branch system.

A key for analyzing the impacts on the water resources of the changed surface water conditions caused by the Gabcikovo plant is an integrated description of the river-aquifer system. The chapter provides details on the modelling of the river branch system and the aquifer system, which is being carried out using a newly developed full coupling between DHI's two modelling systems for rivers and hydrology, MIKE 11 and MIKE SHE, respectively.

CASE STUDY – MODELLING THE INFLUENCES OF GABCIKOVO HYDROPOWER PLANT 235

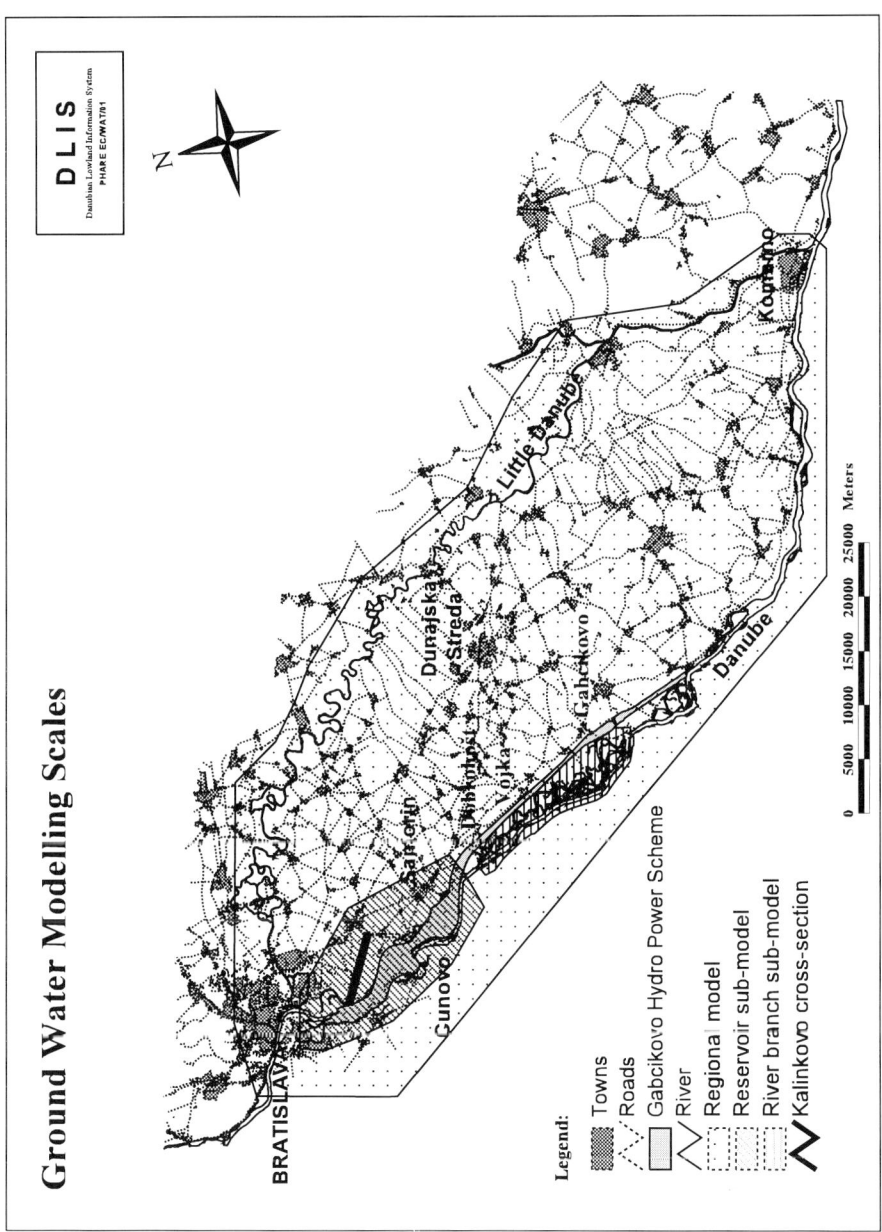

Figure 1. The Danubian Lowland area, layout of the Gabcikovo scheme and the extent of the regional model as well as of local modelling scales.

2. Modelling approach

2.1. INTEGRATED MODELLING SYSTEM

In order to address the problems within the project area an integrated modelling system (Fig. 2) has been established by combining the following existing and well proven mathematical modelling systems:
* *MIKE SHE* (Refsgaard and Storm, 1995) which, on catchment scale, can simulate the major flow and transport processes of the hydrological cycle which are traditionally divided in separate components:
  - 1-D flow and transport in the unsaturated zone
  - 3-D flow and transport in the ground water zone
  - 2-D flow and transport on the ground surface
  - 1-D flow and transport in the river.

  All the above processes are fully coupled allowing for feedbacks and interactions between components. In addition to the above mentioned components, MIKE SHE includes modules for multi-component geochemical and biodegradation reactions in the saturated zone (Engesgaard, Chapter 5).
* *MIKE 11* (Havnø et al., 1995), which is a one-dimensional river modelling system. MIKE 11 is used for hydraulics, sediment transport and morphology, and water quality. MIKE 11 is based on the complete dynamic wave formulation of the Saint Venant partial differential equations. The modules for sediment transport and morphology are able to deal with cohesive and non-cohesive sediment transport, as well as the accompanying morphological changes of the river bed. The non-cohesive model operates on a number of different grain sizes, taking into account shielding effects.
* *MIKE 21* (DHI, 1995), which is a two-dimensional hydrodynamic modelling system. MIKE 21 is used for reservoir modelling, including hydrodynamics, sediment transport and water quality. The sediment transport modules deals with both cohesive and non-cohesive sediment, and the non-cohesive module operates on a number of different grain size fractions.
* MIKE 11 and MIKE 21 include *River/Reservoir Water Quality (WQ) and Eutrophication (EU)* (Havnø et al., 1995; VKI, 1995) modules to describe oxygen, ammonium, nitrate and phosphorus concentrations and oxygen demands as well as eutrophication issues such as bio-mass production and degradation.
* *DAISY* (Hansen et al., 1991) is a one-dimensional root zone model for simulation of soil water dynamics, crop growth and nitrogen dynamics for various agricultural management practices and strategies. The particular processes considered include transformation and transport involving water, heat, carbon and nitrogen.

The integrated modelling system is formed by the exchange of data and the feed-backs between the individual modelling systems. The structure of the integrated modelling system and the exchange of data between the various modelling systems are illustrated in Fig. 2. The interfaces A-E between the various models are briefly described below:
A) MIKE SHE forms the core of the integrated modelling system having interfaces

to all the individual modelling systems. The coupling of MIKE SHE and MIKE 11 is a fully dynamic coupling where data is exchanged after each computational time step. This interface is described in more details below.
The remaining modelling systems are coupled in a more simple manner involving a sequential execution of individual models and subsequently a transfer of boundary conditions from one model to another. Some examples are listed below.

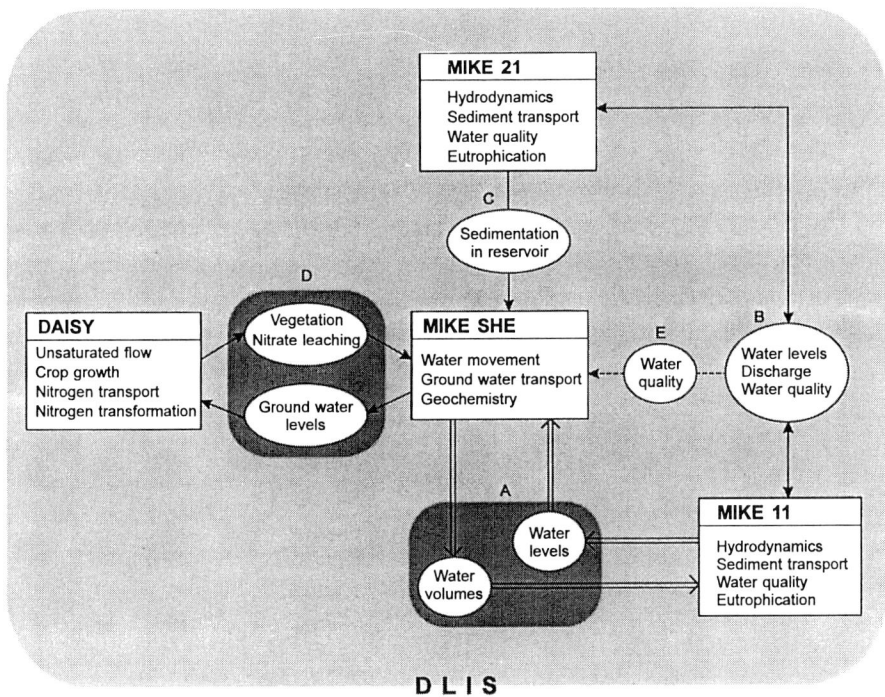

Figure 2. Structure of the integrated modelling system with indication of the interactions between the different models.

B) Results of eutrophication simulations with MIKE 21 in the reservoir are used to estimate the concentration of various water quality parameters in the water that enters the Danube downstream of the reservoir to be used for water quality simulations for the Danube using MIKE 11.

C) Sediment transport simulations in the reservoir with MIKE 21 provide information on the amount of fine sediment on the bottom of the reservoir. This information is used to calculate leakage coefficients which are used in ground water modelling with MIKE SHE to calculate the exchange of water between the reservoir and the aquifer.

D) The DAISY model calculates vegetation parameters which are used in MIKE SHE to calculate the actual evapotranspiration. Ground water levels calculated with MIKE SHE act as lower boundary conditions for DAISY unsaturated zone simulations. Consequently, this process is iterative and requires a few model simulations.

E) Results from water quality simulations with MIKE 11 and MIKE 21 are used to estimate the concentration of various species in the water that infiltrates to the aquifer from the Danube and the reservoir. This is being used in the ground water quality simulations (geochemistry) with MIKE SHE.

The Danubian Lowland Information System (DLIS) is a combined data base and geographical information system which has been developed under this project. The DLIS is based on Informix (database) and Arc/Info (GIS) and provides a framework for data storage, maintenance, processing and presentation. In addition, an interface between DLIS and MIKE SHE allowing import and export of maps and time series files in MIKE SHE file formats has been established.

With regard to simulation of floodplain hydrology and ecology the core of the integrated modelling system is constituted by the MIKE SHE, the MIKE 11 and the newly developed, full coupling of the two systems described below.

2.2. A COUPLING OF MIKE SHE AND MIKE 11

The focus in MIKE SHE lies on catchment processes with a comparatively less advanced description of river processes. In contrary MIKE 11 has a more advanced description of river processes and a simpler catchment description than MIKE SHE. Hence, for cases where full emphasis is needed for both river and catchment processes a coupling of the two modelling systems is required.

A full coupling between MIKE SHE and MIKE 11 has been developed (Fig. 3). In the combined modelling system, the simulation takes place simultaneously in MIKE 11 and MIKE SHE, and data transfer between the two models takes place through shared memory. MIKE 11 calculates water levels in rivers and floodplains. The calculated water levels are transferred to MIKE SHE, where flood depth and areal extent are mapped by comparing the calculated water levels with surface topographic information stored in MIKE SHE. Subsequently, MIKE SHE calculates water fluxes in the remaining part of the hydrological cycle. Exchange of water between MIKE 11 and MIKE SHE may occur due to evaporation from surface water, infiltration, overland flow or river-aquifer exchange. Finally, water fluxes calculated with MIKE SHE are exchanged with MIKE 11 via source/sink terms in the continuity part of the Saint Venant equations in MIKE 11.

The MIKE SHE-MIKE 11 coupling is crucial for a correct description of the dynamics of the river-aquifer interaction. Firstly, the river width is larger than one MIKE SHE grid, in which case the MIKE SHE river-aquifer description is no longer valid. Secondly, the river/reservoir system comprises a large number of hydraulic structures, the operation of which cannot be accounted for in MIKE SHE. Thirdly, the very complex branch system with loops and flood cells needs a very efficient hydrodynamic formulation such as MIKE 11's.

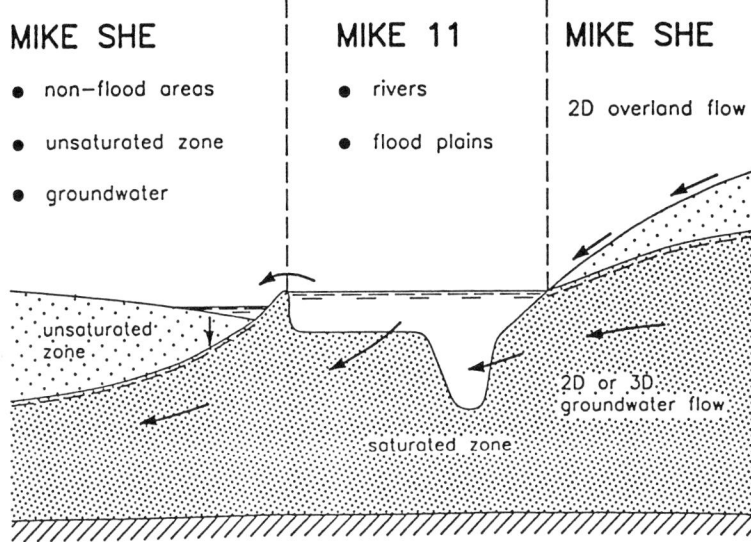

Figure 3. Structure of the MIKE 11 - MIKE SHE coupling.

2.3. MODELLING SCALES

As indicated in Fig. 1 modelling has been carried out at different spatial scales with different objectives. Thus, the following ground water models have been established:
- a regional ground water model for pre-dam conditions,
- a regional ground water model for post-dam conditions,
- a local ground water model for an area surrounding the reservoir (post-dam conditions),
- a local ground water model for the river branch system (both pre-and post-dam conditions), and
- a cross-sectional (vertical profile) model near Kalinkovo.

The regional model area covers about 3000 km^2. It applies a horizontal discretization of 500 m and in the vertical the aquifer has been divided into four geologically determined computational layers. The main objectives of the regional ground water modelling are to study the impacts of the damming of the Danube on the hydrological regime within the project area, in particular in terms of ground water levels and dynamics, and to provide reliable boundary conditions for local ground water models.

The ground water model for the reservoir area is a sub-model of the regional model and the parameter values of the local model are identical to the ones in the regional model with the only change that the local model applies a horizontal discretization of 250 m, and a finer vertical discretization with 7 computational layers. The objectives of the local ground water model around the reservoir are to provide more accurate results for this area than can be done with the more coarse regional model and to provide boundary conditions for the cross-sectional model.

Similarly, the local model for the river branch area is a sub-model of the regional model. This local model has a horizontal discretization of 100m. The objective of the local model for the branch system area is to make detailed predictions of the hydrological regime for alternative water management schemes and thus enable assessments of possible ecological changes in the floodplain area.

A 2 km long cross-sectional model near Kalinkovo with a horizontal discretization of 10 m and 24 vertical layers was established within the area of the local reservoir model in order to provide the hydraulic basis for comprehensive geochemical modelling.

2.4. PROCEDURES FOR MODEL CONSTRUCTION, CALIBRATION, VALIDATION AND APPLICATION

2.4.1 Model construction
All the applied models are based on distributed physically-based model codes. This implies that most of the required data for establishing the model setup and input data can be measured directly in the nature.

A setup of a MIKE 11 hydrodynamic model requires that the river geometry is known involving river cross-sections and various hydraulic structures in the system. A setup of the Hrusov reservoir in MIKE 21 requires that topographical information on the reservoir bottom (bathymetry) is available.

For the MIKE SHE hydrological modelling the same type of data are required, and in addition information on soils and geology is required as well as climatological and vegetation data.

Therefore, a setup for a physically-based model, as a minimum, always involves a geometrical description of the problem and some physical characteristics of the system. This could for instance be hydraulic conductivities in the saturated zone or roughness coefficients in river cross-sections and on the reservoir bottom.

In addition, the value of some state variables has to be known on the model boundaries (boundary condition). For hydrodynamic models the boundary conditions are always a combination of prescribed water levels and discharges. Ground water modelling involves, in principle, the same but described as ground water table or ground water flow. For water quality and geochemical modelling concentration of certain species, for instance nitrate and oxygen, must be known in the water that enters the system.

2.4.2 Model calibration
The calibration of a physically-based model implies that a sequence of simulation runs are carried out and model results are compared with measured data for a certain period

of time.

MIKE 11 was calibrated against measurements of water levels and discharges. MIKE 21 was calibrated against flow velocities and MIKE SHE was calibrated against measured ground water levels, river flows and water levels.

When using physically-based models, the 'amount' of required calibration are reduced the more precise the geometry, the physical characteristics and the boundary conditions of the physical problem are described. In principle, if the model setup exactly reflects the real conditions no model calibration would be required at all. Obviously, this is a hypothetical situation and in practice some model calibration is always needed.

The available amount and quality of data within the project area provided a very good basis for model constructions. Almost all physical characteristics used in the model setup are based on measured data. Hence, the calibration of the various models has been performed by adjusting only a limited number of physical characteristics within a relatively narrow range.

For the calibration of the MIKE 11 hydrodynamic model the roughness of the river bed (the Manning number) was the main calibration parameter. For some structures the precise crest elevations was not known and the exact capacity of culverts was uncertain. Therefore, such geometrical data has been subject to limited calibration, but in general the geometry of the system was not adjusted during the calibration process.

For the MIKE 21 reservoir hydrodynamic model the main calibration parameters were the bottom roughness (Chezy number) and the eddy viscosity.

For the MIKE SHE ground water modelling the main calibration parameter was the hydraulic conductivity in the saturated zone. All other physical characteristics was, in general, kept at the measured values or at experience values from previous studies.

The DAISY unsaturated flow simulations were based on measured soil water retention curves and hydraulic conductivities.

2.4.3 Model validation
Because the calibration process involves some manipulation of parameter values good model results during a calibration process cannot automatically ensure that the model can perform equally well also for other periods. Therefore, model validations on independent data are required.

All the models have been validated by demonstrating the ability to reproduce measured data for a period outside the calibration period. For the MIKE SHE regional ground water flow models the model was even calibrated on pre-dam conditions and validated on post-dam conditions where the flow regime at some locations were significantly altered due to the construction of the reservoir and related hydraulic structures and canals.

2.4.4 Model application - integrated scenario simulations
The validated models were applied in a scenario approach simulating the hydrological conditions resulting from alternative possible operations of the entire system of hydraulic structures (alternative water management regimes). Thus, one historical (pre-

dam) regime and three hypothetical water regimes corresponding to alternative operation schemes for the structures of the Gabcikovo system were simulated. Due to the integration of the overall modelling system (Fig. 2) each scenario simulation involves a sequence of model calculations. Interpretation of results are made for each single step describing the hydraulic/ecological/chemical conditions within a certain field of this study. However, the integrated modelling system is formed by the exchange of data and results between the various models. A typical integrated scenario simulation could involve the following model simulations:

Step 1) MIKE 11/MIKE 21 Hydrodynamic
Hydrodynamic simulations are carried out for the reservoir (MIKE 21) and for the rivers (MIKE 11) given a certain water management regime. Hydrodynamic modelling of the MIKE11 post-dam model provides boundary conditions for the reservoir MIKE 21 model.

Main output: Flow velocities and water levels in the reservoir, in the Danube and in the river branches.

Step 2) MIKE 21/MIKE 11 Water quality/Eutrophication/Sediment transport
Based on the simulated flow fields (step 1) sediment transport, water quality and eutrophication simulations are carried out. Eutrophication modelling of the reservoir provides concentration boundaries for water quality models for the downstream Old Danube and for eutrophication modelling of the downstream river branch system.

Main output: Amount of different sediment grain size fractions on the reservoir bottom and concentrations of oxygen and nitrate distributed in time and space.

Step 3) MIKE SHE/MIKE 11 regional ground water flow simulation
A ground water flow simulation for the entire model area is carried out using the coupled version of MIKE SHE and MIKE 11. The reservoir is included in the MIKE 11 hydrodynamic simulations in a simplified one-dimensional flow description. Based on information on the distribution of sediment on the reservoir bottom (step 2) leakage coefficients are calculated and subsequently used directly in the MIKE SHE ground water flow simulations. Here, the applied leakage coefficients play an important role for the exchange of water between the reservoir and the aquifer.

Main output: Ground water flow and ground water levels distributed in time and space.

Step 4) MIKE SHE/MIKE 11 local reservoir model simulation
Using time varying boundary conditions (ground water levels) from the regional model (step 3) a more detailed model simulation is carried out using the local ground water model for the reservoir area. This model uses detailed information on the sediment layer provided by step 2.

Main output: Detailed three-dimensional ground water flow regime including recharge of water from the reservoir to the ground water.

Step 5) MIKE SHE/MIKE 11 local model for the river branch system on the Slovak floodplain.
Using time varying boundary conditions from the regional ground water model (step 3), a detailed ground water/surface water flow simulation for the river branch system is carried out. The output from this model forms the basis for the description of the ecological conditions in the river branch system.

Main output: Moisture content in the unsaturated zone, ground water levels, infiltration to the ground water and seepage to the Old Danube, depth and areal extent of inundations of flood plains.

Step 6) DAISY crop growth and nitrate leaching
Using time varying ground water levels from the regional ground water model (step 3) DAISY unsaturated flow, crop growth and nitrate leaching simulations are carried out.

Main output: Crop development parameters (leaf area index and root depth), crop yield, nitrate leaching and irrigation requirements distributed in time and space.

Step 7) MIKE SHE geochemical modelling
Based on the flow field from the local reservoir model geochemical modelling is carried out. Results from the sediment transport modelling and results from eutrophication and water quality modelling (step 2) are used to estimate the concentration of various species (oxygen, nitrate, organic matter etc.) in the water that recharges the ground water from the reservoir and the Danube, respectively. Nitrate leaching simulated with DAISY (step 6) are used to estimate nitrate concentrations in the water that percolating to ground water in the remaining parts of the model area.

Main output: Concentration of various species (nitrate, nitrite etc) in the ground water distributed in time and space.

3. Modelling studies in the Danubian Lowland - a few results

Comprehensive modelling studies have been carried out under the PHARE project (Slovak Ministry of Environment, 1995). In the present paper a few selected results are presented with regard to hydrology of the floodplains.

3.1. MODEL CONSTRUCTION FOR RIVER BRANCH SYSTEM

The complexity of the floodplain with its river branch system is indicated in Figs. 4 and 5 for the 20 km reach downstream the reservoir on the Slovakian side where alluvial forest occurs. In order to enable predictions of possible changes in floodplain ecology it is crucial to provide a detailed description of both the surface water and the groundwater systems in this area as well as of their interaction. For this purpose the MIKE SHE-MIKE 11 coupling is required.

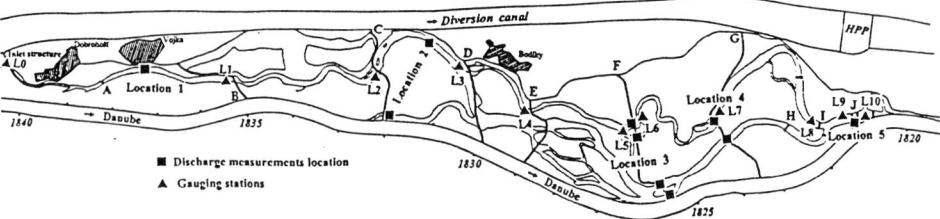

Figure 4. Layout of the river branch system on the Slovakian side of the Danube.

Before the damming of the Danube in 1992 the river branches were connected with the Danube during periods with discharge above average. However, some of the branches were only active during flood situations a few days per year. After the damming the water level in the Old Danube has decreased significantly. Therefore, in order to avoid that most water drained from the river branches to the Old Danube, resulting in totally dry river branches, the connections between the Danube and the river branches have been blocked except for the downstream one at chainage 1820 rkm (see Fig. 4). Instead, the river branch system receives water from an inlet structure in the hydropower canal at Dobrohost (see Fig. 4). This weir has a design capacity of 234 m^3/s and together with the various hydraulic structures in the river branches, it controls the hydraulic, hydrological and ecological regime in the river branches and on the floodplains. The extent of the floodplain model area is indicated in Fig. 1. and a perspective view of the area with the river branch system and floodplains are shown in Fig. 5. The horizontal discretization of the model is 100 m, while the groundwater zone is represented by two layers. Several hundreds of cross-sections and more than 50 hydraulic structures in the river branch system were included in the MIKE 11 setup for the river system.

For the pre-dam model setup the surface water boundary conditions comprise a discharge time series at Bratislava and a water level - discharge relationship at the downstream end (Komarno). For the post-dam model setup the Bratislava discharge time series has been divided into three discharge boundary conditions, namely at Dobrohost (intake from hydropower canal to river branch system), at the inlet to the hydropower canal and at the inlet to Old Danube from the reservoir. For the groundwater system, time varying ground water levels simulated with the regional ground water models act as boundary conditions. The Old Danube river forms an important natural boundary for the area. The Old Danube is included in the model, located on the model boundary, and symmetric ground water flow is assumed under the river. Hence, a zero-flux boundary condition is used as boundary condition for ground water flow below the river.

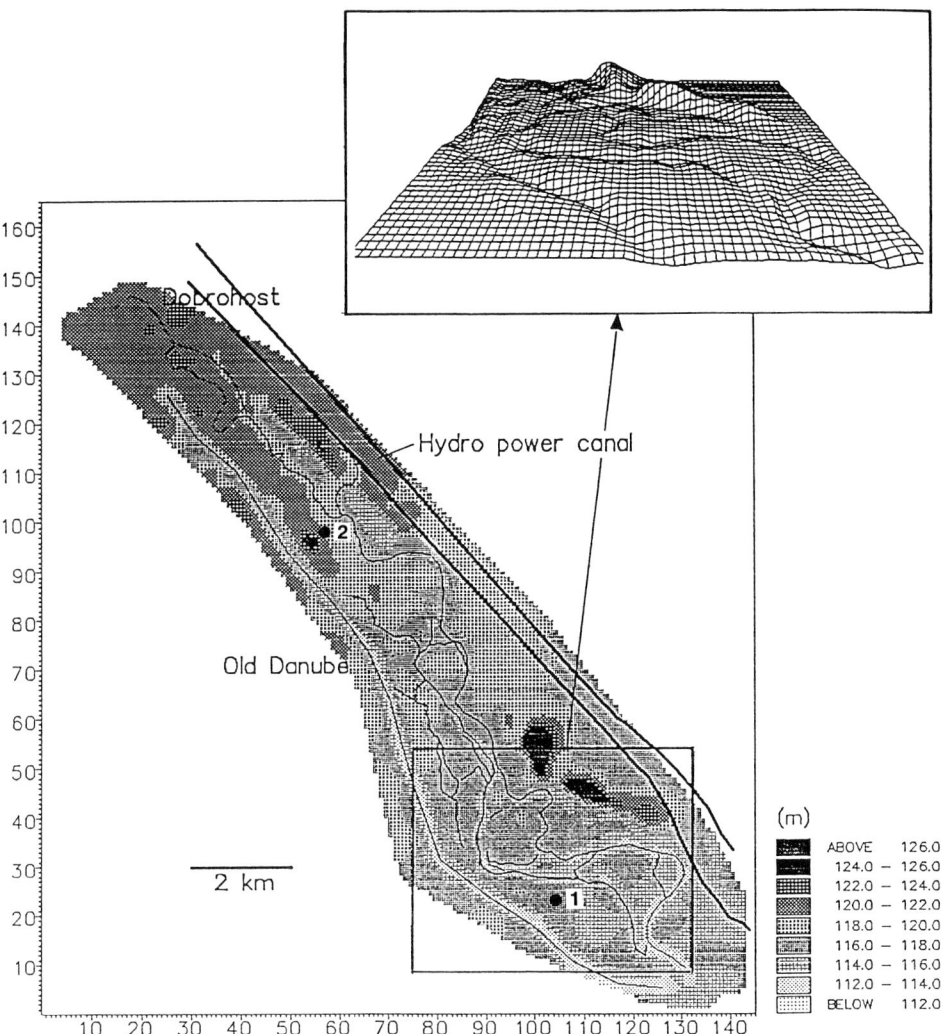

Figure 5. Plan and perspective view of the surface topography, of the river branches and the related flood plains as represented in a model network of 100 m grid squares.

3.2. MODEL CALIBRATION AND VALIDATION

The hydraulic model of the river branch system was calibrated against water level and discharge data measured during a three-week field campaign in June 1994 (Holubova et al., 1994). It appeared that the major part (up to 80%) of the inlet discharge at Dobrohost disappears between the intake structure and the downstream confluence with the Old Danube. The reason for this water loss is infiltration from the river branches to the aquifer system from where a significant part of it discharges the Old Danube which has a lower water table than the river branch channels. The calibration parameters included Manning numbers in the channels, flow capacity for some of the hydraulic structures and leakage coefficients for the channel beds. Although the structure geometry were known from measurements and design the field conditions were some times different due to blocking of culverts by dead trees etc.

The model was validated by testing its ability to reproduce water levels measured during the summer of 1993. Some of the validation results are shown in Fig. 6. It is seen from this figure that the largest deviations between model simulations and observed water levels are at the upstream location (L1). The reason for this is that the model parameter values describing the hydraulic structures correspond to the situation during the calibration period, i.e. 1994, while some modifications, mainly of crest elevations, were made to a few of the hydraulic structures in the upstream part of the river branch system between the 1993 validation period and 1994 . Thus the model setup corresponds to the new crest parameters, while the field data during the 1993 validation period correspond to old ones. Nevertheless, in spite of these minor inconsistencies between model parameters and field situation the simulation results correspond well to the measured water levels both with regard to dynamics and levels.

3.3. MODELLING OF FLOODPLAIN DYNAMICS

To illustrate the complex functioning of the MIKE SHE - MIKE 11 floodplain model and the interaction between the surface and subsurface hydrological processes model simulations for a period in June - July 1993 are shown in Figs. 7 and 8.

Fig. 7 presents the inlet discharges at the upstream point of the river branch system (Dobrohost), while the discharges and water levels at the confluence between the Danube and the hydropower outlet canal downstream of Gabcikovo during the same period are shown in Fig. 8. Fig. 7 shows furthermore the soil moisture conditions for the upper two m below terrain and the water depth on the surface at location 2. Similar information is shown for location 1 in Fig. 8. A soil water content above 0.40 (40 vol. %) implies that the soil is saturated. Location 2 is situated in the upstream part of the river branch system, while location 1 is located in the downstream part (see Fig. 5).

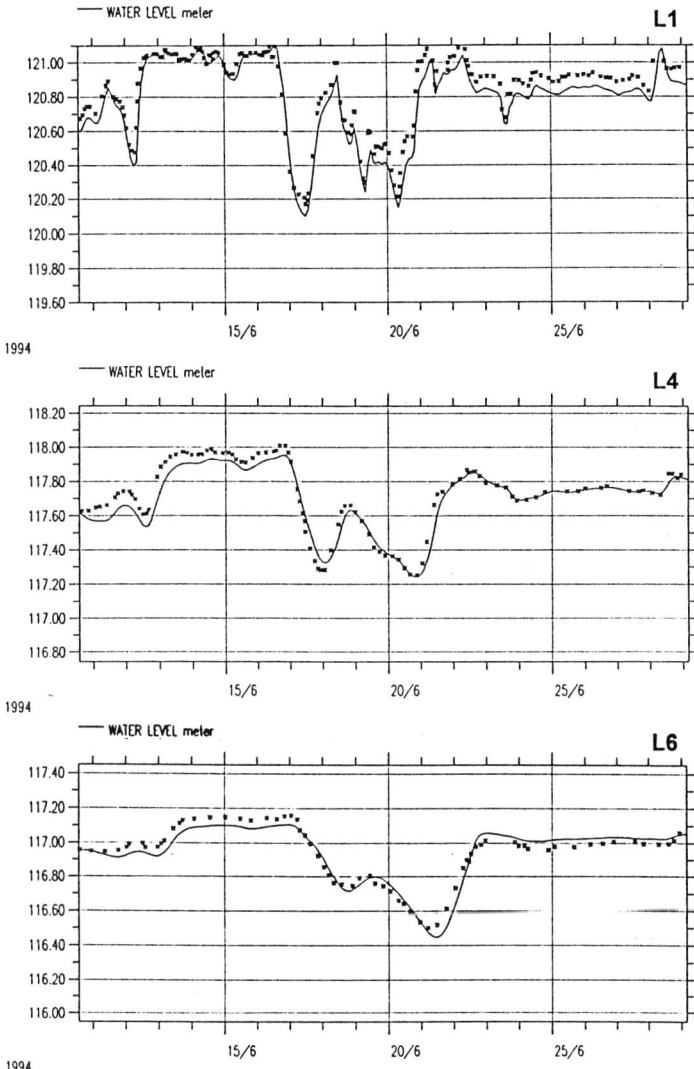

Figure 6. Validation of the river branch model against measured water levels for July-August 1993. The locations of the three sites L1, L4 and L6 can be seen in Fig. 4.

At location 2 (Fig. 7) flooding is seen to occur as a result of river spilling (surface inundation occurs *before* ground water table rises to surface) whenever the inlet discharge exceeds approximately 60 m³/s. The soil moisture content is seen to react relatively fast to the flooding and the soil column becomes saturated. In contrary, full saturation and inundation does not occur in connection with the flood in the Danube in July, but the event is seen in terms of increasing groundwater levels following the temporal pattern of the Danube flood.

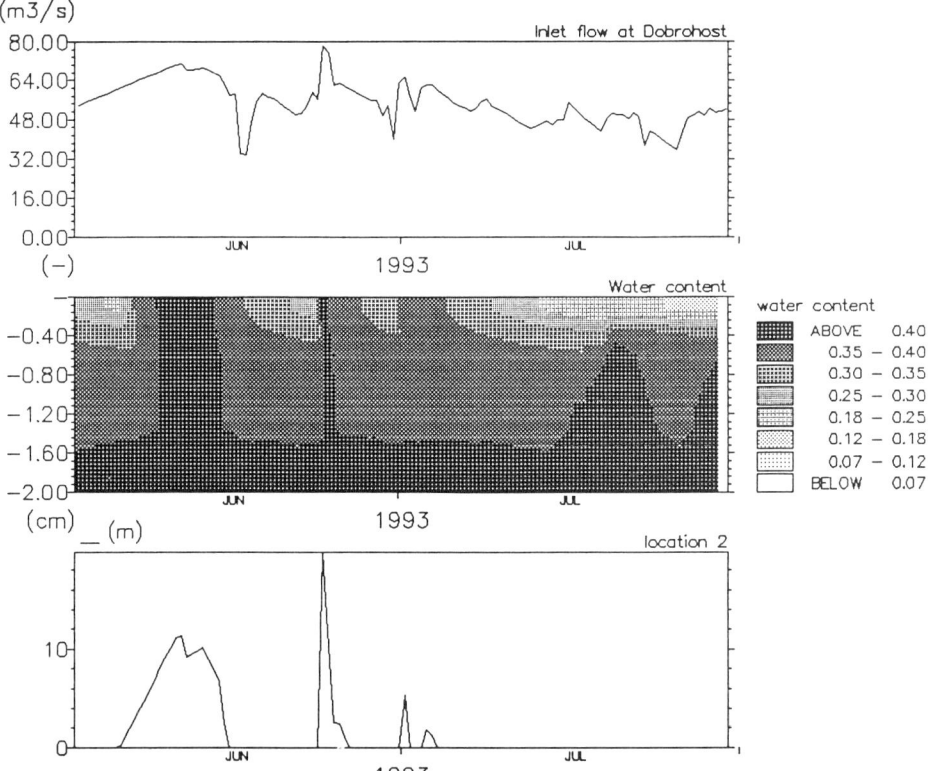

Figure 7. Observed inlet discharge to the river branch system at Dobrohost; simulated moisture contents at the upper two m of the soil profile at location 2 and simulated depths of inundation at location 2 during June-July 1993.

At location 1 (Fig. 8) the conditions are somewhat different. During the simulation period location 1 never becomes inundated due to high inlet flows at Dobrohost. However, during the July flood in Danube inundation at location 1 occurs as a result of increased ground water table caused by higher water levels in river branches due to backwater effects from the Danube. The surface elevation at location 1 is 116.4 m which is 0.4 m below the flood water level shown in Fig. 8 at the confluence (5 km downstream of location 1). It is noticed that the inundation at this location occurs as a result of ground water table rise and not due to spilling of the river (surface inundation occurs *after* the ground water table has reached ground surface).

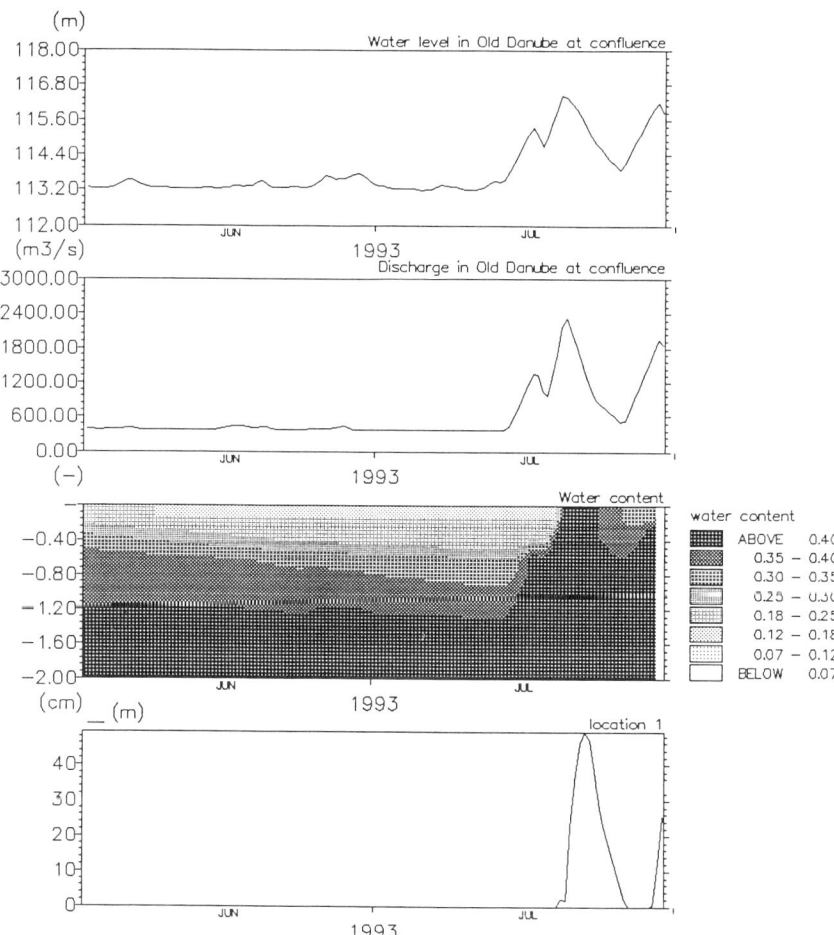

Figure 8. Simulated discharge and water levels in the Danube at the confluence between Old Danube and the outlet canal from the hydropower plant; simulated moisture contents at the upper two m of the soil profile at location 1 and simulated depths of inundation at location 1 in the river branch system during June-July 1993.

3.4. RESULTS OF MODEL APPLICATION

The floodplain model is a management tool which can simulate the operation of the hydraulic structures, enabling an optimization of the hydraulic and ecological conditions for the unique floodplain environment. The floodplain model provides detailed information in time and space about water levels in river branches and on the floodplains, groundwater levels and soil moisture conditions in the unsaturated zone. Such information can directly be compared with quantitatively formulated ecological criteria.

As an example of the results which can be obtained by the floodplain model Fig. 9 shows a characterization of the area according to flooding and depths to groundwater. The map has been processed on the basis of simulations for 1988 for pre-dam conditions. The classes with different ground water depths and flooding have been determined from ecological considerations according to requirements of (semi)terrestrial (floodplain) ecotopes. From the figure the contacts between the main Danube river and the river branch system is clearly seen. Similar computations have been made by alternative water management schemes after damming of the Danube. The results of one of the hypothetical post-dam water management regimes, characterized by average water flows in the power canal, Old Danube and river branch system intake of 1470 m^3/s, 400 m^3/s and 45 m^3/s, respectively, are shown in Fig. 10. By comparing Fig. 9 and Fig. 10 the differences in hydrological conditions can clearly be seen. From such changes in hydrological conditions inferences can be made on possible changes in the floodplain ecosystem.

Further scenarios (not shown here) have, amongst others, investigated the effects of establishing some underwater weirs in the Old Danube and in this way improve the connectivity between the Old Danube and the river branch system.

4. Conclusions

The ecological system of the Danubian Lowland is so complex with so many interactions between the surface and the subsurface water regimes and between physical, chemical and biological changes that a comprehensive mathematical modelling system of the distributed physically-based type is required in order to provide quantitative assessments of environmental impacts.

Such modelling system coupled with a comprehensive data base/GIS system has been developed. The integrated system makes it possible at a quite detailed level to make quantitative predictions of the surface and ground water regime in the floodplain area, including e.g. frequency, magnitude and duration of inundations. Such information constitutes a necessary basis for subsequent analysis of flora and fauna in the floodplain.

In the present chapter some of the capabilities of the modelling system have been illustrated by a few selected results on flood plain hydrology.

CASE STUDY – MODELLING THE INFLUENCES OF GABCIKOVO HYDROPOWER PLANT 251

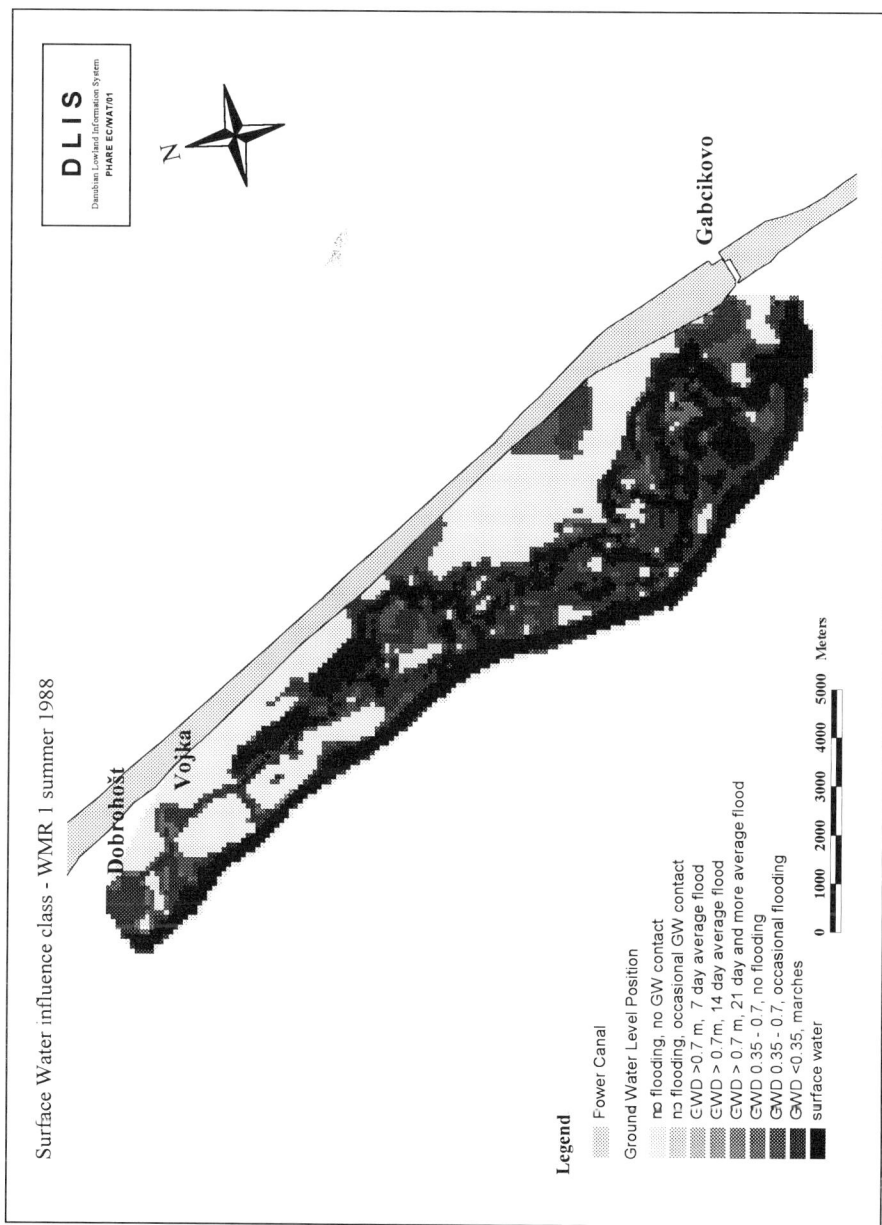

Figure 9. Hydrological regime in the river branch area for 1988 for pre-dam conditions characterized in ecological classes.

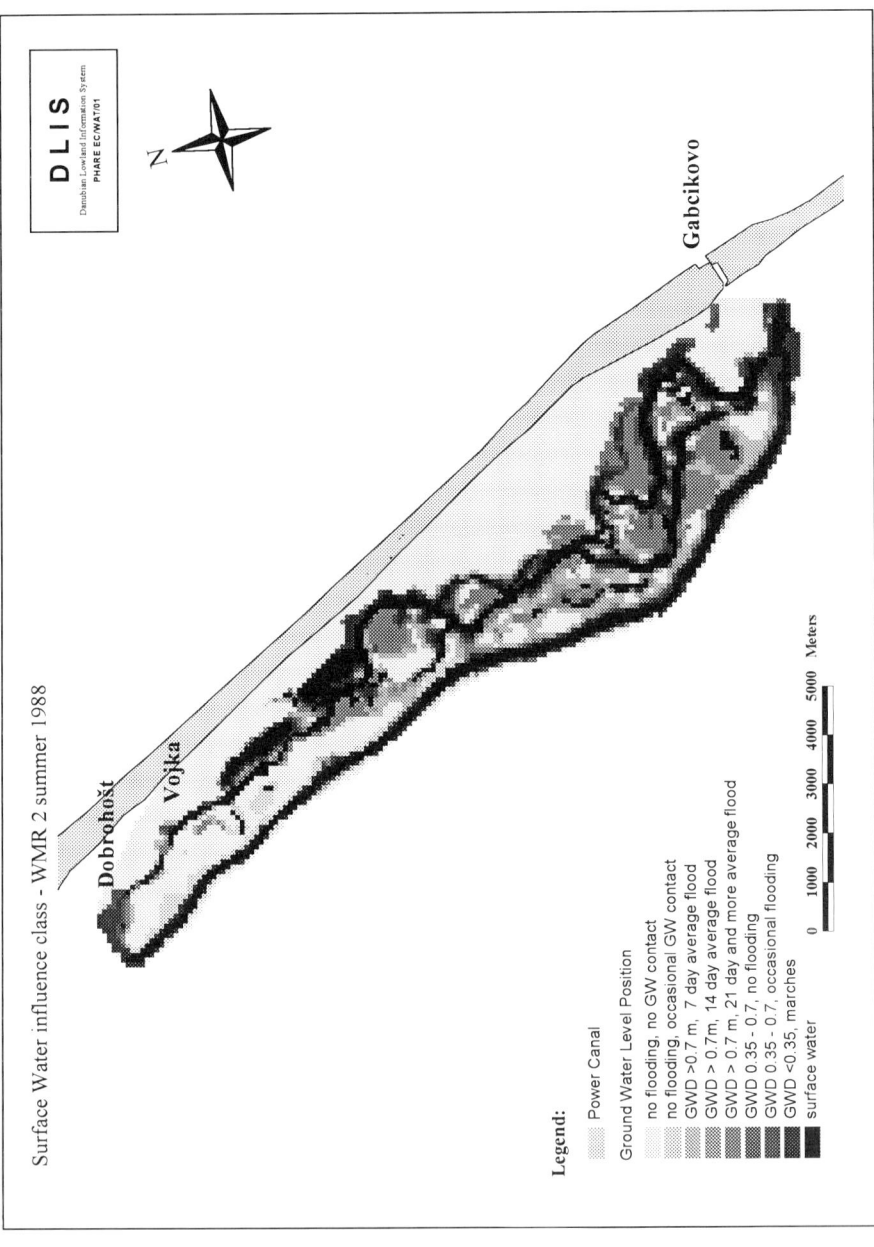

Figure 10. Hydrological regime in the river branch area for a post-dam water management regime characterized in ecological classes. The scenario has been simulated using 1998 observed upstream discharge data and a given hypothetical operation of the hydraulic structures.

5. Acknowledgement

The PHARE project was executed by the Slovak Ministry of the Environment and supported financially by the European Commission. A Danish-Dutch consortium of six organizations was selected as Consultant for the project. The Consultant was headed by Danish Hydraulic Institute (DHI) and comprised the following associated partners: DHV Consultants BV, The Netherlands; TNO-Applied Institute of Geoscience, The Netherlands; Water Quality Institute (VKI), Denmark; I Krüger Consult AS, Denmark; and the Royal Veterinary and Agricultural University, Denmark.

6. References

DHI (1995) MIKE 21 Short Description. Danish Hydraulic Institute, Hørsholm, Denmark.

Hansen, S., Jensen, H.E., Nielsen, N.E., and Svendsen, H. (1991) Simulation of nitrogen dynamics and biomass production in winter wheat using the Danish simulation model DAISY. *Fertilizer Research*, 27, 245-259.

Havnø, K., Madsen, M.N. and Dørge, J. (1995): MIKE 11 - A Generalized River Modelling Package. *In: V.P. Singh (Ed): Computer Models of Watershed Hydrology*, Water Resources Publications, 733-782.

Holubová, K., Capeková, Z., Lukác, M. and Misik, M. (1994) Discharge conditions within the Danube river branch system (Dobrohost-Gabcikovo), Water Research Institute (VUVH), Bratislava.

Mucha, I. (Editor) (1995) Gabcikovo part of the hydroelectric power project environmental impact review. Evaluation based on two years monitoring. Faculty of Natural Sciences, Comenius University, Bratislava. 384 pp.

Refsgaard, J.C. and Storm, B. (1995): MIKE SHE. *In: V.P. Singh (Ed): Computer Models of Watershed Hydrology*, Water Resources Publications, 809-846.

Slovak Ministry of Environment (1995): PHARE project Danubian Lowland - Ground Water Model (EC/WAT/1), Final Report. Bratislava.

VKI (1995) Short Description of water quality and eutrophication modules. Water Quality Institute, Hørsholm, Denmark.

CHAPTER 13A
A DISCUSSION OF DISTRIBUTED HYDROLOGICAL MODELLING

KEITH J BEVEN
Centre for Research on Environmental Systems and Statistics, Institute of Environmental and Biological Sciences, Lancaster University, Lancaster LA1 4YQ

1. Introduction: a friendly critique of distributed modelling in hydrology.

Inspired by the rapidly increasing power of computers and the development of geographical information systems and digital terrain maps, distributed models in hydrology (and other areas such as ecology) have been developing rapidly since the first outline of a physics-based distributed model published by Freeze and Harlan in 1969. Coupled to the opening up of the technological feasibility of making such models possible has been the recognition and development of a demand for more distributed predictions as outlined by Beven and O'Connell (1982), Abbott et al. (1986), Bathurst and O'Connell (1992) and Chapter 1. There are now a number of systems that are being used regularly for practical applications, such as the various versions of the Système Hydrologique Européen (SHE, see Abbott et al., 1986; Bathurst et al., 1995; Refsgaard and Storm, 1995; and Chapters .....), and the Institute of Hydrology Distributed Model (IHDM, see Beven et al. 1987; Calver, 1988; Calver and Wood, 1995). These models claim to be physically-based (in accordance with the outline laid down by Freeze and Harlan, 1969) and to have parameters that are physically measurable in the field, at least in principle.

The qualifier, in principle, is significant here. It has appeared in a number of papers, some of which have been critical of the current formulation of distributed hydrological models (Beven, 1987, 1989a,b, 1993); others of which have been promoting the case for the use of distributed models (e.g. Bathurst, 1986; Bathurst and O'Connell, 1992). It is indicative of a fundamental problem in formulating a distributed mechanistic description of hydrological processes that is realistic. The original blueprint for a physically-based model of Freeze and Harlan (1969) was formulated from process descriptions that in 1969 were thought to describe adequately the flows of water through soil, over the soil surface and in channels under experimental conditions. At that time, very little work had been done in studying the details of hydrological processes in the field. The reality defined in well controlled experimental situations was simply assumed to extend to other situations. This assumption will not necessarily hold in the face of the heterogeneity and complexity of water flows in the field (see discussions of this realist approach in Bhaskar, 1989; Cartwright, 1983)

A good example is the work of Emmett on surface flows which was not widely disseminated until the publication of his chapter in the book *Hillslope Hydrology* (Kirkby, 1978). He showed that the hydraulics of surface flows on rough partly vegetated surfaces was qualitatively different from the parking lots, airfields and uniform grass swards that had been studied by Izzard and others before. These irregular flows were treated in terms of an "equivalent" sheet flow for the purposes of back calculating a roughness coefficient within the theoretical framework then available, but this is not a realistic description. It is however, the description used in all the distributed modelling systems available today, partly, at least, to facilitate the coupling with subsurface flow models at the boundary of the soil surface. More realistic descriptions have been studied (e.g. Dunne et al., 1991) that allow for the irregularity of the surface but, in doing so, necessarily introduce additional parameters.

Similarly, all physically-based subsurface flow descriptions are based on Darcy's law and the Richards' equation for unsaturated flows. This is known to be a good description of flow in laboratory soil columns, where the soil has been well mixed and repacked. It is not a good description of flow in undisturbed soil columns, where it is common to find measured hydraulic gradients that are negative with respect to the known bulk flow direction as a result of differential wetting and preferential flow within the column. It is debatable whether a Darcian description is valid at some local scale within such columns, it is certainly not valid at the column scale or plot scale (see for example Luxmoore et al., 1981; Schulin et al., 1987; Hornberger et al., 1991; Flury et al., 1994) and consequently not at the scale of the element grids used in distributed hydrological models, which may require these equations to be applied with parameters and variables assumed uniform over a spatial scale of tens of meters or even kilometres (see for example Jain et al., 1992).

Thus, it is suggested that the equations on which the current generation of physically-based models are based on the wrong equations to describe hydrological fluxes in the field. Why then do they work? They work (more or less) because although in principle (again) the parameter values of the models can be measured in the field, in practice this is rarely done. Essentially the measurement techniques do not exist to provide parameter values at the scale required by the model. The parameter values must therefore be estimated or calibrated. This is usually done by a comparison of observed and predicted system variables, just as in any other conceptual hydrological model. In fact, it can be plausibly argued that the currently available physically-based distributed models are lumped conceptual models, albeit that they work at the grid scale rather than at the catchment scale of more traditional lumped conceptual models (Beven, 1989).

Distributed models in hydrology are examples of what Morton (1993) calls *mediating models*. The mediate between an underlying theory, which may be partly qualitative or perceptual in nature, and quantitative prediction. As such, they have the general characteristics discussed in Morton's analysis: they have assumptions that are false and *that are known to be false* (for discussions in relation to Darcy's law see Schrader-Frechette, 1989, Hofmann and Hofmann, 1992 and Oreskes et al., 1994); they tend to be purpose specific with parameters and auxiliary conditions that are different for different purposes; they reflect physical intuition but may contain some more or less arbitrary elements; they have real explanatory power but may never (nor should they be expected to) develop into full theoretical structures. They also have a history, in that successful model structures are copied and refined by later models. However, the predictions and parameters of such models should not be taken as valid outside of the context of the particular model structure being used.

We should expect, of course, that the models will be improved as our understanding of hydrological processes improves. It is not currently clear whether such an expectation is justified. There are many things about hydrological systems that are essentially unknowable, especially the nature of flow processes below the ground in structured soils. In principle (again) a full physical description is possible and we could solve the Navier-Stokes equations for flow through a system of pores, given sufficient knowledge of the geometry of the pore structure (even a statistical description might suffice) and the parameters controlling factors such as the effect of different organic coatings on the wetting angle and the roughness of pore surfaces. Other processes such as air pressure effects, microbiological effects, soil surface sealing and other temporal changes could also be parameterised, at least for simple cases. Even ignoring the problem of the computer power that such simulations would require, such descriptions merely move the problem of unknowability to smaller time and space scales.

A more practical alternative strategy would be to apply the Darcian equations at a scale to which they are appropriate, the scale of the "representative elementary volume" (*REV*). This is the scale at which the pore water potentials can be assumed to be locally in equilibrium so that a continuum description involving potential gradients might be acceptable. This would need to be much smaller than the grid scales of current distributed models, although no-one to my knowledge has ever determined the scale of the *REV* for a real soil. The Darcy-Richards equation will then be a description compatible with the scale at which it is being applied. This scale will, however, be much smaller than the grid scale used by current distributed models implying that, given the heterogeneity of the soil, many more parameter values will be need to be specified and that considerably more computer time will be needed for solutions at the *REV* scale. It would not, of course, be possible to measure all the parameter values required without destroying the system of interest, since current measurement techniques are, for the most part destructive. Thus, the parameters will necessarily remain essentially unknowable, even for a stationary system. Changes over time will only exacerbate the problem. Distributed hydrological modelling is thus essentially a *transcientific* problem (Weinberg, 1972; Philip, 1978).

For subsurface flow, the best that might be expected is some statistical sample of small scale measurements. Given such a sample, some theory is needed to move from the statistical description at the *REV* scale to the grid scale at which predictions are required. Such scaling theories have been the subject of considerable research effort in groundwater systems, where it has been shown that the "effective" parameter values required at larger grid scales can be derived from statistical models of the small scale heterogeneity of Darcian parameters under a variety of assumptions, including fractal porous media (e.g. Dagan, 1986; Wheatcraft et al., 1990). The results suggest that the effective parameters will be scale dependent and subject to uncertainty.

Much less has been achieved for unsaturated systems, and even then only for what might be called "Darcian" heterogeneity, without consideration of the types of soil structure leading to preferential flow (e.g. Mantoglou and Gelhar, 1987; Nicholson et al., 1989; Yeh et al., 1985; Jensen and Mantoglou, 1992). Models have been developed for flow in structured systems involving two or more flow components. All are theoretical constructs derived from prior assumptions and introduce more parameters, without a clear indication of how those parameters can be determined at larger scales (see for example Beven and Germann, 1981; Hoogmoed et al., 1980; Beven and Clarke, 1986; Steenhuis et al., 1990; Jarvis et al., 1991; Booltink, 1994).

There is also the problem of coupled hydrological processes. The study of Binley et al. (1989) suggests that in the case of coupled surface and subsurface flows in a heterogeneous domain, it may not be possible to find effective parameter values that are generally applicable. The heterogeneity is important in controlling the coupled response. In fact, Binley et al. (1989) did not take account of the effects of "run-on" infiltration, which might exacerbate the problem.

And yet there have been a number of studies published which suggest that the application of these physically-based models has been successful, despite all the problems outlined above. How is this possible? Probably, because most of the measures of assessment of model performance have been those of predicting discharges and the discharge response of hydrological systems at the catchment scale is relatively simple to model (which is why the unit hydrograph technique still survives as an engineering tool after half a century of use and criticism as an inadequate representation). Following a rainfall, the stream discharge rises and falls with a certain characteristic timing, reflecting the integrated effect of all the complex processes occurring within the catchment. The volume of runoff for individual storms is more difficult to predict, partly because of a complex nonlinear relationship with the pattern of heterogeneity of catchment characteristics and antecedent conditions and partly because the available input data may be of limited accuracy.

Thus, even a minimal hydrological model requires some acceptable way of predicting the relationship between the runoff coefficient and antecedent conditions and some acceptable way of routing or distributing the resulting runoff through time. This can be achieved to reasonable accuracy using relatively simple non-distributed models based on data analysis techniques (see the studies of Jakeman and Hornberger, 1993 and Young and Beven, 1994 noted above). Physically-based distributed models also provide both functions with sufficient parameter values that can be calibrated to fit observed discharges. The essential difference about distributed models is that they also make predictions of internal state variables at the grid element scale, but there have been very few tests of these predictions against observed variables. In fact there is a similar problem with such measurements as with the model parameters, since the measurable state variables are also at a much smaller scale than the grid element scale.

In summary, it cannot be assured that distributed models are based on the correct equations to describe hydrological reality at the grid element scale; it is very difficult (if possible at all) to estimate effective model parameter values for those equations at the element scale; and, while they are overparameterised for the purposes of estimating discharges, they have not been properly tested in terms of simulating the internal state variables that is their principal advantage over catchment scale models. These problems have been discussed more extensively elsewhere (see Beven, 1985, 1989, 1993; Grayson et al., 1992). In what follows here I wish to consider various responses to the problems outlined above that might provide a way ahead for distributed modelling, for there is no doubt that a practical capability of making distributed predictions of flow pathways would be extremely valuable (as discussed in Chapter 1).

2. Is there an example of a successfully validated distributed model at the catchment scale?

One response to these problems is that they are not as bad as might first appear. As already noted there have been a number of published applications of models such as SHE and IHDM that have been declared successful in predicting discharges at the catchment scale, suggesting that these are valid, or at least acceptable, hydrological predictors (e.g. Bathurst, 1986; Calver, 1988). There are only a handful of studies that have tried to validate such rainfall-runoff models in the prediction of internal hydrological state variables. The process of model validation has attracted a lot of attention in the hydrological literature recently, primarily in the groundwater literature, where both research programmes, public inquiries and litigation concerning waste disposal sites and groundwater contamination has focused attention on our abilities to model the subsurface. Two well respected groundwater hydrologists have suggested that groundwater models cannot be validated (Konikow and Bredehoeft, 1992), while a number of post-audit studies of groundwater modelling exercises have revealed serious deficiencies in the original predictions (Anderson and Woessner, 1992). The debate has also attracted the attention of the philosophical community. Oreskes et al. (1994) for example discuss the meaning and use (and misuse) of the terms validation, verification, confirmation and falsification in the groundwater modelling context. The practitioner, on the other hand, knows a good modelling study when he sees it and can persuade a jury to accept it in court (Bair, 1994).

A similar debate has been continuing in the catchment modelling literature, initiated by the discussions in Beven (1989) and Grayson et al. (1992), who highlighted the limitations of the current distributed models and the need to resort to calibration strategies to make them work. Responses have been mixed. Smith et al. (1994) for example suggest that it is important to distinguish between a mathematical abstraction of observed natural behaviour based on fundamental hydrodynamic concepts (their type 1 model) that is verifiable under controlled experimental conditions and a computer implementation of those concepts with numerical approximations and possibilities of errors in boundary and other ancillary conditions and "ever-possible coding mistakes" (their type 2 model). They point out that difficulties and failure of the type 2 model in the face of heterogeneities and uncertainties should not be confused with failure of the type 1 model. They also suggest that a type 2 model should not be expected to be more accurate in its predictions than the reproducibility of nature itself and point to the study of Hjelmfelt and Burwell (1984) which revealed great heterogeneity in responses from 40 adjacent 0.01 ha plots with coefficients of variation of runoff volumes varying from 0.071 to 1.09 for storms ranging from 6 to 96 mm. These plot responses, presumably varying in response to different soil conditions and rainfall inputs on the individual plots, appear virtually unpredictable at the level of soil and rainfall information and Smith et al. (1994) ask "why should we expect more of a type 2 model than we expect from nature?" Why indeed?

There is probably enough evidence that many of the type 1 models found in the literature are not adequate descriptions of reality, but even if they were it is not of that much use to know that there is a type 1 model that might (or might not) apply in principle (again) if the parameters of that model are essentially unknowable without costly (and possibly destructive) experimentation at all locations in a catchment. Thus, if the type 2 approximation to this model, with parameters calibrated in some way, cannot be expected to accurately mimic the variability in the measurements of the real responses given the available estimates of the inputs and other boundary conditions, then those models cannot be validated at any detailed level and should, in fact, be expected to be in error.

This aspect of model validation was recognised very early on by Stephenson and Freeze (1974) who found that, albeit having tried only a few runs of a hillslope scale finite difference model due to limitations in computer power at that time, they could only achieve relatively poor fits to both discharge and internal water table measurements, partly because there was great uncertainty about the initial and boundary conditions for the slope as well as for the parameter values. Any attempt at validation, they noted, would also be dependent on the definition of initial and boundary conditions and since these would always be to some extent unknown there could therefore never be any true validation of such a model.

A more successful test of a distributed model has been reported by Jensen et al. (1993) in a simulation of a tracer test in a sandy aquifer in Jutland, Denmark, using the three-dimensional subsurface flow and transport component of MIKE-SHE. Hydraulic conductivities were estimated from slug tests in boreholes on the site, dispersivities were estimated by calibration. One run was made using a stochastically generated field of hydraulic conductivity values but most of the runs used a three layer representation of the flow domain with homogeneous "effective" hydraulic conductivity values in each layer. This is one of the still very few studies published in which model predictions are compared with internal state measurements within the flow domain. They found a very good reproduction of the water table over a period of several months using recharge estimates derived from a one-dimensional Darcian unsaturated flow model (it is not clear how the parameters of the unsaturated zone model were derived). Results from the transport model were also good, although fitted longitudinal and transverse dispersivity values varied by a factor of 10 at different wells. The general spread of the plume was reproduced using homogeneous dispersivity values but while the breakthrough curves at some wells were simulated well there were marked discrepancies at others. The small values of dispersivity used suggest that this is a relatively homogeneous medium.

These results contrast with those of Binley and Beven (1991) who compared the results from a three-dimensionally saturated-unsaturated finite element model of a small hillslope hollow with heterogeneous soil depths and hydraulic conductivities with a representation of the same area using homogeneous two-dimensional hillslope planes. The subsurface solution was the same in both cases, the simpler model using an approximate geometry and "effective" parameter values. In this case, the approximations led to relatively poor reproduction of the three-dimensional model results, perhaps because a greater degree of heterogeneity had been assumed than for the real Jutland site of Jensen et al. (1993).

These examples suggest that there is no clear answer to the problem of model validation,. confirmation or acceptability. It is intrinsically bound up with issues of specifying model discretisations, parameter estimation and calibration and the definition of initial and boundary conditions. A wider view also introduces other issues. Beck et al. (1995) identify two other aspects to model validation in addition to the intrinsic properties of the model itself. These are the nature of the predictive task to be performed and the magnitude of the risk of making a wrong decision based on the model predictions. They suggest that if the task to be performed is similar to one that has been studied before and the risk of making a wrong decision is low then making a judgement on model validity (or acceptability) should be relatively straightforward. If, on the other hand, the task is novel and the cost of a wrong decision high, then the tests should be more stringent and a decision on acceptability will be much harder. More stringent tests may reduce the chance of accepting a false model but will also make it more likely that good or acceptable models will be rejected. The decision might then depend on more qualitative information such as peer review of the model structure and conceptual model of the flow domain. An example of such a situation is in the assessment of potential deep radioactive waste disposal sites such as at Yucca mountain in the USA and at Sellafield in the UK. Expert opinions can differ widely about the adequacy of various modelling approaches, and recent studies of the way in which experts use different levels of information in constructing a conceptual model of the flow domain for site assessment is not too reassuring in this respect (Mackay, 1995).

3. Are improved process parameterisations possible?

If, as has been suggested above, the small ("*REV*") scale equations used in current "physically-based" distributed models are not easily scaled up in heterogeneous and structured flow domains, is it possible that improved parameterisations might be found? It is important to remember that there are (at least) two problems here. One is that associated with the problem of change of scale and heterogeneity of parameters, even if the small scale equations were correct at the local scale. If this were the only problem then it might be possible to derive a theory of scaling, that would allow scale dependent parameters to be developed based on knowledge of a statistical model of the heterogeneity (see for example Dagan, 1986). However, it is not the only problem. The second problem is that the small scale equations may not be correct at the local (profile to plot) scale.

One example, as noted in section 1 above, is the use of Darcy's law to describe flow in structured soil where flow may be responding to quite different potential gradients and may not have a simple linear relationship between flux rate and gradient (i.e. a problem of the second type). More realistic descriptions are also needed for runoff on irregular and vegetated surfaces; the controls of soil water on water use by plants, (or of water use by plants on soil moisture when the roots grow faster than water can move by Darcian flow towards them and a fundamental control on evapotranspiration appears to be levels of abcissic acid (ABA) in the plant); and of the effects of soil layering on downslope flows. Current models of transport processes and geochemical processes also need improvement.

In most cases, as with flow in structured soil, improved, or at least more complex, descriptions are already available, often have been for some time but are difficult to apply. Think of the simple modification to Darcy's law to allow hysteretic soil moisture characteristics (e.g. Mualem, 1976), still quite rarely implemented in distributed models. These descriptions share some common features, regardless of the process to which they refer. Being more complex they take more computer time and almost certainly require more parameters and state variables. Those parameter values may not be readily estimated, except by calibration against some data. Calibration of these models is already a problem, adding parameters will only make the problem worse.

Thus, it would appear that there is a fundamental dilemma in "physically-based" distributed modelling. If the limitations of the descriptive equations are recognised, it is necessary to introduce more complex parameters with more parameters. These parameters may not be easily measured and may also suffer from the problem of heterogeneity, thereby necessitating large number of measurements to assess the degree and importance of spatial variability. And yet, the response to rainfall is not that complex. A rainfall occurs, surface and subsurface flow rates increase, a hydrograph results. Reproducing the dominant modes of the integrated response at the catchment scale is not that difficult and requires a relatively simple model with few parameters. How can this lesson be adapted for the case of distributed models? What would be the minimal model to describe the distributed response? Would it actually be simpler than today's distributed models?

I believe the answer to this last question to be yes, but difficult to prove without appropriate distributed measurements from which to develop such a description. Certainly the resolution of the dilemma identified above does not lie in ever more detailed descriptions and ever more demanding requirements for data collection for parameters and state variables. The resolution lies in the search for simpler distributed models, based on data collected at appropriate scales. In that the water table integrates the distributed hydrological responses upslope and around a point, a fundamental building block along this path might be the imaging of water table responses over time. Even given such data, however, it is not clear how best to proceed to develop an appropriate model, since every additional measurement point that is added should allow for some further local calibration of the model to take account of the local heterogeneities of the system. What is not clear is whether enough information is added by such internal measurements to permit such local calibration, especially where the measurement scale is less (sometimes much less) than the model grid elements, and also whether model structural errors can be separated from local calibration errors. The availability of such data, reflecting the local responses of the system, might allow a disaggregation approach towards distributed modelling. This will eventually prove to be a more promising approach than the aggregation of small scale parameter information that underlies much of today's hydrological theorising.

4. Is a disaggregation approach to sub-catchment scale parameterisations possible?

A disaggregation approach to distributed modelling has been discussed by Beven (1995) who has outlined a set of objectives and principles on which to base such an approach. The objectives are simple. All that is required is to define the input fluxes and other boundary conditions correctly at the chosen element scale (subcatchment, hillslope, patch... the problem becomes more difficult as the scale decreases, but the subcatchment scale may often be adequate); and to get the partitioning between discharges, storage and latent heat fluxes correct, however this is done. There may also be some subsidiary objectives for some problems. In some cases the timing of discharges and evapotranspiration fluxes may be important, in other cases flow pathways may be important, especially for transport and quality prediction. A final objective should be to get an estimate of the uncertainty in the element scale predictions correct.

The principles on which to base an appropriate parameterisation include the satisfaction of mass balance; that flow will follow (but not necessarily be linearly proportional to) the dominant potential gradient, which at larger scales will often be gravity; that the equations of a parameterisation appropriate at one scale will not be the same as at smaller or larger scales; that the form of the parameterisation used may depend on the data support and may vary with the environment and purpose of the application; that preferential flow may exist at all scales; that the most important nonlinearities in hydrological systems are those due to the coupling of different processes; that at any scale the extremes of the sub-element distribution of responses may be important in controlling the overall water balance partitioning (both for discharges and evapotranspiration fluxes); and finally (but not exclusively) that the current (and future) response of the element may be conditioned on its history. Other principles may need to be added for specific purposes but, as far as I am aware, there is no current hydrological model that reflects these principles in sufficient detail to be considered an appropriate parameterisation (or, consequently, hydrological model).

There are some that are working along the right lines by attempting to take the distribution of responses into account, rather than using a lumped description at the element scale. This presupposes, however, that some data are available to allow this. TOPMODEL (Beven et al., 1995), for example, uses topographic data, now widely available at the 50 m scale or better, to allow a distributed description based on a topographic index. However, not only are there different ways of processing the topographic data to derive the index (see Moore and Grayson, 1991; Quinn et al., 1995; Costa-Cabral and Burges, 1994; Wolock and Price, 1994, Bruneau et al., 1995) but it is clear that in some drier environments the topographic index needs to be dynamic (Barling et al., 1994; Durand et al., 1993).

The distribution function approach to disaggregation is also the subject of research efforts aimed at improving the land surface parameterisation of Global Circulation Models (GCMs) which have previously used lumped descriptions of land surface hydrology within elements of up to 300 km in scale. A number of models are now available that represent the variability in the hydrological response within the element as a number of patches (see for example, Avissar, 1992; Koster and Suarez, 1992; Blyth et al., 1993; Quinn et al., 1994). Macroscale hydrological models are also starting to take advantage of these developments by linking together such element distributions by channel routing algorithms (e.g. Jolley and Wheater, 1993). How far these descriptions can be considered physically-based depends on one's individual working definition of "physically-based" but most hydrologists might agree that they are at least more physically-based than the lumped descriptions used hitherto.

Again, as for any sub-element parameterisation of this type, the acceptability of the description depends critically on the data that supports the description. The role of the availability of data here is critical. Consider the TOPMODEL concepts, for example. The idea of using a topographic index as a measure of hydrological similarity predates the widespread availability of digital terrain models (Kirkby, 1975; Beven and Kirkby, 1979) but the current popularity of this approach has more to do with the availability of digital terrain data than with any fundamental belief in (or even desire to test) the similarity concepts on which these models are based.

Another example of the interaction between data available at a given scale and an appropriate parameterisation is given by the transport of solutes in rivers. In this case it is actually relatively simple to carry out tracer experiments to determine the reach scale mixing characteristics. Such experiments can also provide the necessary data to develop and test parameterisations. In many cases the data suggest that a description based on simple Fickian dispersion, leading to the advection-dispersion equation (ADE), on which most one-dimensional distributed models of transport are based, is not an adequate description of the data since it will not produce the long tails characteristically seen in most tracer experiments. These long tails are due to the effects of "dead zones" or imperfect or inefficient mixing of the flow even over long distances. This means that the effective "mixing length" for the transport process is apparently much longer than the reach scale such that the ADE, valid only beyond the mixing length, should not be used (see discussion in Young and Wallis, 1993). The ADE can be modified to fit the data by incorporating a dead zone component (e.g. Bencala and Walters, 1983; Hart, 1995) but this introduces additional storage and mixing coefficients that are not easily determined except by fitting the tracer data. Extension to two- or three-dimensional descriptions would require both more computer time and more geometric and parametric inputs. Yet, most tracer curves are easily described by a much simpler model based on a simple advective time lag and linear store dispersion as incorporated into the aggregated dead zone model (ADZ, see Wallis et al., 1989; Green et al., 1994) which assumes, in effect, that the dispersion is dominated by the dead zones, rather than by Fickian shear dispersion. The ADE is the description that is found in the text books, but is wrong. Is the simpler ADZ description, which works, any less physically-based at the scale for which it is intended?

There is a direct analogy between the ADZ model and the catchment scale models used for example by Jakeman and Hornberger (1993). The parameterisations used are based directly on data at the scale of the modelling. The inference is that improved parameterisations of distributed hydrological models must be developed from distributed data, collected at a scale appropriate to the prediction scale required. If, for example it was possible to measure directly the volume of water stored in a block or hillslope of soil, then hydrological theories might well be developed in terms of storage-discharge relations at that scale rather than moisture content-hydraulic conductivity relations at the *REV* scale. It would be expected, of course, that such larger scale relationships might demonstrate hysteretic differences in wetting and drying. This is already deemed theoretically acceptable at the *REV* scale, why not at the larger scale (if the distributed measurement techniques were available to allow it)?

One strong criticism of such a databased mechanistic approach is that it is purely empirical in the sense that, while the relationships developed might provide a good representation of current conditions, it would be difficult to extrapolate behaviour to other, perhaps more extreme conditions or speculate about the effects of change in the system. However, again, it is not clear how the more "physically-based" parameters have any advantage in this respect. How would one speculate about the changes in grid element effective parameter values to be expected as a result of field underdrainage, or different land use in a changed climate, or the mechanical operations associated with deforestation or afforestation. The potential for inaccurate prediction is clearly great for both approaches and, at the very least, these predictions should be associated with an estimate of predictive uncertainty.

5. Can uncertainty in the predictions of distributed models be assessed?

The answer to this question is, given enough computer time, yes; but there are two basically different approaches to the problem of assessing predictive uncertainty. The more traditional approach assumes that the uncertainty derives from uncertainty in parameter values (and sometimes boundary conditions) around some "true" parameter values. This is the approach used in classical statistical estimation theory. Maximum likelihood or some other parameter estimation technique is used to find optimal and preferably unbiased estimates of the parameters, , given some sample of data, **Y**. The parameters are, however, considered to be stochastic variables and by assuming that they are normally distributed then, at least for linear models, the uncertainty in the predictions can be estimated by integrating over the expected parameter variability or, in calibration, as a function of the Jacobian of the likelihood matrix. Extension to nonlinear models is also possible and example applications of this type for distributed hydrological models are given by Jensen and Mantoglou (1992), Destouni et al. (1994) and Zhang et al. (1994).

However, as pointed out above, distributed hydrological models are neither linear nor do they have an optimal set of parameter values. If the concept of equifinality of parameter sets (and models) is accepted then some other approach to estimating model uncertainty must be explored. This problem is being recognised in a number of fields (e.g. O'Neill et al., 1982; Freeze et al., 1990; van Straten and Keesman, 1991; Patwardhan and Small, 1992; Dilks et al., 1992; Klepper and Hendrix, 1994; Brooks et al., 1994), even amongst a sample of statisticians (e.g. Draper, 1995). A framework for the estimation of uncertainty in such a situation has been provided in the Generalised Likelihood Uncertainty Estimation (GLUE) methodology outlined in Beven and Binley (1992). GLUE is based on a Monte Carlo simulation, using randomly chosen sets of parameter values. It extends the type of Generalised Sensitivity Analysis (GSA) of Hornberger and Spear (1991) by evaluating the simulation results for each parameter set against some observed data by calculating a likelihood value. Those parameter sets which give the best fit to the data have the highest likelihood values. Those that are considered as "nonbehavioural" are given a likelihood of zero.

One parameter set will give the highest likelihood value. This would be the optimal set found by a more traditional search algorithm (or at least close to it). There will be many other parameter sets, however, that have likelihood values nearly as high, perhaps from different parts of the parameter space. It is also almost certain that given another period of observed data the ranking of these good fitting models will change. The best fitting parameters for the first period will not have the highest likelihood for the second period. A further period might yield another parameter set with the highest likelihood. This should be expected given the errors in the model conceptualisation and the errors in both boundary condition data and observed data. Essentially the distribution of likelihood values will depend on the calibration period.

This is not a problem within the GLUE methodology. The idea is to use the distribution of likelihood values as a weighting function to condition the range of predicted variables from all the "behavioural" parameter sets to determine uncertainty quantiles. Thus, the predictions of the model for a parameter set with the highest likelihood are given the greatest weight in assessing the distribution of predictions. It is then a relatively simple matter to update the likelihood distribution as more observed data is available using Bayes equation. Since each parameter set is treated only as a set , Bayes equation can be applied independently to the associated likelihood values in the form

$$L_p(_i|Y) = L_o(_i) \, L_y(_i|Y_j)$$

where $L_o(_i)$ is the prior likelihood associated with parameter set $_i$; $L_y(_i|Y_j)$ is the likelihood calculated for the simulations using parameter set $_i$ given the observed data

Y_j of period j and $L_p(_i|Y)$ is the posterior likelihood value.

Many different definitions of the likelihood measure are possible, including traditional goodness of fit indices used in hydrology (see Beven and Binley, 1992; Beven, 1993; Freer et al., 1996). Functions that take more explicit account of a model for the residuals from each simulation are also possible (e.g. Romanowicz et al., 1994) as are fuzzy likelihood or possibility measures (van Straten and Keesman, 1991). Likelihood measures for different types of predicted variables can also be combined (Binley and Beven, 1991).

A number of applications of GLUE and similar Monte Carlo based estimation of uncertainty for distributed models have been published. Binley et al. (1991) used multiple realisations (but without conditioning against observed data) to evaluate the uncertainty of predictions of the effects of land use change for the Institute of Hydrology Distributed Model (IHDM). A particularly pertinent study, Binley and Beven (1991) showed how a finite element hillslope segment model, representing a catchment as two-dimensional vertical slices of variable width, provided highly uncertain predictions of internal water tables depths, even after conditioning on both discharge and observed water table levels. The "observed" data for this study came from a model: a fully three-dimensional solution of the same descriptive equations (Richards' equation, without hysteresis), with the same finite element solution algorithm, but with heterogeneous fields of soil properties and depths.

The GLUE methodology has also been applied to the quasi-distributed model, TOPMODEL (Beven et al., 1995) by Beven (1993), Romanowicz et al. (1994) and Freer et al. (1996). The latter paper explores the effect of using different likelihood measures on the resultant uncertainty estimates for predicted discharges. In all the cases, however, no internal state measurements have been used in conditioning the predictions. This is now being explored within the Lancaster TOPMODEL group.

There are clearly constraints on the type of model to which this procedure can be applied, merely because of the computational demands of Monte Carlo simulation for distributed models. It is, however, readily implemented on parallel computer systems and is a very flexible and readily understandable procedure that should be developed further in the future. For example, it would be quite possible to extend the methodology to consider competing models as well as competing parameter sets, and to design likelihood measures (and measurements) specifically to discriminate between models or parameter sets (see next section).

6. The future of distributed modelling: on the value of data

To summarise briefly the conclusions of the arguments above, I suggest that the current approach to "physically based" distributed modelling, in which the equations of small scale physics are used at larger scales with an assumption that the change of scales can be accommodated by the use of "effective" parameter values, must be replaced by an approach that recognises much more explicitly the limitations of the modelling process. One way of doing this is to work within an uncertainty framework, but one which like the GLUE methodology allows for the inherent equifinality in model matches to the limited observations available. This implies that many models must be considered and consequently that the computer power (and concomitant disk storage for the results) required will be significantly greater.

The computer power available to the modeller has, of course, increased dramatically since Freeze and Harlan outlined the basis for a physically based approach to modelling in 1969. The recent availability of cheap parallel systems will allow further enhancements of workstation power, both in the size of model that can be run and the number of runs that can be made easily. There are choices to be made here: such power can be used to do the same type of distributed modelling as before but with finer and finer grid scales and longer and longer runs for larger and larger catchments. This is an obvious way to bring the scale of the model elements more into line with the scale for which the theory was developed. However, it must be recognised that this increases the number of parameters which must be specified without increasing the data available for their calibration and validation: these do not depend on computer power. The problem of overparameterisation is consequently greater.

Is there an alternative? Clearly increased computer power can be used to make more runs of a simpler conceptual model to evaluate model uncertainty, rather than longer and more detailed runs of a single model. If it is agreed that all distributed models, even the most "physically-based", are lumped conceptual models, there is an implication that there may be other choices of conceptualisation that are parametrically more parsimonious and yet still retain an acceptable description of reality in the sense of some specified model validation criterion pertinent to the problem at hand. Such models could be made explicitly scale dependent, reflecting directly the changing controls on hydrological responses with changing scale and the changing data available for conditioning of the model parameters.

In essence, such conceptualisations would aim to reflect the hydrological functionality of different parts of the landscape or catchment (Beven, 1995). "Physically-based" models aim to do this in a structured way, but there may be other more efficient ways of doing so if measurements or indices of differences in spatial hydrological behaviour could be made observable. To some extent this is the case already with measures derived from remote sensing (e.g. Bastiaanssen et al., 1994; Moran et al., 1994; Goodrich et al., 1994). In this sense, the development of such conceptualisations is dependent on inference from data rather than theory (noting that this is also true for the theory we have available now except that the inferences were made at small scales (and long ago) and even then are only approximations to the data, see Davis et al., 1992, for a discussion of Darcy's law in this context).

The recognition of model equifinality and the evaluation of models within the framework of predictive uncertainty allows for a different approach to the relationship between model conceptualisations, parameters and data. If, as in the GLUE procedure, the criterion of performance enters directly into the assessment of model uncertainty, there is the possibility of inverting the normal process of model calibration and validation and, rather than regarding model evaluation as a process of confirmation, pose the problem as one of falsification. Data is then used to reject "non-behavioural" models.

Consider the following hypothetical future scenario. The availability of massively parallel workstations means that computer time is essentially unlimited. Storage of results is not too great a problem since runs can be regenerated if necessary. It is proposed to model an ungauged catchment. A GIS database of soils and land use types is available, together with some radar rainfall, raingauge and meteorological data, and sample Landsat, SPOT and spaceborne radar images. A priori, all distributed hydrological models, and all parameter sets within each model conceptualisation, are potential simulators of the catchment. For quality assurance reasons, it is no longer considered acceptable by the Bureau of International Hydrological Standards to give one's own model a prior likelihood of one, and all other models a prior likelihood of zero. Prior judgement and professional expertise can be used to restrict the model and parameter search space, but such a priori rejection of possibly acceptable models must be carefully justified and documented. Numerous runs of the models are used with different parameter sets and compared with the evaluation data available according to predefined performance criteria or likelihood measures.

The most important measures are then those that allow model rejection as "non-behavioural" (there are analogies here with "evolutionary" or "genetic" approaches to model development being studied in a variety of fields, including hydroinformatics). The remaining models are retained as "behavioural" in some sense, although some may be more behavioural than others in having higher likelihood values. Further observations may lead to further rejections and a "smaller" behavioural set. The value of data lies then in its ability to discriminate between behavioural and non-behavioural models. This conditioning of the range of model possibilities is, of course, a form of model calibration but with two important advantages. One is that it recognises the impossibility of ultimately differentiating between many "behavioural" possibilities compatible with the available data (but which will all give different predictions), and secondly it focuses attention on the data and allows data to be used in a hypothesis testing framework. Given a range of behavioural models, what data should be collected in order to most effectively reduce that range by showing that some models or parameter sets are, in fact, non-behavioural.

Returning to the data available for this ungauged catchment, there is in fact very little data to provide such a test. Soil and vegetation types can be used to specify ranges of parameters appropriately. Hopefully the cataloguing of parameters such as the profile of downslope transmissivity and plant parameters controlling responses to water stress will have improved in the future, but such ranges may still be relatively imprecise. Rainfall and meteorological data are used as inputs. Active or passive microwave data may be able to provide an index of surface soil moisture, but only a relatively fuzzy measure of hydrological behaviour. Clearly, in this case discrimination between models will be poor and the range of predictions of future behaviour might be high. Thus, if the project justifies some expenditure, what measurements should be taken to allow the evaluation of the possible models? What data will have the most discriminatory power in the rejection of models that are really non-behavioural?

The answer to these questions is not that clear. It is certainly not simply a question of defining a programme of parameter measurements (especially if these are still small scale measurements) for the purposes of model "calibration". The answer must depend in part on the model conceptualisations and on the modelled hydrological responses and on the purpose of the modelling. One set of models may predict that there are areas of saturated soil acting as contributing and return flow areas in the catchment. This is easily and relatively cheaply checked. It is also relatively cheap, and in discriminating between models probably very effective, to install one or more stream level gauges and carry out a number of check gaugings over one or more hydrographs (perhaps, in the future, using ultrasonic tomography of the velocity field in a natural cross-section).

How well would the current generation of distributed hydrological models stand up in this procedure of analysis, relative to other, parametrically simpler, conceptualisations. There would probably be equivalent "behavioural" simulations in both, unless the data collected and tests used were too discriminating; and therein lies a possible difficulty in following this scientifically respectable hypothesis testing approach. It allows that if the criteria of "behavioural" performance are too severe all the models and parameter sets might be rejected as non-behavioural. In fact, at the present time the tests would not have to be too severe for this to be the case. Hydrological models are not alone in this. All the current generation of global circulation models would have to be rejected if a criterion of accuracy of regional scale predictions of current climate was used (and therefore their predictions of future global warming). Most environmental, geochemical and ecological models are vulnerable in this respect (see for example, Beck, 1987; van Straten and Keesman, 1991). Of course, future developments (and bigger and better computers) will improve the accuracy of such models as process understanding feeds in to improved parameterisations.....or is it the understanding of the modelling process itself that needs to be improved?

Model rejection should, however, not necessarily be seen as a problem within the framework suggested. If predictions are required now, then the criteria for rejection must be relaxed to allow some models to be considered behavioural *within the limitations of current understanding*. The range of predictions should be accordingly wide to allow for the limitations of current understanding. The reasons for rejection can also be examined with a view to providing insights that can be used in the development of new conceptualisations, leading to an appropriately symbiotic relationship between modelling and data collection. Indeed, the intriguing possibility of using computer power to *generate* new parameterisations on the basis of learning algorithms evaluating model performance measures is already being investigated (Babovic and Minns, 1994.). Such developments, however, keep the emphasis firmly on the data, and appropriate data gathering techniques, rather than on *a priori* theorising.

In fact, I have argued before (Beven, 1995) that hydrological science and the development of hydrological theory is awaiting the development of new measurement techniques, especially a method for the rapid assessment of a measure of the heterogeneity of hydrological responses in space, somewhat better than those available from current remote sensing capability. The definition of appropriate hypotheses to be tested might be one mechanism for encouraging the development of such techniques.

The approach being suggested here stems directly from the notion of equifinality or undecidability between model descriptions. It appears to sit somewhat uneasily between a number of philosophical frameworks for a scientific approach to modelling and theory generation. Most hydrologists will agree that model or theory confirmation is a matter of degree of empirical adequacy (van Fraasen, 1980; Oreskes et al., 1994). It appears necessary to accept, however, that there are many different descriptions that are "adequate" in some sense, where adequacy itself may be limited or conditional, requiring further tuning or modification of ancillary conditions as more information becomes available. The idea of ranking the available models in terms of some likelihood measure would appear to result in a purely relativist attitude to the problem of modelling hydrological systems, in keeping with the views of Feyerabend (1975) on the development of scientific thought (see Beven, 1987 for a discussion in relation to hydrology) where there is no requirement for a necessary or strong correspondence between theory or reality. This is, however, unacceptable to many hydrological scientists.

Alternatively, the rejection of "non-behavioural" models is also in the tradition of both logical empiricist and critical rationalist philosophical stances. Neither have traditionally, however, allowed for the possibility of multiple descriptions that remain "acceptable" even after a process of rejection based on goodness-of-fit, hypothesis testing or the rational appraisal of the underlying theory. The difficulties associated with rejection or falsification have been well aired in the philosophical literature (see for example Bhaskar, 1989) but it would appear to provide a useful way to proceed faced with the inherent difficulties of hydrological modelling. The difficulties, as outlined above, suggest that a purely critical rationalist approach to distributed modelling is untenable, since it would lead to the rejection of all current models. Even a logical empiricist approach will require some fairly relaxed criteria of acceptability, particularly in respect of predictions of internal states of the system, to allow that some models can be retained as "acceptable".

If, however, acceptability is to be assessed in terms of some quantitative measure of performance or "likelihood" as in the Bayesian GLUE procedure described briefly above, it is clear that acceptability is necessarily a relative measure. Figure 1 shows a number of plots of likelihood value versus parameter value for an application of TOPMODEL to the small Ringelbach catchment in the Vosges, France. Each point on these plots represents the results of one run of the model for different randomly selected parameter sets, each parameter being chosen from a uniform distribution within the a priori chosen ranges shown. These plots are similar to many others calculated at Lancaster for different types of models in different applications. It is common to find that any particular parameter value can be associated with both good and bad likelihood values (depending on the values of the other model parameters); that high likelihood values may be found across a very wide range for each parameter; and that there is no clear

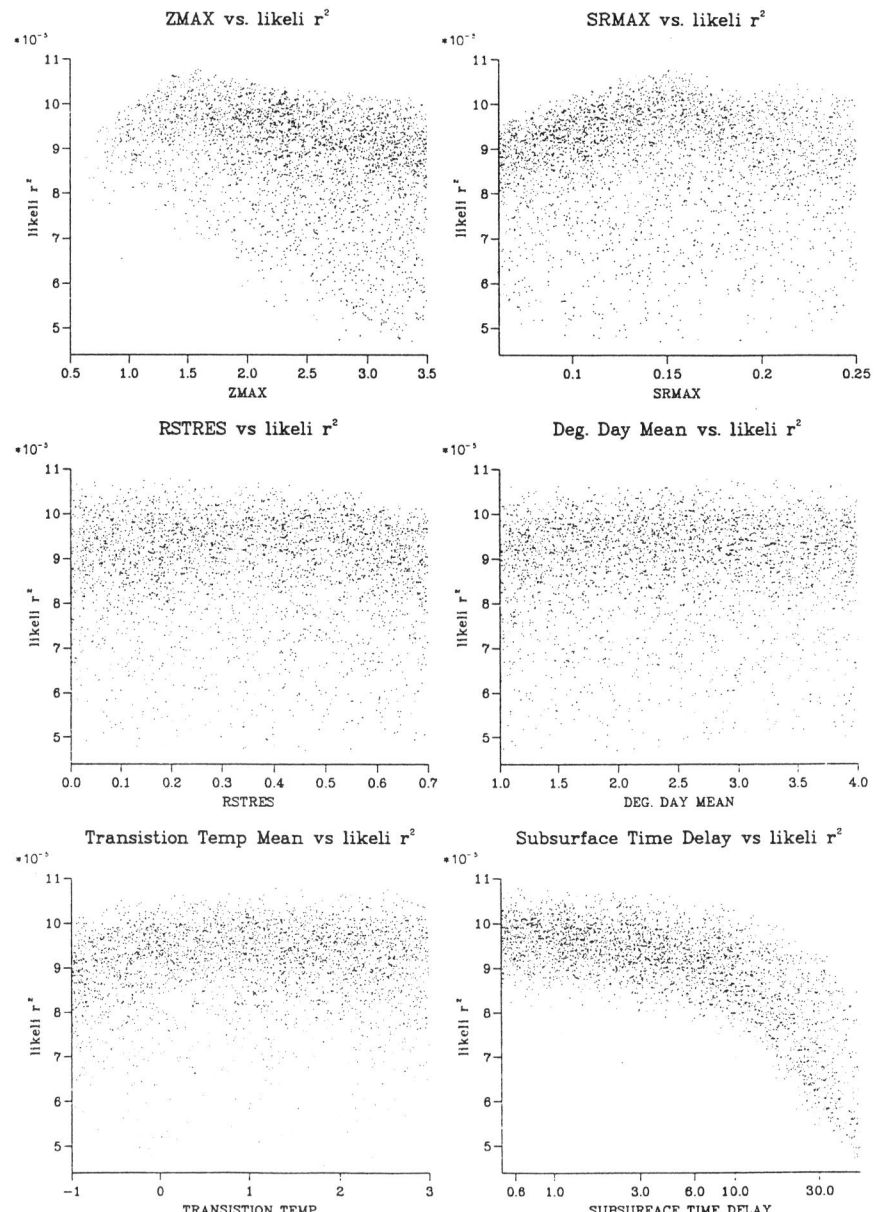

Figure 1. Results from Monte Carlo experiments using a version of TOPMODEL plotted as a goodness of fit or likelihood index (in this case based on a coefficient of determination for a one year simulation period) against parameter value for 6 different parameters in the model. Each point represents a run with different randomly chosen parameter values. Each parameter was sampled uniformly across the range shown (after Freer et al., 1996).

demarcation between good simulations and poor simulations. The choice of a point of acceptability is therefore a subjective one and acceptability, however assessed, will be a relative measure

There is an argument that if it is not possible to distinguish between models or theories on the basis of their predictions then *they are simply poor models*. It should then be the aim of the hydrological scientist to develop processes of observation and hypothesis testing that would allow competing models or hypotheses or parameter sets to be accepted or rejected (see, for example, Haines-Young and Petch, 1983). Such a view is not incompatible with the approach described above in which rejection is a fundamental process in constraining predictive uncertainty and improving model structure. I believe, however, that equifinality will prove to be inherent in modelling environmental systems as a result of the unknowability and uncertainty in defining the system of interest and that the hydrologist will have to find ways of coping with it.

7. Conclusions

There is a continuing need for distributed predictions in hydrology but a primary theme of the analysis presented here is that distributed modelling should be approached with some circumspection. It has been shown that the process descriptions used in current models may not be appropriate; that the use of effective grid scale parameter values may not always be acceptable; that the appropriate effective parameter values may vary with grid scale; that techniques for parameter estimation are often at inappropriate scales; and that there is sufficient uncertainty in model structure and spatial discretisation in practical applications that these models are very difficult (if not impossible) to validate. As a result of these problems, the modeller should expect that there will be multiple acceptable models to represent the catchment system of interest, possibly differing markedly in their model structures or parameter values, but all reproducing any observations of the catchment behaviour to some acceptable level. This *equifinality* of models will result in uncertainty in the model predictions that can be assessed by techniques such as the GLUE procedure.

Distinguishing between these multiple models requires data that will allow some models to be rejected as unacceptable. Some data gathering techniques, reflecting the large scale distributed responses of the system may be of greater value in this respect than many measurements at scales much smaller than the model grid elements.

Finally, a speculative suggestion has been made that the future development of distributed models lies more in developing sub-grid scale parameterisations based directly on large scale measurements than on the improvement of the aggregation of small scale theory and parameter values that underlies all of today's distributed models. Progress, in this scenario, will be towards simpler, more robust, and more easily calibrated representations of distributed hydrology rather than introducing ever more complexity and ever more parameters to be defined. This analysis then leads to the conclusion that significant progress in distributed modelling will be dependent on the development of new measurement techniques that properly reflect the variability in responses at the scale of interest, leading to new theoretical formulations.

Acknowledgements

The results summarised in Figure 1 were calculated by Jim Freer and were based on data provided for the Ringelbach Catchment by Bruno Ambroise.

References

Abbott, M B, Bathurst, J C, Cunge, J A, O'Connell, P E and Rasmussen, J, 1986, An introduction to the European Hydrological System-Système Hydrologique Européen (SHE), *J. Hydrol.* 87, 45-59.

Anderson, M P and Woessner, W W, 1992, The role of the post-audit in model validation, *Adv. Wat. Resour.*, 15, 167-173.

Avissar, R, 1992, Conceptual aspects of a statistical-dynamic approach to represent landscape sub-grid scale heterogeneities in atmospeheric models, *J. Geophys. Res.*, **97**(D3), 2729-2742.

Babovic, V and Minns, A W, 1994, Use of computational adaptive methodologies in hydroinformatics, in V Verwey, A W Minns, V Babovic and C Maksimovic (Eds.), *Proc. 1st International Conference on Hydroinformatics*, Blakema, Rotterdam, 201-210.

Bastiaanssen, W G M, Hoekman, D H and Roebling, R A, 1994, A methodology for the assessment of surface resistance and soil water storage variability at mesoscale based on remote sensing measurements, *IAHS Special Publication* 2, IAHS Press, Wallingford, UK. 66pp

Bair, E S, 1994, Model (in)validation - a view from the courtroom, *Ground Water*, 32(4), 530-531.

Barling, R D, Moore, I D and Grayson, R B, 1994, A quasi-dynamic wetness index for characterising the spatial distribution of zones of surface saturation and soil water content, *Wat. Resour. Res.*, 30, 1029-1044.

Bathurst, J C, 1986, Physically-based distributed modelling of an upland catchment using the Système Hydrologique Européen, *J. Hydrol.*, 87, 79-102.

Bathurst, J C and O'Connell, P E, 1992, The future of distributed modelling: the Système Hydrologique Européen, *Hydrol. Process.*, 6, 265-277.

Bathurst, J C, Wicks, J M and O'Connell, P E, The SHE/SHESED basin scale water flow and sediment transport modelling system, in V P Singh (Ed.), *Computer Models of Watershed Hydrology*, Water Resource Publications, Colorado, 563-594.

Beck, M B, 1987, Water quality modelling: a review of the analysis of uncertainty, *Wat. Resour. Res.*, 23, 1393-1442.

Beck, M B, Mulkey, L A, Barnwell, T O, and Ravetz, J R, 1995, Model validation for predictive exposure assessments, preprint.

Bencala, K E and Walters, R, 1983, Simulation of solute transport in a mountain pool and riffle stream: a transient storage model, *Wat. Resour. Res.*, 19, 732-738.

Beven, K J, 1985, Distributed Models, in M G Anderson and T P Burt (Eds.), *Hydrological Forecasting*, Wiley, Chichester.

Beven, K J, 1987, Towards a new paradigm in hydrology, in *Water for the Future: Hydrology in Perspective*, IASH Pubn. No. 164, 393-403.

Beven, K J, 1989a, Changing ideas in hydrology: the case of physically-based models. *J. Hydrol.*, 105, 157-172

Beven, K J, 1989b, Interflow, in H J Morel-Seytoux (Ed.), *Unsaturated Flow in Hydrologic Modelling*, Proc. NATO ARW, Arles, France, Reidel, Dordrecht, 191-219.

Beven, K J, 1993, Prophecy, reality and uncertainty in distributed hydrological modelling, *Advances in Water Resources*, 16, 41-51.

Beven, K J, 1995, Linking parameters across scales: sugrid parameterisations and scale dependent hydrological models, *Hydrol. Process.*, 9, 507-525.

Beven, K J, and Binley, A M, 1992, The future of distributed models: model calibration and uncertainty prediction, *Hydrol. Process.* 6, 279-298.

Beven, K J and Clarke, R T, 1986, On the variation of infiltration into a homogeneous soil matrix containing a population of macropores, *Wat. Resour. Res.*, 22, 383-388.

Beven, K J, Calver, A and Morris, E M, 1987, The Institute of Hydrology Distributed Model, Institute of Hydrology Report 98, Wallingford, UK.

Beven K, J and Germann, P F, 1981, Water flow in soil macropores. II. A combined flow model. *J. Soil Sci.*, 32, 15-29.

Beven, K J and Kirkby, M J, 1979, A physically based variable contributing area model of catchment hydrology, *Hydrol. Sci. Bull.*, 24, 43-69.

Beven, K J, Lamb, R, Quinn, P, Romanowicz, R and Freer, J, 1995, TOPMODEL,in V P Singh (Ed.), *Computer Models of Watershed Hydrology*, Water Resource Publications, Colorado, 627-668.

Bhaskar, R, 1989, *Reclaiming Reality*, Verso, London

Binley, A M, Beven, K J and Elgy, J, 1989, A physically based model of heterogeneous hillslopes. 2. Effective hydraulic conductivities, *Wat. Resour. Res.*, 25(6), 1227-1233.

Binley, A M and Beven, K J, 1991, Physically-based modelling of catchment hydrology: a likelihood approach to reducing predictive uncertainty, in D G Farmer and M J Rycroft (Eds) *Computer Modelling in the Environmental Sciences*, IMA Conference Series, Clarendon Press, Oxford, 75-88.

Binley, A M, Beven, K J, Calver, A and Watts, L G, 1991, Changing responses in hydrology: assessing the uncertainty in physically-based model predictions, *Wat. Resour. Res.*, 27, 1253-1261.

Blyth, E M, Dolman, A J and Wood, N, 1993, Effective resistance of sensible and latent heat flux in heterogenous terrain, *Q. J. Roy. Meteorol. Soc.*, 119, 423-442.

Booltink, H W G, 1994, Field scale distributed modelling of bypass flow in a heavily textured clay soil, *J. Hydrol.*, 163, 65-84.

Brooks, R J, Lerner, D N and Tobias, A M, 1994, Determining the range of predictions of a groundwater model which arises from alternative calibrations, *Wat. Resour. Res.*, 30, 2993-3000.

Bruneau, P, Gascuel-Odoux, C, Robin, P, Merot, Ph, and Beven, K J, 1995, The sensitivity to space and time scales of a hydrological model using digital elevation data, *Hydrol.Process.*, 9, 69-82.

Calver, A, 1988, Calibration, sensitivity and validation of a physically-based rainfall-runoff model, *J. Hydrology*, 103, 103-115.

Calver, A and Wood, W L, 1995, The Institute of Hydrology Distributed Model, in V P Singh (Ed.), *Computer Models of Watershed Hydrology*, Water Resource Publications, Colorado, 595-626.

Cartwright, N, 1983, *How the Laws of Physics Lie*, Clarendon Press, Oxford.

Costa-Cabral, M C and S J Burges, 1994, Digital elevation model networks (DEMON): a model of flow over hillslopes for computation of contributing and dispersal areas, *Wat. Resour. Res.*, 30, 1681-1692.

Dagan, G, 1986, Statistical theory of groundwater flow and transport: pore to laboratory, laboratory to formation and formation to regional scale, *Wat. Resour. Res.*, 22(9), 120s-134s.

Davis, P A, Olague, N E abd Goodrich, M T, 1992, Application of a validation strategy to qDarcy's experiment, *Adv. Wat. Resour.*, 15, 175-180.

Destouni, G, Sassner, M and Jensen, K H, 1994, Chloride migration in heterogeneous soil. 2. Stochastic modelling, *Wat. Resour. Res.*, 30(3), 747-758.

Dilks, D W, Canale, R P and Meier, P G, 1992, Development of Bayesian Monte Carlo techniques for water quality model uncertainty, *Ecological Modelling*, 62, 149-162.

Draper, D, 1995, Assessment and propogation of model uncertainty, *J. Roy. Statist. Soc.* B37, 45-98.

Durand, P, Robson, A and Neal, C, 1992, Modelling the hydrology of submediterranean montain catchments (Mont Lozère, France) using TOPMODEL: initial results. *J. Hydrol.*, 139, 114.

Dunne, T, Zhang, W and Aubrey, B F, 1991, Effects of rainfall, vegetation and microtopography on infiltration and runoff, *Wat. Resour. Res.*, 27(9), 2271-2286.

Feyerabend, P K, 1975, *Against Method: Outline of an Anarchist Theory of Knowledge*, New Left Books, London.
Flury, M, Flühler, H, Jury, W A and Leuenberger, J, 1994, Susceptibility of soils to preferential flow of water: a field study, *Wat. Resour. Res.*, 30, 1945-1954.
Freer, J, Beven K J and Ambroise, B, 1996, Bayesian estimation of uncertainty in runoff prediction and the value of data: an application of the GLUE approach, *Water Resour. Res.*, in press.
Freeze, R A, and Harlan, R L, 1969, Blueprint for a physically-based digitally simulated hydrologic response model, *J. Hydrol.*, 9, 237-258.
Freeze, R A, Massmann, J, Snith, L, Sperling, T and James, B, 1990, Hydrogeological decision analysis. 1. A framework. *Ground Water*, 28, 738-766.
Goodrich, D C, Schmugge, T J, Jackson, T J, Unkrich, C L, Keefer, T O, Parry, R, Bach, L B and Amer, S A, 1994, Runoff simulation sensitivity to remotely sensed initial soil water content, *Wat. Resour. Res.*, 30, 1393-1405.
Grayson, R B, Moore, I D and McMahon, T A, 1992, Physically-based hydrologic modeling. 2. Is the concept realistic?, *Wat. Resour. Res.*, 28(10), 2659-2666.
Green, H M, Beven, K J, Buckley, K and PC Young, Pollution incident prediction with uncertainty, in K J Beven, P C Chatwin and J Millbank (Eds), *Mixing and Transport in the Environment*, Wiley, Chichester, 113-137.
Haines-Young, R H and J R Petch, 1983, Multiple working hypotheses: equifinality and the study of landforms, *Trans. Instn. Brit. Geogr.* 8, 458-466.
Hart, D R, 1995, Parameter estimation and stochastic interpretation of the transient storage model for solute transport in streams, *Wat. Res. Res.*, 31, 323-328.
Hjelmfelt, A T and Burwell, R E, 1984, Spatial variability of runoff, *J. Irrig. Drain. Eng. ASCE*, 110(1), 46-54, 1984.
Hofmann, J R and Hofmann, P A, 1992, Darcy's law and structural explanation in hydrology, in D Hull, M Forbes and K Okruhlik (Eds) *PSA 92*, 1, 23-35, Philosophy of Science Association, East Lansing, Mich.
Hoogmoed, W D and Bouma, J, 1980, A simulation model for predicting infiltration into cracked soil, *Soil Sci. Soc. Amer. J.*, 44, 458-461.
Hornberger, G M and Spear, R, 1981, An approach to the preliminary analysis of environmental systems, *J. Environ. Manag.*,12, 7-18.
Hornberger, G M, P F Germann and K J Beven, 1991, Throughflow and solute transport in an isolated sloping soil block in a forested catchment, *J. Hydrol.* 124, 81-99.
Jain, S K, Storm, B, Bathurst, J C, Refsgaard, J-C and Singh, R D, 1992, Application of the SHE to catchments in India. 2. Field experiments and simulation studies on the Kolar subcatchment of the Narmada river, *J. Hydrology*, 140, 25-47.
Jakeman, A J and Hornberger, G M, 1993, How much complexity is warranted in a rainfall-runoff model, *Wat. Resour. Res.*, 29, 2637-2650.
Jarvis, N J Jansson, P-E, Dik, P E and Messing, I, 1991, Modelling water and solute transport in macroporous soil. 1. Model description and sensitivity analysis. *J. Soil Sci.*, 42,59-70.
Jensen, K H and Mantoglou, A, 1992, Application of stochastic unsaturated flow theory: numerical simulations and comparisons to field observations, *Wat. Resour. Res.*, 28(1), 269-284.
Jensen, K H, Bitsch, K and P L Bjerg, 1993, Large scale dispersion experiments in a sandy aquifer in Denmark: observed tracer movements and numerical analysis. *Wat. Resour.Res.*, 29, 673-696.
Jolley, T and Wheater, H S, 1993, Macromodelling of the River Severn, in W B Wilkinson (Ed.), *Macroscale Modelling of the Hydrosphere*, IAHS Publn. No. 214, 91-100.
Kirkby, M J, 1975, Hydrograph modelling strategies, in R Peel, M Chisholm and P Haggett (Eds), *Process in Physical and Human Geography*, Heinemann, 69-90.
Kirkby, M J (Ed.),1978, *Hillslope Hydrology,* Wiley, Chichester.
Klepper, O and Hendrix, E M T, 1994, A method for robust calibration of ecological models under

different types of uncertainty, *Ecol. Modell.* 74, 161-182.

Konikow, L F and Bredehoeft, J D, 1992, Groundwater models cannot be validated, *Adv. Wat. Resour.*, 15, 75-83.

Koster, R D and Suarez, M J, 1992, Modelling the land surface boundary in climate models as a composite of independent vegetation stands, *J. Geophys. Res.*, 97(D3), 2697-2715.

Luis, S J and McLaughlin, D B, 1992,A stochastic approach to model validation, *Advances in Wat. Resources*, 15, 15-32.

Luxmoore, R J, Grizzard, T and Patterson, M R, 1981, Hydraulic poperties of Fullerton Cherty Silt Loam, *Soil Sci. Soc. Amer. J.*, 45, 692-698.

Mackay, R, 1995, A study of the effect of the extent of site investigation on the estimation of radiological performance, *Report Doe/HMIP/RR/93.053*, Her Majesty's Inspectorate of Pollution, London.

Mantoglou, A and L W Gelhar, 1987, Capilary head variance, mean soil moisture content and effective specific soil moisture capacity of transient unsaturated flow in stratified soils, *Wat. Resour. Res.*, 23, 47-56

Moore, I D and Grayson, R B, 1991, Terrain based prediction of runoff with vector elevation data, *Wat. Resour. Res.*, 27, 1177-1191.

Moran, M S, Clarke, T R, Kustas, W P, Weltz, M, and Amer, S A, 1994, Evaluation of hydrologic parameters in a semiarid reangeland using remotely sensed spectral data, *Wat. Resour. Res.,* 30, 1287-1297.

Morton, A, 1993, Mathematical models: questions of trustworthiness, *Brit. J. Phil. Sci.*, 44, 659-674.

Mualem, Y, 1976, Hysteretical models for prediction of the hydraulic conductivity of unsaturated porous media, *Wat. Resour. Res.,* 13, 657-664.

Nicholson, T J, P J Weirenga, G W Gee, E A Jacobson, D J Polmann, D B McLaughlin and L W Gelhar, 1989, Validation of stochastic flow and transport models for unsaturated soils: field study and preliminary results, in B E Buxton (Ed.) *Geostatistical, Sensitivity and Uncertainty Methods for Groundwater Flow and Radionuclide Transport Modelling*, Batelle Press, 1989.

O'Neill, R V, Gardner, R H and Carney, J H, 1982, Parameter constraints in a stream ecosystem model: incorporation of a priori information in Monte Carlo error analysis, *Ecol. Modell.* 16, 51-65.

Oreskes, N, Schrader-Frechette, K, Belitz, K, 1994, Verification, validation and confirmation of numerical models in the Earth Sciences, *Science*, 263, 641-646.

Patwardhan, A and M J Small, 1992, Bayesian methods for model uncertainty analysis with application to future sea level rise, *Risk Analysis*, 12, 513-523.

Philip, J R, 1978, Some remarks on science and catchment prediction in T G Chapman (Ed.), Prediction in Catchment Hydrology,

Quinn, P F, Beven, K J and Culf, A. 1994, The introduction of macroscale hydrological complexity into land surface-atmosphere transfer function models and the effect on planetary boundary layer development, *J. Hydrology*, 166, 421-444

Quinn, P F, Beven, K J and Lamb, R, 1995, The ln(a/tan) index: how to calculate it and how to use it in the TOPMODEL framework, *Hydrol. Process.*, 9, 161-182

Refsgaard, J-C and Storm, B, 1995, MIKE SHE, in V P Singh (Ed.), *Computer Models of Watershed Hydrology*, Water Resource Publications, Colorado, 809-846.

Romanowicz, R, Beven, K J and Tawn, J, 1994, Evaluation of predictive uncertainty in nonlinear hydrological models using a Bayesian approach, in V Barnett and K F Turkman (Eds.), *Statistics for the Environment 2: Water Related Issues* Wiley, Chichester, 297-317.

Schrader-Frechette, K S, 1989, Idealised laws, antirealism and applied science: a case in hydrogeology, *Synthese*, 81, 329-352.

Schulin, R, van Genuchten, M Th, Fluhler, H and Ferlin, P, 1987, An experimental study of solute transport in a stony field soil, *Wat. Resour. Res.*, 23, 1785-1794.

Smith, R E, Goodrich, D R, Woolhiser, D A and Simanton, J R, 1994, Comment on "Physically-based hydrologic modeling. 2. Is the concept realistic?" by R B Grayson, I D Moore and T A McMahon,

Wat. Resour. Res., 30, 851-854.

Steenhuis, T S, Parlange, J-Y and Andreini, M S, 1990, A numerical model for preferential solute movement in structured soils, *Geoderma*, 46, 193-208.

Stephenson, G R and Freeze, R A, 1974, Mathematical simulation of subsurface flow contributions to snowmelt runoff, Reynolds Creek watershed, Idaho, *Wat. Resour. Res.*, 10, 284-298.

van Fraasen, B C, 1980, *The Scientific Image,* Clarendon, Oxford, 235pp.

van Straten, G and Keesman, K J, 1991, Uncertainty propagation and speculation in projective forecasts and environmental change: a lake-eutrophication example, *J. Forecast.*, 10, 163-190.

Wallis, S G, Young, P C and Beven, K J, 1989, Experimental investigation of the aggregated dead zone model for longitudinal solute transport in stream channels, *Proc. Inst. Civ. Engr.* Part 2, 87, 1-22.

Weinberg, A M, 1972, Science and trans-science, *Minerva*, 10, 209-222.

Wheatcraft, S W, Sharp, G A and Tyler, S W, 1990, Fluid flow and solute transport in fractal heterogeneous porous media, in *Dynamics of Fluids in Hierarchical Porous Media*, J H Cushman (Ed.), Academic Press, 305-326.

Wolock, D M and Price, C V, 1994, Effects of digital elevation model map scale and data resolution on a topography-based watershed model, *Wat. Resour. Res.*, 30(11), 3041-3052.

Yeh, T-C J, Gelhar, L W and Gutjahr, A L, 1985, Stochastic analysis of unsaturated flow in heterogeneous soils. 3. Observations and applications, *Wat. Resour. Res.*, 21, 465-471.

Young, P C and Beven, K J, 1994, Data-based mechanistic modelling and the rainfall-flow nonlinearity, *Econometrics*, 5, 335-363.

Young, P C, and Wallis, S G, 1993, Solute transport and dispersion in stream channels, in K J Beven and M J Kirkby (Eds.) *Channel Network Hydrology*, Wiley, Chichester, 129-174.

Zhang, H, Haan, C T and Nofziger, D L, 1993, An approach to estimating uncertainties in modelling transport of solutes through soils, *J. Contam. Hydrol.*, 12, 35-50.

CHAPTER 13B
COMMENT ON 'A DISCUSSION OF DISTRIBUTED HYDROLOGICAL MODELLING' BY K. BEVEN

J.C. REFSGAARD[1], B. STORM[1] AND M.B. ABBOTT[2]
[1] *Danish Hydraulic Institute, Hørsholm, Denmark*
[2] *International Institute for Infrastructure, Hydraulic and Environmental Engineering, Delft, The Netherlands*

1. Introduction: Terminology and Content of Comment

Before, commenting on Keith Beven's questions we should like to emphasize two issues relating to our terminology and our assessments of the key objectives of distributed hydrological modelling. Indeed, since the terminology used by Beven differs in some respects significantly from the terminology used in some of the other chapters of this book, we should like first to define some key terms as we have used them in the present discussion. We begin by distinguishing between a model and a modelling system. A *model* is defined as a particular hydrological model established for a particular catchment. A *modelling system* or a *model code*, on the other hand, is defined as a generalized software package which can be used without program changes to establish a model with a range of generic basic types of equations (but allowing different parameter values) for different catchments. We then define *model validation* as the validation of a site-specific model, while *code verification* refers to the testing of algorithms etc. Our terminology is defined and discussed in more details in this book by Refsgaard (Chapter 2) and Refsgaard and Storm (Chapter 3).

With respect to the objectives of distributed hydrological modelling we see four different major types of objectives, namely:
(a) Simulation of discharges under stationary catchment conditions.
(b) Simulation of the effects of catchment changes due to human interference, such as land use change, groundwater development and irrigation.
(c) Water quality and soil erosion modelling.
(d) Research on hydrological processes.

Whereas objective (a) can be equally well addressed by simpler hydrological models, such as the lumped conceptual models, we see no real alternatives to distributed models with respect to objectives (b) and (c). Furthermore, distributed physically-based models are the most suitable for many research purposes, since they directly allow the user to incorporate new hypotheses on process descriptions for testing. Most of the criticisms made during the past decade against distributed models as being unnecessarily complex, have been made in relation to discharge simulations (objective (a)). Hence, although we agree with Beven as to many of the fundamental problems outlined in Chapter 13A, we believe it is important to realize that for many important modelling purposes there are at present no realistic alternatives to the distributed physically-based approach.

With our background as being involved both in engineering hydrology (consultancy) and in research it might be expected that the engineering part of our activities should require simple approaches (which is common in consulting work). However, our experience it that, although we have the full range of model codes available in-house, including the simple traditional ones, many of the environmental projects, in which we have been involved in fact required distributed modelling. Thus, we have as modellers chosen to use distributed catchment models because we believed that these tools, in spite of the scientific simplifications, errors and limitations, can provide the best possible information to support decisions in water resources management.

2. The basic problem with distributed hydrological modelling

Beven argues that the currently available physically-based distributed model codes are in fact lumped conceptual model codes. The justifications for this statement are the lack of physical realism of certain process descriptions such as those of overland flow and preferential flow paths in the unsaturated zone.

To our mind, however, there are still clear fundamental differences in the way a typical lumped conceptual code, such as the Sacramento, and a typical distributed physically-based code, such as the MIKE SHE, operate with regard to process descriptions, spatial variability of hydrological variables, physical realism of parameter values, and applicability. As examples of models which cannot, meaningfully, be classified as lumped conceptual, consider the geochemical model in Engesgaard (Chapter 4) or the coupled MIKE SHE - MIKE 11 floodplain model in Sørensen et al. (Chapter 12). To state that there are no differences between distributed physically-based and lumped conceptual model types adds more to the confusion than it does to the clarification of these issues.

We agree that in many applications the distributed physically-based codes have been used to construct hydrological site-specific models that may rather be classified as lumped conceptual, albeit very detailed ones. This can for example be argued with respect to a SHE application to catchments in India, where grid sizes of 1- 4 km were used (Refsgaard et al., 1992). However, this does not imply that the code in general cannot be used to construct distributed physically-based models.

We acknowledge that the present knowledge on hydrological processes suggests the need to construct much more complex hydrological models than were foreseen in the blueprint for a physically-based model of Freeze and Harlan (1969), and this process is already well under way. The very complex soil erosion processes described by Lørup and Styczen (Chapter 6) is a good example in this respect. However, this does not make the need for such models less; it 'just' makes it more difficult to establish a generally applicable code. We regard this development of improved hydrological codes as a continuous, on-going and apparently never-ending, process, whereby new knowledge about processes and new capabilities of accommodating new data types have regularly to be added. Indeed, from this point of view one of the chief merits of the development of such codes is that they provide a framework for the accumulation and integration of new knowledge, introduced from many different disciplines by a very considerable number of individual experts. A modelling system of this kind then also provides the technical means for *consensus building* across the widest possible range of disciplines in hydrology.

Thus, although we agree to a large extent with Beven's summary statement: "it cannot be assured that distributed models are based on the correct equations to describe hydrological reality at the grid scale", we consider it an expression of a rather pessimistic view. What is 'correct' and what is 'reality', and how 'correctly' do we have to describe 'reality' in different situations? (For the definition of such terms as 'reality' and 'truth' as used in hydroinformatics, see Abbott, 1994). Certainly, one can always specify performance criteria that are so strict that no model will ever pass a validation test. We consider it a more fruitful challenge to hydrological modelling to develop model codes and construct models on the basis of the best available knowledge, knowing that this basis can be improved in the future, and to use this at any time as the best *current* basis for decision making.

Furthermore, Beven argues that "while they (the distributed models) are overparameterised for the purposes of estimating discharges, they have not been properly tested in terms of simulating the internal state variables". While we agree that this may often be the case in some particular applications, we do not see how Beven can make such induction from a few examples to a universal statement. The important fact for us is that distributed physically-based codes enable model validation against internal state variables, and furthermore allow rigorous and transparent parameterisation schemes to be used. Whether models are used in such a responsible way or rather everything is confused and unjustified claims are made is the responsibility of the individual model users and should not be confused with the applicability of and general statements about the codes.

3. Examples of Successfully Validated Distributed Models at the Catchment Scale

In accordance with the terminology defined in Section 1, we associate model validation with site-specific models. A generic model code can be verified, implying the successful testing of the mathematical algorithms and the program code. Furthermore, a given model validation should always be related to pre-specified performance criteria. Within this framework, many examples of successful model validations exist, although the performance criteria most often have not been pre-specified explicitly. Refsgaard (1996) reports successful validations of lumped and distributed models from a study on Zimbabwean catchments, where rigorous test schemes were used. Jensen and Jørgensen (1988) describes a successful post-audit study of the distributed groundwater/surface water model for the 1000 km^2 Danish Suså catchment, developed and calibrated 10 years earlier (Refsgaard and Hansen, 1982). Originally, the Suså model was calibrated against soil moisture data from four plots, groundwater heads from about 40 observation wells and discharges data from 6 stations. During the post-audit period, 1981-87, the groundwater abstractions were slightly different from those in the original study, 1951-80. The post-audit comprised validation against data from 38 groundwater observation wells and 4 discharges stations, and the model predictions were found to match the values to the same degree of accuracy as during the calibration period.

In Beven's terminology, model validation refers to a kind of general validity of the model code. We agree that a model code can, in principle, never be documented as universally valid. Each model application, with its inherent assumptions and assessments made in connection with construction and possibly calibration of the model, must be validated separately. Because one model user succeeds in constructing and validating a model for given conditions in the case of one catchment, this does not imply that another user can automatically construct a valid model for another catchment using the same generic code.

The question of model validation applies equally well to all model types. Thus no lumped conceptual model code can be claimed to be universally valid either. Instead of the validity of a model code, we talk about an apparently less strict term: credibility of a given model code. By this we mean that a generic code, which has been used to construct many successfully-validated models covering a wide range of specific types of application gains a higher credibility than another code which has not been through such a chain of experience accumulation through incremental development.

In our opinion, the issues of model validation and code verification deserve much more effort in the future, and there is certainly a strong need for a common terminology (see Refsgaard, Chapter 2; Refsgaard and Storm, Chapter 3).

4. Are improved Process Parameterisations possible ?

Beven argues that the two main problems related to process descriptions (parameterisations) are (1): "the problem of change of scale and heterogeneity of parameters, even if the small scale equations were correct at the local scale" and (2): "the small scale equations may not be correct at the local (profile to plot) scale". We agree to this general statement as well as to the examples given by Beven. In fact, on the basis of our past 20 years experience with SHE and MIKE SHE we agree to the fundamental dilemma in distributed physically-based modelling outlined by Beven, namely that when limitations of the descriptive equations are recognised, it is necessary to introduce more complex parameterisations with (often) more parameters, and these may not always be easily measured and may also introduce new problems of heterogeneity.

Again, however, although we agree to Beven's problem assessment, our conclusions with regard to the modelling strategy that needs to be followed in order to accommodate these problems are quite different.

From a research point of view, we do not necessarily consider it a problem, but rather an inconvenience that more knowledge about processes and a greater access to field data lead to more complex models. If the objective of research is to achieve improved understanding of the hydrological processes in nature, new knowledge will inevitably have to be incorporated in models in order to imitate nature and test new theories. If we take the route of working with simpler models, we would not be able to utilize all available information on hydrological data and, not knowing beforehand which new data contains significant information for the problem at study, we would risk stopping further progress in the deeper and more detailed understanding of hydrological processes.

From a practical model application point of view, we can agree with Beven that if the sole objective is to model the rainfall-runoff process and predict discharges at the outlet of a catchment, then simpler models are adequate. According to our experience from several studies, such as the intercomparative study (Refsgaard, 1996) using a lumped conceptual code (NAM), a distributed physically-based code (MIKE SHE) and a semi-distributed conceptual/physically-based code (WATBAL) on Zimbabwean catchments, lumped conceptual and semi-distributed models with relatively simple process descriptions are generally just as good as more complex models such as MIKE SHE.

However, as outlined above there are many modelling objectives, such as prediction of effects of land use changes or ground water abstractions, and simulation of water quality and soil erosion, for which we see no alternatives to even more complex model codes than the existing ones.

5. Disaggregation and Scale Problems

We agree that the aggregation and scale problems are very fundamental, and indeed are of a fundamental nature. Thus, process equations and effective parameter values which are valid at one scale may not necessarily be valid at larger or smaller scales. The fact that data collection is carried out at a large range of scales does not make the problem easier. Thus, we agree to the need for identifying appropriate scaling procedures.

Comprehensive researches have been carried out in a stochastic framework for certain aspects of subsurface hydrology, such as groundwater transport (Gelhar, 1986) and unsaturated zone flow (Jensen and Mantoglou, 1992). A few attempts have also been made in the case of rainfall-runoff processes on hillslopes (Freeze, 1980). In recent years this issue has also received considerable attention in connection with simulations of the interaction between the land surface and the atmosphere (Entekhabi and Eagleson, 1989; Famiglietti and Wood, 1994). Much of the same kind of problem has been extensively studied in other fields, and notably in the theory of turbulence (e.g. Leslie and Quarini, 1979; Leonard, 1974), where the way in which in which it leads to the introduction of additional higher order terms in existing continuum equations has been rather fully analyzed.

All the reported studies on scaling in hydrology, however, have highlighted the problem only for particular cases and have, at best, provided theoretically scaling methodologies for such cases. Thus, no universal methodology has yet been developed, nor appears to be within sight in the short term.

In addition to the above research areas, which can be characterized as local scale problems, we often encounter scale problems when applying distributed models for several hundred km^2 size catchments (see Storm and Refsgaard, Chapter 4). In engineering applications of distributed models, the scale problems are, in practice, often circumvented through specific model calibration. In this way parameter values are indirectly fitted to the particular scale, and it is fully accepted that these may not necessarily be valid at other scales. Thus, if models are not calibrated, or, in particular, if models after calibration and validation are used with different spatial discretizations, then the use of appropriate scaling procedures becomes important.

6. Assessment of Uncertainty of Model Predictions

Beven emphasizes the importance of associating model predictions with estimates of predictive uncertainty. We agree that this is a very important issue which requires more attention in the future.

The most common method today of assessing model prediction uncertainties is sensitivity analysis, but this is most often carried out in a rather qualitative, unsystematic manner. More comprehensive and rigorous methodologies require a joint stochastic-deterministic modelling approach, such as state space formulations or Monte Carlo techniques. As demonstrated for the Sacramento model code (Kitanidis and Bras, 1978) lumped conceptual models can easily be reformulated in a state space form and imbedded in a Kalman filtering framework enabling predictions of uncertainty bands to be placed upon discharges. Refsgaard et al. (1983) and Storm et al. (1988), using the NAM lumped conceptual code in a Kalman filtering framework, made analyses of the propagation of uncertainties due to both the uncertainty of rainfall input and the uncertainty of certain key model parameters. Joint stochastic-deterministic modelling, including the prediction of uncertainty bands due to incomplete knowledge of the spatial variability of hydraulic parameter values, is also common in subsurface hydrological modelling (Gelhar, 1986; Kros et al., 1993; Zhang et al., 1993; Jensen and Mantoglou, 1994).

Although joint stochastic-deterministic modelling is not yet common in distributed hydrological modelling, we agree with Beven that it is feasible, at least in the form of a Monte Carlo approach. Assessments of model prediction uncertainties will be useful in connection with the following types of applications of distributed models:
* Use of model results for supporting management decisions.
* Updating (data assimilation) by use of point data from traditional monitoring networks and spatial data from remote sensing.
* Inverse modelling in much more general contexts.

Among the methodologies reported in literature the Generalised Likelihood Uncertainty Estimation (GLUE) methodology appears so far as the most comprehensive approach, and one which we look very much forward to see further developed and applied.

7. The Future of Distributed Modelling: on the Value of Data

As discussed by Refsgaard and Abbott (Chapter 1) distributed models have in general had less data available than they could have used. Thus, we agree with Beven that hydrological science is awaiting the development of new measurement techniques, especially with respect to spatial data and their heterogeneity.

However, it appears to us that several developments in these years indicate that this situation is likely to become significantly improved within the coming years. Firstly, new data sources are emerging, such as remote sensing data from new active sensor systems and geophysical data from new sensor types as supplements to geological data, while new in-situ water quality sensors are rapidly being developed, and weather radar data are gradually becoming more reliable. Secondly, the widespread use of GIS and other data base systems are gradually making the existing data, which previously in practice were not accessible in large volumes, much more easily available and applicable.

In today's water resources management systems, data bases are being developed in many places and models are being used to some extent; but very seldom is all available data used on a routine basis together with distributed models. We foresee that distributed models in the future will be integrated with permanent comprehensive data collection systems and data bases as decision support tools in water resources management.

Beven argues that with more available data "the problem of overparameterisation is consequently greater". As discussed by Refsgaard and Storm (Chapter 3) and Storm and Refsgaard (Chapters 4) we believe that this is not necessarily correct. A key issue in this respect is how the parameterisation is carried out. We advocate avoiding make too many degrees of freedom in connection with calibration procedure. Hence, almost all distributed data should be data which are not subject to calibration.

8. Conclusions

Basically, we agree with Keith Beven's listing of the problems existing in the present generation of distributed physically-based model codes with respect to general (in)adequacy of local scale process descriptions, heterogeneity and scaling problems, and the need to make assessments of uncertainties in model predictions. Furthermore, we agree that many non-documented claims have been made about the capabilities and overall validity of generic model codes and, particular, individual models. We agree that these issues deserve much further attention in terms of fundamental and applied research.

However, although we basically agree with Beven's assessment of these problem areas, our conclusions with regard to the future of distributed hydrological modelling are in most respects quite different.

Whereas we agree that for runoff prediction the complex distributed physically-based models are generally unnecessary, we disagree with Beven's more general conclusion that there is no benefit in using the comprehensive distributed physically-based codes, and that distributed models should therefore be made simpler. In our view the main justification for the distributed physically-based codes are the demands for prediction of effects of such human intervention as land use change, groundwater abstractions, wetland management, irrigation and drainage and climate change as well as for subsequent simulations of water quality and soil erosion. For these important purposes, we see no alternative to further enhancements of the distributed physically-based model codes, and we believe that the necessary codes in this respect will be much more comprehensive and complex than the presently existing ones.

9. References

Abbott, M.B. (1994) Hydroinformatics: a Copernian revolution in hydraulics, *Hydroinformatics, Guest Issue of Journal of Hydraulic Research*, 11, 3-14.

Entekhabi, D. and P.S. Eagleson (1989) Land surface hydrology parameterization for atmospheric general circulation models including subgrid scale spatial variability, *J. Climate*, 2, 816-831.

Famiglietti, J.S. and E.F. Wood (1994) I: Multiscale modelling of spatially variable water and energy balance processes. II: Application of multiscale water and energy balance models on a tallgrass prairie, *Water Resources Research*, 30(11), 3061-3093.

Freeze, R.A. (1980) A stochastic-conceptual analysis of rainfall-runoff processes on a hillslope. *Water Resources Research* 16(2), 391-408.

Freeze, R.A. and Harlan, R.L. (1969) Blueprint for a physically-based digitally-simulated hydrological response model. *Journal of Hydrology*, 9, 237-258.

Gelhar, L.W. (1986) Stochastic subsurface hydrology. From theory to applications. *Water Resources Research*, 22(9), 135-145.

Jensen, K.H. and Mantoglou, A. (1992) Application of stochastic unsaturated flow theory, numerical simulations and comparison to field observations. *Water Resources Research*, 28(1), 269-284.

Jensen, R.A. and Jørgensen, G.H. (1988) Hydrological surface water/groundwater model (in Danish). Technical report prepared by Danish Hydraulic Institute for the County of Storstrøm and the County of Vestsjælland.

Kitanidis, P.K. and Bras, R.L. (1978) Real time forecasting of river flows. Technical Report 235. Ralph M. Parson's Laboratory for Water Resources and Hydrodynamics, MIT, Cambridge, Massachusetts.

Kros, J., DeVries, W., Janssen, P.H.M. and Bak, C.I. (1993) The uncertainty in forecasting trends of forest soil acidification. *Water, Air and Soil Pollution*, 66, 29-58.

Leonard, A. (1974) Energy cascade in large eddy simulation of turbulent fluid flows, *Advances in Geophysics*, 18A, 237-248.

Leslie. D.C. and Quarini, G.L. (1979) The application of turbulence theory to the formulation of subgrid modelling procedures, *Journal of Fluid Mechanics*, 97 (Part 1), 65-91.

Refsgaard, J.C. and Hansen, E. (1982) A distributed groundwater/surface water model for the Suså catchment, Part I: model description. *Nordic Hydrology*, 13, 299-310.

Refsgaard, J.C., Rosbjerg, D., and Markussen, L.M. (1983) Application of the Kalman filter to real-time operation and to uncertainty analyses in hydrological modelling. *Proceedings of the Hamburg Symposium, Scientific Procedures Applied to the Planning, Design and Management of Water Resources Systems, August 1983*. IAHS Publication 147, 273-282.

Refsgaard, J.C., Seth, S.M., Bathurst, J.C., Erlich, M., Storm, B., Jørgensen, G.H., and Chandra, S. (1992) Application of the SHE to Catchments in India - Part 1: General Results. Journal of Hydrology, 140, 1-23.

Refsgaard, J.C. (1996) Model and data requirements for simulation of runoff and land surface processes, in S. Sorooshian and V.K. Gupta (Ed), *Proceedings from NATO Advanced Research Workshop*

"Global Environmental Change and Land Surface Processes in Hydrology: The Trials and Tribulations of Modelling and Measuring, Tucson, May 17-21, 1993. Springer-Verlag, (in press).

Storm, B., Jensen, K.H., and Refsgaard, J.C. (1988) Estimation of catchment rainfall uncertainty and its influence on runoff prediction. *Nordic Hydrology*, 19, 77-88.

Zhang, H., Haan, C.T., Nofziger, D.L. (1993) An approach to estimating uncertainties in modelling transport of solutes through soils. *Journal of Contaminant Hydrology*, 12, 35-50.

CHAPTER 13C
RESPONSE TO COMMENTS ON 'A DISCUSSION OF DISTRIBUTED HYDROLOGICAL MODELLING' BY J C REFSGAARD ET AL.

KEITH J BEVEN
Centre for Research on Environmental Systems and Statistics, Institute of Environmental and Biological Sciences, Lancaster University, Lancaster LA1 4YQ, UK

On model formulations

It is good to see this developing dialogue on this very important topic of the issues involved in distributed models. In response to the comments of Chapter 13B, I must re-emphasise that my discussion in was always intended as a friendly critique. I do not deny the need for distributed predictions. I am not suggesting that lumped models or simpler distributed models will necessarily provide adequate predictions. I myself have been involved in distributed modelling and in trying to do it in some sense "properly" for more than 20 years. I do feel, however, that the limitations of making distributed predictions need to be made more explicit with a view to moving the science ahead, perhaps in alternative ways.

One comment seems to have provoked a particularly strong response is my suggestion that physically-based distributed models are, at some elemental scale, lumped conceptual models. I think there may have been some misunderstanding here, but I do feel that this point needs reinforcing. I do not suggest that "there are no differences between distributed physically-based and lumped conceptual model types" (Chapter 13B, section 2). Clearly there are differences. Distributed models make distributed predictions at scales larger than the element scale. But why obscure that fact that the equations used at that elemental scale may be essentially lumped and conceptual. Manning's "law", widely used in distributed models, is an empiricism that was rejected by Manning (1891) in his comparison of a variety of different open channel flow equations in favour of a more complex formulation. The hydraulic radius exponent of 0.667 was a compromise value: experimental data at the time suggested values of between 0.65 to 0.84 (Chow, 1959). It is a steady state formulation that is widely used as an approximation for gradually varied flows. It is also widely used floodplain and overland flows for which the only justification is based on roughness coefficients that have been back-calculated from experimental data at a variety of scales. The resulting coefficients may be scale dependent, (even for channels e.g. Beven and Carling, 1992) but there is no adequate theory of scaling of the coefficients from the cross-sectional "point" measurements to the reach or element scale required by the models.

I also continue to stress that Darcy's "law" and the Richards' equation are conceptual when used for unsaturated flow in structured soil at the element scale. For such soil, there is no adequate theory of scaling of the coefficients from the point scale to the element scale required by the model. The "effective" parameters of these models at the element scale are lumped conceptual entities. I have chosen the word lumped quite deliberately, to emphasise the mismatch between the element scale at which the theory is being applied and the "point" scale at which the theory can be more properly (but in many cases not entirely) justified. I do not agree that this adds confusion if it makes the user think about what is actually being done.

Refsgaard et al. (Chapter 13B) commend the geochemical model of Engesgaard (Chapter 4) as a model that cannot be regarded as a lumped conceptual model. It is certainly a model that makes distributed predictions but geochemical models are also notoriously conceptual when applied to environmental (as opposed to well controlled laboratory) fluxes. It is common, for example, to include a Gibbsite reaction in equilibrium models of aqueous geochemistry to control the aluminium/pH concentrations. Models such as PROFILE (Warfinge and Sverdrup, 1992), that include such a Gibbsite reaction, are used widely in western Europe in assessing critical loads for acid deposition - despite the fact that Gibbsite is essentially absent from the soil in most of Europe. Worse, the Gibbsite reaction coefficient appears to be the most sensitive parameter in PROFILE (see Jonsson et al., 1995; Zak et al., 1996). It is used as an analogue reaction, lacking better knowledge. There is even some evidence that the _form_ of the model that is derived from equilibrium geochemical theory may not be appropriate when applied to the integrated effects of interactions with heterogeneous materials in the field (see discussion in Neal et al., 1990) - geochemical models merely assume that the form should be the same in the same way as the hydrologist assumes that the form of Darcy's law should be the same at the element scale, despite the fact that heterogeneous soils mean that there may be highly variable hydraulic gradients at the sub-element scale.

We all recognise that as knowledge and understanding improve, so process descriptions will change. I am not yet convinced, however, that this necessarily means that they should become more complex, as Refsgaard et al. (Chapter 13B, section 2) suggest. Model complexity may well be required for improving and structuring the understanding gained from detailed experimentation but also implies that more parameters will need to be specified. Complexity will only be advantageous in applications if it means that the parameters can be more easily estimated on the basis of readily available characteristics or can be related to "universal" constants. There is no evidence that this may be so, indeed it seems unlikely given the complexity of the geometry and boundary conditions of the flow domains that we are interested in. It is more likely that there will be more parameters requiring calibration for application at a particular site and the calibration problem will then be made worse by additional model complexity. This *may* outweigh any advantage of the improved process description when it is necessary to apply that description in prediction at other sites and in other circumstances. I stress the *may* because this whole area requires much further work. A simpler (possibly fuzzy) description may have as much predictive value as a more complex but overparameterised description in incorporating the knowledge from past experimental work (from which there is undoubtedly still much to be learned by reinterpretation) and in predicting the effects of a distributed land use change.

On model validations

Refsgaard et al. (Chapter 13B) distinguish the processes of validating the code of a model system, validating the modelling system concepts and validating the application of that system to a particular problem. It is true that in different parts of Chapter 13A I was addressing these different issues without clear distinction between them. Validation of the code itself is the only case where validation is indeed possible. I did not discuss this, but assumed that all modellers at least attempt to ensure that their code is properly validated. This may not be a good assumption; developments in formal code validation and implementations of quality assurance techniques have not penetrated to most of the research modelling community. It is also well known that even the largest software houses can release code with remaining bugs even after extensive beta testing, but let as assume that continued use and testing of new numerical algorithms results in a modelling system that is consistent with the equations on which it is based.

However, my main concern in Chapter 13A was with the scientific aspects of validation in terms of both model concepts and particular applications. I continue to feel that there has been inadequate validation of distributed catchment models in both respects. The previous section revisits some of the discussion on the validation of model concepts. I would refer the reader again to the analysis of mediating models in Morton (1993). I do not consider it unhealthy to recognise the limitations of current physical descriptions of hydrological processes and to state that we <u>know</u> that they are wrong. Quite the contrary; only in this way will the science progress.

This does, however, limit confidence in the general validity of the current distributed modelling systems. This does not mean that particular model applications developed from those systems cannot give valuable results, as pointed out in Chapter 13B, particularly after some calibration against observed data. The number of validated models remains, however, small, particularly in respect of validation against internal state data. My own experience suggests that reproducing internal state variables is quite difficult, for reasons both fundamental (as discussed in section 13A) and practical (Lamb et al., 1996). Post-hoc analyses in the groundwater literature suggest that such difficulties are not uncommon (e.g. Anderson and Woessner, 1992) , although it is clearly difficult to differentiate between the limitations of the modelling system and the limitations of the process of application of such a system, perhaps by a variety of users with different levels of expertise.

What is needed here are some comparative studies based on particular data sets, both field measurments and hypothetical data, similar to the INTRAVAL data sets in groundwater transport (Larsson, 1992). Experience in INTRAVAL suggests that this will result in a more circumspect attitude to the predictive capabilities of modelling systems. In catchment hydrology, arguably an even more complex problem because of the interaction of processes and nonlinearities of the unsaturated zone, there are very few such datasets with adequate internal state measurements. Contenders include the artificial HYDROHILL catchment in China, the artificial sprinkler experiments carried out at the Shale Hills catchment in Pennsylvania (Corbett et al., 1975) (which include some experiments with some interesting different patterns of inputs); the small MINIFELT catchment in Norway with 105 piezometers in an area of 7500 m2 (Erichson and Myrabø; Lamb et al., 1996); the UK Nirex supported experiments in the Slapton Wood catchment, Devon, UK (Ragab and Cooper, 1993); and, in a semi-arid environment, the Lucky Hills watershed in Arizona (Goodrich et al., 1994; Faurès et al., 1995). Although none of these are ideal, those that are still current could be improved with some additional effort to make worthwhile validation and comparison data sets.

The Slapton Wood catchment is currently being used to test the hydrological component of the SHETRAN UK modelling system in a "blind" validation test (i.e. without prior access to the observed discharge and water table variables to be predicted) with prior specification of the criteria for success or failure to be met. A previous, more limited, blind validation has already been carried out for the Rimbaud sub-catchment of the Réal Collobrier in southern France (Parkin et al., 1996) with reasonable success in discharge prediction. Note that these studies make use of multiple simulations based on estimated ranges of parameters values and produce ranges of the simulated variables.

If some observed data are available, however, it would be expected that predictions of both discharges and internal state variables would be improved by modifying parameter values in some calibration process. As discussed in Chapter 13A this is not a well posed problem because of the sheer number of parameters that could be modified. Availablility of local data will allow some local calibration but it is not clear how best to make use of such data, which is often at the "point" scale. Indeed, one problem appears to be that the more local data are available, the more the limitations of the models are revealed. Much further work is required to understand the value of different types of data in this respect to guide measurement programmes in the future. Again, the availability of some specific well documented data sets would be very helpful in this repect.

On the value of data

Yes, distributed data are becoming much more widely available and GIS systems are being integrated with modelling systems, but much of this data is only indirectly hydrologically relevant. The hydrological value of information on vegetation type or soil type or remotely sensed images depends on the use of intermediate interpretive models, which have their own parameters and simplifying assumptions. The difficulties of deriving estimates of surface soil moisture from remote sensing images, for example, are clear from the studies of Wang et al. (1992), Merot et al. (1994), and Lin et al. (1994).

Models for relating soil hydraulic parameters to soil texture variables are also being increasingly used (e.g. the regression relationships of Rawls and Brakensiek, 1989; see also Ragab and Cooper, 1993b) since soil texture is readily available from soil maps and soil hydraulic characteristics are expensive and time consuming to measure. Two things must be remembered in using such models: the models can only be as good as the data on which they are based (the hydraulic conductivity data in the USDA database for example is based on "fist-sized" samples and takes no account of structural porosity) and that the estimates are associated with a standard error of estimation that may be large. We must move towards a modelling practice in which such uncertainties are routinely taken into account; only then will a proper research programme develop addressing the best economic means of uncertainty constraint.

We really know very little about the value of different kinds of data for distributed hydrological modelling. This is another good reason for pursuing the possibility of creating some test catchment datasets, including intrenal measurements of dynamic water storages and parameter values. This is relatively easy to do for hypothetical catchments, much more difficult and expensive for real catchments. However, only in this way will it be possible to assess the relative value of different strategies for investment in data collection.

To the reader.......

..... if you have managed to persevere this far, I hope you will agree that this has been a valuable exchange of views and that it will give you cause for thought in the future development and application of distributed models. I would not have characterised the general conclusion of my earlier discussion that "there is *no* benefit in using the comprehensive distributed physically-based codes" as suggested by Refsgaard et al. in Chapter 13B (my emphasis). I would rather suggest that such models should be treated as contenders in the type of modelling process I have advocated that recognises the possible equifinality of modelling systems and parameter sets. It is quite possible that in some situations (perhaps the prediction of land use change impacts) such models may be the only contenders that are not rejected within some criteria of fit-for-purpose and acceptability in reproducing the available data. However, the limitations of the current generation of models must be recognised, notwithstanding the high tech integration of sophisticated GIS and graphical user interfaces. The possibilities of other possible contending modelling systems (such as a methodology for assessing land use impacts based on fuzzy logic) should also be considered. And, to the field hydrologist, it would be really extremely valuable to have the type of well documented distributed data sets discussed above to test some of the various possiblities.

If I have argued for simpler process representations, it is because they may both allow the user to think more carefully about the way in which the model is representing the processes in a catchment (see Beven et al., 1995), but also, and more importantly, that within current computing constraints simpler models allow predictive uncertainty to be assessed more easily (e.g. Freer et al., 1996). I have no doubt at all that if we did have different, larger scale, measurement techniques available - for example, a technique that could rapidly, accurately and continuously measure total water storage or actual evapotranspiration fluxes at the hillslope or element scale - then our "physically-based" theory in hydrology would look very different. Failing such developments, which today seem a long way off, we should approach distributed predictions with circumspection and careful thought.

References

Anderson, M P and Woessner, W W, 1992, The role of the post-audit in model validation, *Adv. Wat. Resour.*, 15, 167-173.

Beven, K J and Carling, P, 1992, Velocities roughness and dispersion in the lowland river Severn. In: Petts, G.F. & Carling, P. (Eds.) *Lowland Rivers: Geomorphological Perspectives*, pp. 71-93, Wiley, London.

Beven, K J, Lamb, R, Quinn, P, Romanowicz, R and Freer, J, TOPMODEL in V P Singh (ed.), *Computer Models of Watershed Hydrology*, Water Resource Publications, Colorado, 627-668.

Chow, V T, 1959, *Open Channel Hydraulics*, McGraw-Hill, New York (footnote, pp99-100).

Corbett, E S, Sopper, W E and Lynch, J A, 1975, Watershed response to partial area applications of simulated rainfall. in *Hydrological Characteristics of River Basins and the Effects on these Characteristics of Better Water Management, IASH Publ. No. 117*, 63-73.

Erichsen, B and Myrabø, S, 1990; Studies of the relationship between soil moisture and topography in a small catchment, in G Gambolati (ed.), *Proc. 8th Int. Conf. on Computational Methods in Water Resources*, Venice.

Faurès, J-M, Goodrich, D C, Woolhiser, D A and Sorooshian, S, 1995, Inpact of small-scale spatial rainfall variability on runoff-modelling, *J. Hydrol.*, 173, 309-326.

Freer, J, Beven, K J and Ambroise, B, 1996, Bayesian estimation of uncertainty in runoff prediction and the value of data: an application of the GLUE approach, *Wat. Resour. Res.*, in press.

Goodrich, D C, Schmugge, T J, Jackson, T J, Unkrich, C L, Keefer, T O, Parry, R, Bach, R B and Amer, S A, 1994, Runoff simulation sensitivity to remotely sensed initial soil moisture, *Wat. Resour. Res.*, 30, 1393-1405.

Jonsson, C, Warfvinge, P and Sverdrup, H, 1995, Uncertainty in predicting weathering rate and environmental stress factors with the PROFILE model, *Water, Air and Soil Pollution, 81, 1*

Lamb, R, Beven, K J and Myrabø, S, 1996, A generalised topographic-soils hydrological index, in *Landform Monitoring, Modelling and Analysis, Proc. BGRG Annual Conference 1995*, Wiley, in press.

Larsson, A, 1992, The International Projects, INTRACON, HYDROCON and INTRAVAL, *Adv. Wat. Resour.*, 15, 85-87.

Lin, D S, Wood, E F, Beven, K J and Saatchi, S, 1994, Soil moisture estimation over grass covered areas using AIRSAR, *Int. J. Remote Sensing*, 15(11), 2323-2343.

Manning, R, 1891, On the flow of water in open channels and pipes, *Trans. Instn. Civ. Engrs. Ireland*, 20, 161-207; suplement, 24, 179-207, 1895.

Merot, Ph., Crave, A, Gascuel-Odoux, C and Louhala, S, 1994, Effect of saturated areas on backscattering coefficient of the ERS1 synthetic aperture radar: first results, *Water Resour. Res.*, 30, 175-179.

Morton, A, 1993, Mathematical models: questions of trustworthiness, *Brit. J. Phil. Sci.*, 44, 659-674.

Neal, C, Reynolds, B, Stevens, P A, Hornung, M, and Brown, S G, 1990, Dissolved inorganic aluminium in acidic stream and soil waters in Wales, in R W Edwards (ed.), *Acid Waters in Wales*, Kluwer, Dordrecht, 173-188.

Parkin, G, O'Donnell, G, Ewen, J, Bathurst, J C, O'Connell, P E and Lavabre, J, 1996, Validation of catchment models for predicting land use and climate change impacts: 2. Case study for a Mediterannean catchment, *J. Hydrol.*, in press.

Ragab, R and Cooper, JD, 1993a, Variability of unsaturated zone water transport parameters: implications for hydrological modelling. 1. In situ measurements, *J. Hydrol.*, 148, 109-131.

Ragab, R and Cooper, JD, 1993b, Variability of unsaturated zone water transport parameters: implications for hydrological modelling, 2. Predicted vs. in situ measurements and evaluation of methods. *J. Hydrol.*, 148, 109-131.

Rawls, W J and Brakensiek, D L, 1989, Estimation of soil water retention and hydraulic properties, in H J Morel-Seytoux (ed.), *Unsaturated Flow in Hydrologic Modeling: Theory and Practice*, Kluwer NATO ASI Series C, 275, 275-300, Dordrecht.

Wang, J R, Gogineni, S P and Ampe, 1992, Active and passive microwave measurement of soil moisture in FIFE, *J. Geophys. Res.*, 97(D17), 18979-18985.

Warfinge, P and Sverdrup, H, 1992, Calculating critical loads of acid deposition with PROFILE - a steady state soil chemistry model, *Water, Air and Soil Pollution*, 63, 119-143.

Zak, S, Beven, K J and Reynolds, B, 1996, Uncertainty in the estimation of critical loads: a practical methodology, submitted to *Water, Air, and Soil Pollution*.

CHAPTER 14
HYDROLOGICAL MODELLING IN A HYDROINFORMATICS CONTEXT

A.W. MINNS
International Institute for Infrastructural, Hydraulic and Environmental Engineering (IHE)
P.O. Box 3015, 2601 DA, Delft, The Netherlands
V. BABOVIC
Danish Hydraulic Institute
Agern Allé 5, 2970 Hørsholm, Denmark

1 Introduction to Hydroinformatics

The informational revolution of the last 30 years has fundamentally altered the traditional planning, modelling and decision-making methodologies of the water-related sciences and technologies. Information technology (IT) now plays an essential role in the sustainable development of water resources and the responsible management of the aquatic environment. The general availability of sophisticated computers with ever-expanding capabilities has given rise to an increasing complexity in terms of computational ability and in the storage, retrieval and manipulation of information flows. Hydroinformatics is the field of study of the flow of information and its processing by knowledge as applied to the flow of fluids and all that they transport.

From one point of view, hydroinformatics can be seen as having emerged from the well-established technology of computational hydraulics that utilises numerical modelling techniques to describe physical systems with sets of numbers and simulates the laws acting upon these systems with sets of operations on these numbers. The introduction of computational hydraulics some 20 years ago initiated a correspondingly significant revolution in classical hydraulics and a thorough reformulation of laws and concepts in order to accommodate the new possibilities represented by the discrete, sequential and recursive processes of digital computation. Now, with the introduction of hydroinformatics, more fundamental changes again are taking place.

In addition to water quantity and quality data, the information necessary to describe and assess the state of any given body of water must also include a plethora of social, legal and environmental factors. In this context, the typical information to be incorporated into a hydroinformatics system must include such variables as international and national laws, local bye-laws - either temporary or permanent - and any applicable physical, chemical and biological parameters. Added to this, flows of water, sediment, chemicals and other water-borne substances must be calculated and measured, and the sites and water quality parameters of the area's water users identified. Lastly, the locations and production rates of heat, chemical and biological pollution must be introduced into the representation, as well as the presence, position and capacity of control elements in the area, such as pumps, retention basins and treatment plants.

An important feature of a hydroinformatics system is that it allows the use of those numerical simulations that are subject to constraints expressed in natural language (such as applicable legislation, contracts and agreements). However, producing an accurate impact prediction requires a wealth of knowledge, much of which can only be obtained by studying previous experiences under similar circumstances. Hydroinformatics facilitates this assessment process by encapsulating expert knowledge and experience, and by making this knowledge available in informational form to hydro-scientists and engineers, thereby raising the level of their professional performance.

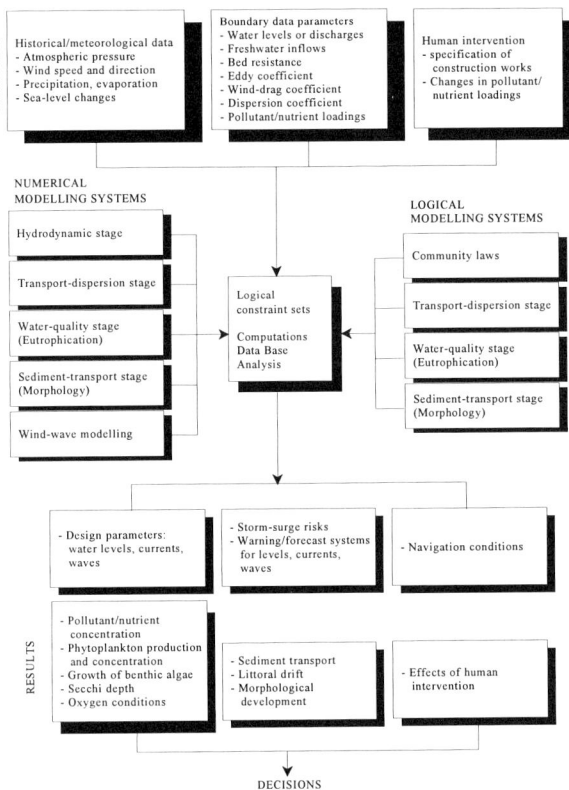

Figure 1   A prototype of a relatively complete hydroinformatics system for the Venice Lagoon, Italy.

In order to operate effectively, a hydroinformatics system may therefore be connected to measuring equipment, through a SCADA (supervisory control and data acquisition) system; it may contain numerical models to quantify the movements and changes within a body of water; it may use graphical interfaces to present the results of computations in a form which is understandable to a wide audience; it may assist in its own instantiations and in the interpretation of the results that it provides through the use of expert-advice systems; and it may store this information in data- and knowledge-bases. The size and complexity of such undertakings is well illustrated in Figure 1 (reproduced from Abbott, 1991, p. 33) that shows the schematisation of a relatively complete prototype for a hydroinformatics system of the Venice Lagoon in Italy as conceived already in 1989. Such systems have now entered service for the real-time control of urban drainage systems, they are being realised for coastal management systems and are being prepared for such applications as river-basin management and real-time control of irrigation systems.

The very development of hydroinformatics and the corresponding value that its integrating function adds to each of its components separately, leads in its turn to an accelerated development of measuring equipment, to much more sophisticated SCADA systems, to new modelling capabilities, to new means to relate measurements and models through data assimilation, automatic constitutive equation generation, automatic calibration procedures and other such applications of inverse and adjoint methods, to new data base technologies, to new user interfaces and indeed to any number of other such developments.

Despite the new ground that has already been broken by hydroinformatics, these systems are far from fulfilling their potential. Hydroinformatics research does not remain limited to the fields of hydraulics and hydrology alone, but has recourse to the latest IT developments in the fields of artificial intelligence (including machine learning, evolutionary algorithms and artificial neural networks), artificial life, cellular or finite-state automata and other, previously unrelated sciences and technologies.

Through studying and exploiting elements of these seemingly unrelated sciences, hydroinformatics is producing new and innovative solutions to hydraulic and hydrological problems, as represented by real-time control and diagnosis, real-time forecasting, calibration of numerical models, data analysis and parameter estimation. (*see* Verwey *et al*, 1994; IAHR, 1994; Babović, 1995). In particular, these new approaches can be used to generate important components of physically-based, distributed hydrological modelling systems by *inducing* models or sub-models of individual physical processes based only upon measured data. These (sub)models may then replace whole systems of complex, non-linear, differential equations that would otherwise require great skills from the modeller to calibrate and powerful computing devices to solve.

This chapter then introduces some research results with which the authors are familiar covering a range of applications varying from the modelling of certain aspects of the hydrological cycle to the analysis and control of complete water resources systems. The examples used are by no means exhaustive of the total range of applications of hydroinformatics in hydrology, but are indicative of the effectiveness of these new approaches in finding solutions to some long-standing problems in hydrological modelling.

2   Symbolic and Sub-symbolic Paradigms

In order to appreciate more fully the power of the new approaches described in this chapter, it is necessary to introduce a fundamental notion that expresses the essential difference between these approaches and the more traditional modelling approaches. This is the notion of the differentiation between symbols and signs and thus between symbol manipulation and sign manipulation. Symbols are an artefact of our beliefs about the natural world. These symbols are tokens that *stand in the place* of the objects that they represent. A collection of symbols, however, does not constitute a model. It is only when we interpret a collection of symbols, thereby giving them 'meaning' or semantic content, that we say that this collection of symbols becomes a *sign* that *points towards* a certain natural phenomenon. The set of symbols that we recognise as the one or the other of the Richards' equations for unsaturated flow then constitute a hydrological model because each such equation constitutes 'a collection of signs that serves as a sign' (Abbott, 1992). There can only be a finite number of signs in this world created by the modeller and the potential infinity of details in the physical-world that cannot be described within this limited sign vocabulary are often gathered together in the form of assumptions and simplifications that have to be applied in order to read a meaning into the sequence of symbols that is the differential equation. If the modeller accepts the limitations imposed upon the model by the assumptions and simplifications, then he or she accepts that this sign vocabulary, or language, is the best available description of the physical processes being considered. Natural systems, however, rarely conform to these assumptions. Subsequently, this symbolic language is commonly very restrictive for research into novel and innovative approaches.

The inherent limitations of the traditional modelling approach and its associated language are exemplified in the *Philosophical Investigations* of Wittgenstein (Part I, §§ 2-3);

> "Let us imagine a language ... (that is) ... meant to serve for communication between a builder A and an assistant B. A is building with building-stones: there are blocks, pillars, slabs and beams. B has to pass the stones, and that in the order in which A needs them. For this purpose they use a language consisting of the words 'block', 'pillar', 'slab' and 'beam'. A calls them out; - B brings the stone which he has *learnt* to bring at such-and-such a call. - Conceive this as a complete primitive language.
>
> "[On the other hand,] Augustine, we might say, does describe a system of communication; only not everything that we call language is this system. And one has to say this in many cases where the question arises 'Is this an appropriate description or not?' The answer is: 'Yes, it is appropriate, *but only for this narrowly circumscribed region, not for the whole of what you were claiming to describe.*' " (emphasis added)

In this simple example, the universe of discourse consists only of the words 'block', 'pillar', 'slab' and 'beam'. It would be impossible for the characters in this 'language game' ever to talk about 'doors', 'windows' or 'roofs' - let alone an entire house! Similarly, the hydrological modeller is restricted in his or her description of a hydrological catchment by the limited language of computational hydrology. The description of this catchment can only be as detailed as the model that is to be used to simulate the catchment processes. No amount of extra measured data will ever change the basic structure of the underlying differential equations of the model, but may only be used to adjust certain calibration parameters in order to bring the results of model simulations closer to the observed and measured phenomena. Since functional similarity to the natural system is supposed to be comprehended by the equations themselves, it is the calibration parameters that must then capture the correspondence between the model and the real world. These parameters serve in effect as *error compensation devices* that artificially adjust the model results to compensate for the fundamental discrepancies that exist between the real world and its differential equation representation in the model.

Calibration parameters are, however, usually not at all well-defined in nature. One may even ask 'What is the physical meaning of these parameters - how well are they grounded, and indeed are they grounded at all?'. We may indeed be able to read a certain 'physical meaning' into our calibration parameters, but they do not exist-as-such and are thus 'disconnected' in a fundamental way from the world that they are supposed to model.

The differential equations and the calibration parameters constitute the language of the hydrological modeller. The traditional approach to hydrological modelling is one of simply manipulating and adjusting these symbols in order to arrive at the best possible correspondence between model output and measured data. Nowadays we even recognise the process of symbol manipulation in the many commercial packages that claim to do just that (e.g. *Mathematica*, *Matlab*, etc.). Rarely is it possible for the modeller to create and incorporate new symbols and their associated signs into this language of discourse. The symbolic approach suffers from a rather thoroughly intractable problem of 'symbol grounding' (Harnad, 1990).

One of the greatest strengths of the new approaches described in this chapter is their ability to identify relationships and to induce models of measured data without requiring a detailed knowledge of physical hydrological characteristics *a priori*. One of the reasons for this is that many of these approaches manipulate the data at the level of the computer representation of the numbers. That is, the data are represented in our digital computers as *bits* and the operations upon this data then take place upon these individual bits. The modeller in this case has no direct influence upon the bits that convey the information. After translating the data that describe the symbols of our natural world into bit strings, the original symbols are then further irrelevant for the subsequent manipulations of bits. The algorithm operates at the level of the bits and is referred to as a sub-symbolic approach.

The computer is free to manipulate the bit strings, cutting them and rejoining them again at different places, flipping bits either randomly or in a controlled way. During this process the overall performance of the system can be observed and evaluated. This observation involves translating the bit information back into data that then acts once more as a sign which can be interpreted as a solution to a natural world phenomenon. This search for a solution takes place at the level of the bits and is unrestricted by the limitations of the language of our symbolic world. The computer itself, however, does not interpret the results of these recombinations at this sub-symbolic level.

The most important influence of the modeller in this process then is the translation or interpretation of the results being produced by the computer. These results should somehow 'make sense' to the modeller. The advantage of the sub-symbolic approach is that solutions may emerge whose signs point to other objects in the natural world that were not included in the original collection of symbols contained in the original model.

2.1 A SYMBOLIC APPROACH

It is possible to process strings of symbols which obey the rules of some formal system and which are interpreted by humans as ideas or concepts. It was the hope of early artificial intelligence researchers that all knowledge could be formalised in this way. However, the success of this approach relies very heavily upon selecting a sufficient set of symbols governed by a set of rules that is large enough to comprehend all possible conditions and this all linked together by some algorithm that is simple enough to program on the available, digital computing devices. It soon becomes obvious that even the simplest of tasks are extremely difficult to formalise in this way.

Expert or 'rule-based' systems represent one obvious example of a symbolic system. An expert system consists essentially of only two major components: a knowledge-representation component and an inference engine. Knowledge representation is the key component that 'translates' the utilities of the real-world into a finite collection of symbolic structures. In expert-systems this knowledge takes the form of rules. It is these rules that fundamentally discretise the world and reduce it to a finite number of configurations that are arranged in a tree, as exemplified in Figure 2.

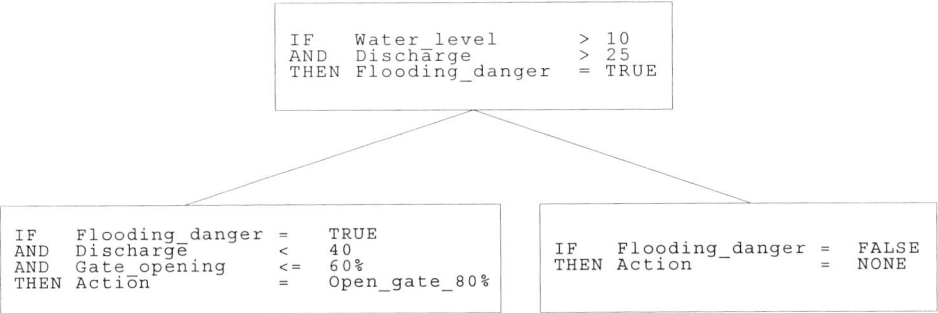

Figure 2   Knowledge representation in tree form in a simple expert system.

Given the represented knowledge in such a formalism, the task of the inference engine is then to find an instance that is most appropriate for a given situation. Correspondingly, the objectives of the inference engine are similar to those of any search problem. Most expert systems use either breadth-first or depth-first search strategies, as exemplified in Figures 3a and 3b, respectively. This figure illustrates the order in which an inference engine would 'visit' particular rules in a knowledge-base in order to locate the one that is most appropriate for a given input.

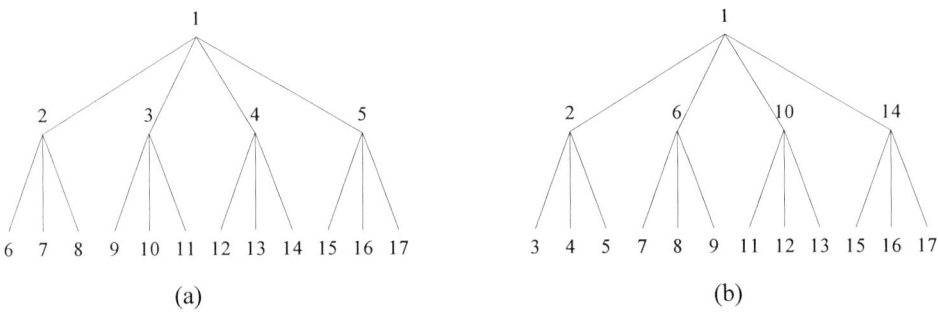

Figure 3   Search strategies in an expert system showing the order of searching the rules in the tree structure for (a) a breadth-first and (b) a depth-first search strategy.

From a cognitive point of view, it is argued that an expert system is the most appropriate programming environment for modelling logical human mental activities. It provides an automation of the reasoning process, with knowledge about the domain (knowledge base) clearly separated from the set of mental operations (inference engine) to be performed on the knowledge base. Babović (1991, pp. 12-16) described several features of expert systems that make them especially useful in complex systems. These features include:
- knowledge is presented in a highly declarative way. The symbolic appearance of the model can be interpreted by almost any user, allowing the logics of the model to be

understood and permitting the alteration of the contents of the model in order to improve its performance;
- in addition to formal knowledge and established theories, fragmentary, ill-structured, approximate, incomplete, uncertain, heuristic and judgemental knowledge can be encoded and used;
- non-deterministic control strategies can be implemented. Modules may not be executed in a predefined sequence, thus enabling the use of various reasoning paradigms;
- the system is transparent, *i.e.* it can provide natural explanations and justifications of its specific lines of thought and reasoning;
- it provides flexibility through incremental creation, debugging and updating of very large knowledge bases.

On the other hand, expert systems suffer from several short-comings which, in addition to the symbol grounding problem, include the so-called *completeness* issue. Using the example in Figure 2, a situation in which `Water_level > 10` and `Discharge = 55` is not represented in the knowledge tree. This would result in the failure of the inference engine to find an appropriate instance for `Action` and to draw the corresponding conclusions. Due to this problem, expert systems are said to be *brittle* systems, in the sense that as long as every question has an explicitly coded answer they will perform well, but as soon as a situation is not explicitly represented in the knowledge-base they fail quite suddenly, in a brittle fashion.

Applications of symbolic approaches can be exemplified by several recent publications such as Almeida and Schilling (1993) in which the construction of `IF...THEN...ELSE` rules in a knowledge base of a hydroinformatics system is described, while the most recent proceedings of the International Conference on Urban Drainage describe the application of expert systems to problems of urban hydrology quite extensively (see, for example, Ahmad *et al*, 1987; Delleur and Baffaut, 1990; Khelil *et al*, 1993; Martin-Garcia, 1995).

2.2 SOME SUB-SYMBOLIC APPROACHES

2.2.1 *Artificial neural networks*

Systems investigations, which Amorocho and Hart (1964) regarded as being concerned with the direct solution of technological problems subject only to the constraints imposed by the available data, and so not subject to 'physical' considerations, has recently undergone something of a renaissance, largely through the adaptation of artificial intelligence techniques, such as Artificial Neural Networks (ANNs) and Evolutionary Algorithms (EAs) (e.g. Babović and Minns, 1994). The particular advantage of the ANN is that, even if the 'exact' relationship between sets of input and output data is unknown - but is still acknowledged to exist - the network can be trained to learn that relationship.

The ability of the brain to perform difficult operations and to recognise complex patterns, even if these patterns are distorted by 'noise', has formed the subject matter of the discipline of cognitive psychology that has in turn strongly influenced the study of artificial intelligence (AI).The particular ability of the brain to learn from experience without a predefined knowledge of underlying physical relationships makes it an exceptionally flexible and powerful calculating device that AI researchers have long tried to mimic.

At the same time, other researchers have been devoted to reproducing, or modelling, physical phenomena by making use of electronic computational machines to solve ever-increasingly complex partial differential equations and related empirical relationships. These researchers are supported by a rapid increase in the computational capacity of modern computers and an emerging recognition of the advantages of massively parallel computation (parallel distributed processing) that performs the required calculations with ever-increasing speed. However, although the design and construction of the hardware for parallel computation is relatively straightforward, the software required for creating algorithms to utilise this parallel architecture for solving partial differential and other such equations efficiently is still quite limited.

These two groups of researchers, pursuing what appear to be quite different goals, have found a common ground in the field of artificial neural networks. One of the major applications of ANNs is in pattern recognition and classification or, more generally, system identification. In brief, an ANN consists of layers of processing units (representing biological neurons - *see* Hopfield, 1994) where each processing unit in each layer is connected to all processing units in the adjacent layers (representing biological synapses and dendrites). Many publications describe in much greater detail the architecture of various types of ANNs (for example, Beale and Jackson, 1990; Aleksander and Morton, 1990; Hertz *et al*, 1991). Figure 4 shows a schematisation of a typical multi-layer, feed-forward ANN.

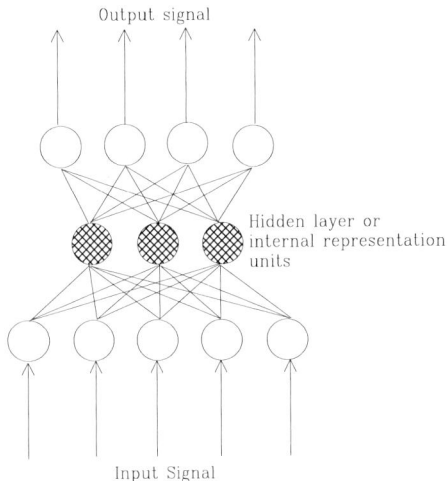

Figure 4   Representation of a multi-layer, feed-forward artificial neural network (ANN)

The working of an ANN can best be described by following the operations involved during training and computation. An input signal, consisting of an array of numbers x_i is introduced to the input layer of processing units or nodes. The signals are carried along connections to each of the nodes in the adjacent layer and can be amplified or inhibited through weights associated with each connection. The nodes in the adjacent layer act as summation devices for the incoming (weighted) signals (Figure 5). The incoming signal is transformed into an output signal O_j within the processing units by passing it through a threshold function. A common threshold function for the ANN is the sigmoid function that is depicted in Figure 4 and defined as:

$$f(x) = \frac{1}{1 + e^{-x}} \tag{1}$$

which provides an output in the range $0 < f(x) < 1$.

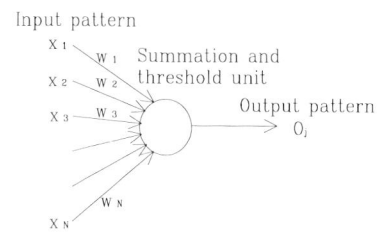

Figure 5   A typical ANN node

The output from the processing unit is then:

$$O_j = \frac{1}{1 + e^{-\sum x_i w_i}}, \quad 0 < O_j < 1, \quad \forall j \tag{2}$$

This output signal is subsequently carried along the weighted connections to the following layer of nodes and the process is repeated until the signal reaches the output layer. The one or more layers of processing units located between the input and output layers have no direct connections to the outside world and are referred to as *hidden* layers. The output signal can then be interpreted as the *response* of the ANN to the given input *stimulus*.

The ANN can be *trained* to produce known or desired output responses for given input stimuli. The ANN is first initialised by assigning random numbers to the interconnection weights. An input signal is then introduced to the input layer and the resulting output signal is compared to the desired output signal. An error or 'energy' function is then computed that represents the amount by which these two signals differ. This error function is defined as:

$$E = \tfrac{1}{2} \sum (D_j - O_j)^2 \tag{3}$$

where O_j is the network output and D_j is the desired output. The interconnection weights are then adjusted to minimise the error. This process is repeated many times with many different input/output tuples until a sufficient accuracy for all data sets has been obtained. A learning rule, known as the *generalised delta rule*, adjusts the weights associated with each connection by an amount proportional to the strength of the signal in the connection and the total measure of the error (*see* Rumelhart and McClelland, 1986). The total error at the output layer is then reduced by redistributing this error value *backwards* through the hidden layers until the input layer is reached. For this reason, this method is referred to as *error back-propagation*. The next input/output tuple is then applied and the connection weights readjusted to minimise this new error. This procedure is repeated until all training data sets have been applied. The whole process is then repeated again, starting from the first data set once more and continuing until the total error for all data sets is sufficiently small and subsequent adjustments to the weights are inconsequential. The generalised delta rule provides in fact a form of *gradient descent* method, where the energy function (3) is calculated and changes are made in the steepest downward direction. Although this method does not guarantee convergence to an optimal solution since local minima may exist, it appears in practice that the back-propagation method leads to solutions in almost every case (Rumelhart *et al*, 1994). In fact, standard multi-layer, feed-forward networks, with only one hidden layer have been found capable of approximating any measurable function to any desired degree of accuracy (Hornik *et al*, 1989). Errors in representation would appear to arise only from having insufficient hidden units or the relationships themselves being insufficiently deterministic.

2.2.2 *Evolutionary algorithms*

Evolutionary algorithms (EAs) are simulation engines of grossly simplified processes occurring in nature and implemented in artificial media such as computers. The fundamental idea is indeed the one of plagiarising natural processes. Darwinian theory of evolution depicts the adaptation of species to its environment as one of natural selection (Darwin, 1859). Perceived in this way, all species currently inhabiting our planet (and for that matter, all species that have ever lived on this planet) are actually the result of this process of adaptation.

Evolutionary algorithms in effect depict an alternative approach to problem solving, in which solutions to the problem are evolved rather than problems being solved directly. The family of evolutionary algorithms may be characterised by four main streams: Evolution Strategies (Schwefel, 1981), Evolutionary Programming (Fogel *et al*, 1966), Genetic Algorithms (Holland, 1975) and Genetic Programming (Koza, 1992).

Although different and applied for different purposes, all EAs share a common conceptual base. In principle, an initial population of individuals is created in a computer and allowed to evolve using the principles of *inheritance* (so that offspring resemble parents), *variability* (the process of offspring creation is not perfect - some mutations occur) and *selection* (more fit, or 'better', individuals are allowed to reproduce more often and less fit individuals less often so that the 'genealogical' trees of the latter will 'die out' with time).

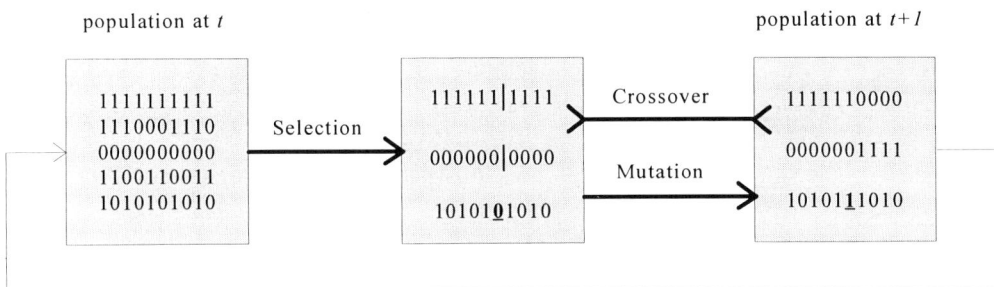

Figure 6   Schematic illustration of an evolutionary algorithm.

Figure 6 depicts the main processes that make up an evolutionary algorithm. From an initial, typically randomly generated, population of individuals the fittest entities are selected to be altered by genetic operators exemplified by *crossover* (corresponding to sexual reproduction) and *mutation*. Selection is performed on the basis of a certain fitness criterion in which the fitter individuals are selected more often. Crossover combines two genotypes by exchanging substrings around a randomly selected point. Mutation simply flips a randomly selected bit.

Similar to the processes of nature, one should distinguish between the evolving entity's genotype and its phenotype. The genotype is essentially a code to be executed (such as a code in the DNA strand in humans), and the phenotype represents the result of the execution of this code (such as a living person). The information exchange between evolving entities (parents) occurs at the level of the genotypes; however, it is the phenotypes in which we are really interested.

The phenotype is in effect an interpretation of a genotype in a problem domain. This interpretation can take the form of any feasible mapping. One of the main advantages of EAs is their domain independence. EAs can evolve almost anything, given an appropriate representation of the evolving structures. For example, for optimisation and constraint satisfaction purposes, genotypes are typically interpreted as independent variables of a function to be optimised. Several applications of genetic algorithms (GAs) that make use of this kind of mapping and with specific emphasis on water resources are described by Babović (1993).

In so-called *learning classifier systems (LCS)*, as introduced by Holland (1986), phenotypes take the appearance of rules in evolving knowledge-bases. LCSs are actually built on the top of ordinary GAs, and continuously augmented the knowledge-base with new and better-performing rules, thus avoiding a rigid and static tree structure. LCSs thus open avenues towards automatic model enhancement through the process of machine learning (*see* Wilson, 1994).

In genetic programming (GP), the evolutionary force is directed towards the creation of models that take a symbolic form. In this evolutionary paradigm, evolving entities are presented with a collection of data, and the evolutionary process is expected to result in a closed-form symbolic expression that describes the data. In principle, GP evolves tree structures representing symbolic expressions in Reverse Polish Notation. The nodes in this tree structure are user-defined. This means that they can be algebraic operators, such as *sin, log, +, -,* etc., or can take a form of *if-then-else* rules, making use of logical operators such as *OR, AND,* etc. (*see* Walker *et al*, 1993).

It is extremely difficult ,if not impossible, to describe the full potential of EAs and their applications in such a limited space. The reader is therefore referred to the original texts that describe the inner workings of EAs and their applications in much more detail. Here, however, we would like to highlight two essential properties of EAs:

- Evolutionary Algorithms are sub-symbolic models of computation. As was suggested before, the exchange of information between evolving entities occurs at the level of the genotypes. The phenotypes represent or contain the *meaning* encoded in the genotypes. This meaning (or semantic interpretation) is acquired through both a *mapping function* (from genotype to phenotype) and an interaction of the phenotypes with their environment. This applies for the entire EA family. GP in its most rudimentary form can be understood as a method for evolving trees which acquire meaning only when they are confronted with the problem domain;

- The most important phenomenon in relation to EA performance is that it attains its knowledge about its environment through interaction with this environment. The knowledge about a problem that is being solved does not explicitly exist within the EA-based problem-solver before the problem-solving (*i.e.* evolutionary) process is initiated. This knowledge is acquired through the process of survival of the fittest. The consequence of this is that the process of solving problems actually transforms to one of adequately describing the problem and then letting the solution to the problem evolve itself.

3 Some Applications

3.1 RAINFALL-RUNOFF MODELLING

For rainfall-runoff modelling it is supposed that, subject to given antecedent conditions, there is an explicit relationship between the depth of rain falling on a catchment and the magnitude of the streamflow emerging from that catchment. Hall and Minns (1993) confirmed that for simple laboratory catchments and small sewered areas, an ANN is capable of learning the relationship between rainfall and runoff to a very high degree of accuracy even in the very simple case of having inputs restricted to antecedent rainfall depths and antecedent flow ordinates Figure 7 shows the results of an ANN model that was trained on actual data from a small urban catchment in the UK. The results of the ANN model have also been compared to the results of the single, conceptual, non-linear reservoir model called RORB.

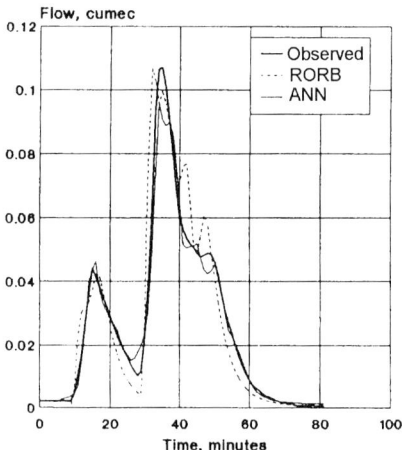

Figure 7   Comparison of ANN with conceptual model RORB

Minns and Hall (1995) continued these investigations into applications on more complex theoretical catchments exhibiting a range of behaviour patterns varying from the linear to the highly (in hydrological terms) non-linear and having inputs restricted to antecedent rainfall depths and antecedent flow ordinates (*see* Figures 8 and 9). The ANN model provides these exceptional results unhindered by constraints of volume continuity in the input and output data and, in fact, the units of the data are chosen simply for convenience of measurement and representation. Furthermore, simple, non-hydrological parameters like the percentage impervious area, may be easily incorporated into the model at the discretion of the modeller. These types of parameters may be derived from simple measurements or may even be highly intuitive, and are likewise unrestricted in terms of conditions of dimension or hydrological-physical consistency (*see* Minns, 1996).

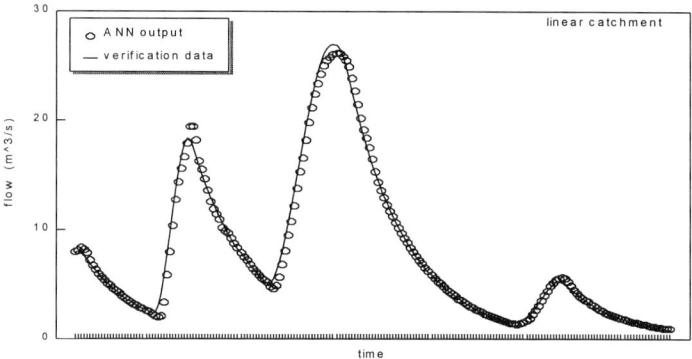

Figure 8   Verification of a 3-layer ANN for a linear catchment

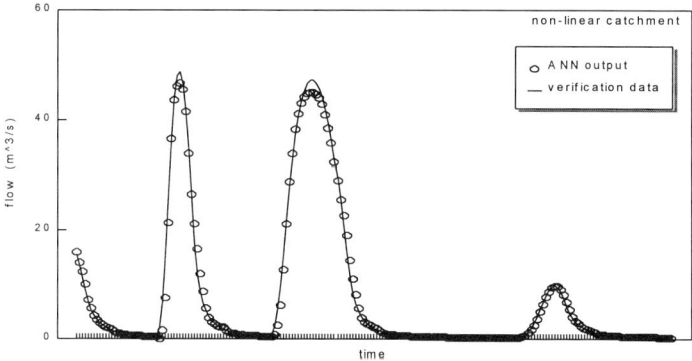

Figure 9   Verification of a 3-layer ANN for a non-linear catchment

Babović (1995, pp. 208-216) applied genetic programming techniques to induce symbolic expressions from the data used by Minns and Hall (1995). The best-performing expression for the linear catchment model (Figure 8) was:

$$q[t] = q[t-1] + 0.3\,r[t-1] - 0.29\sqrt{r[t-14]} \qquad (4)$$

and for the non-linear catchment model:

$$q[t] = q[t-1] + r[t-2] - r[t-8] \qquad (5)$$

where :

$q[t]$ denotes runoff in m^3/s at time t
$r[t]$ denotes rainfall in mm/hr at time t
t denotes time in hours.

Eqs. (4) and (5) performed with a similar accuracy to the ANN models of the same data. Although no attempt was made to interpret these equations physically, it is immediately obvious that the relevant variables emerging from the GP induced expressions might have something to do with the lag-time of the catchment. Extracting a semantic content from the ANN models is by no means as obvious however.

The potential role of ANN and GP models in hydrological modelling in general is manyfold. At the simplest level they may function as a flexible, easy-to-implement, lumped-conceptual models that relate rainfall data to runoff data for *individual catchments*. At the other end of the spectrum, they may be used to generate important components of physically-based, distributed hydrological modelling systems, whereby a sub-model of *individual physical processes* (e.g. unsaturated zone flow dynamics) is induced based only upon measured data. Such a sub-model may then replace whole systems of complex, non-linear, differential equations that would otherwise require great skills from the modeller to calibrate and powerful computing devices to solve.

3.2 MODEL CALIBRATION AND SYSTEM OPTIMISATION

With traditional conceptual hydrological modelling techniques the modeller applies his or her measured data together with some hydrological insight in order to adjust modelling parameters and equations manually and so eventually to calibrate the model. Babović *et al* (1994) describe the application of a genetic algorithm to the problem of model calibration, in which the genotypes of the GA are interpreted as roughness coefficients in a free surface pipe flow simulation and the evolution is directed towards the minimisation of discrepancies of model output and measured water level and discharge values - thus resulting in an automatic calibration of the roughness coefficients in the hydrodynamic model.

Solomatine (1995) explains that the process of calibrating a hydrological model is, in fact, a form of optimisation problem in which the objective function to be minimised is the difference (error) between computed output variables and the corresponding values measured in the physical system. The independent variables in the optimisation problem are the unknown model parameters. He further compares the performance of a GA to that of more traditional optimisation techniques and confirms the power of this methodology in global optimisation.

Rapid selection of the optimal control strategy in a multi-objective water resources system is of primary importance for the real-time control of these systems. Masood-Ul-Hassan and Wilson (1995) describe how both ANNs and GAs can be used to improve system performance and respond to processes in the real world in the face of real-time constraints. They explain how an ANN can be trained off-line to replicate optimised gate-settings in a flood-control scheme. The generation of the optimised gate-settings with which to train the ANN was carried out using a numerical optimiser employing a traditional gradient descent method. This generation of the optimal control strategy data and the training of the ANN with this data is quite time consuming; however, after training, the ANN can be used instantaneously to recall the optimal gate settings corresponding to any given system state for which it was trained This paper further describes the implementation of a learning classifier system to the real-time control of a sewerage network. In a classifier system, actions in response to a given system state are obtained from a rule-based system. The learning classifier system improves its performance with time by generating new rules based on experience. A GA is used to generate the new rules through recombination of the best-performing classifiers that replace the low-performance classifiers.

4 References

Abbott, M.B. (1991), *Hydroinformatics - Information Technology and the Aquatic Environment*, Avebury Technical, Aldershot, U.K.

Abbott, M.B. (1992), The theory of the hydrologic model, or: The struggle for the soul of hydrology, in *Topics in Theoretical Hydrology: A Tribute to Jim Dooge*, ed O'Kane, J.P., Elsevier, Amsterdam, pp. 237-259.

Ahmad, K., Hornsby, C.P.W. and Langdon, A.J. (1987), Expert systems in urban storm drainage, *Proc. 4th Int. Conf. on Urban Storm Drainage*, Lausanne, pp. 297-302.

Aleksander, I. and Morton, H. (1990), *An Introduction to Neural Computing*, Chapman and Hall, London.

Almeida, M. and Schilling, W. (1993), Derivation of IF-THEN-ELSE rules from optimised strategies for sewer systems under real-time control, *Proc. 6th Int. Conf. on Urban Storm Drainage*, Niagara, pp. 1525-1530.

Amorocho, J. and Hart, W.E. (1964), A critique of current methods in hydrologic systems investigation. *Trans. Amer. Geophys. Union*, Vol 45, pp. 307-321.

Babović, V. (1991), *Applied Hydroinformatics: A Control and Advisory System for Real-time Applications*, M.Sc. Thesis, IHE Report Series 26, IHE-Delft, The Netherlands.

Babović, V. (1993), Evolutionary algorithms as a theme in water resources, in *Scientific Presentations of AIO Meeting. '93: AIO Network Hydrology*, Boekelman, R.H. (ed.) Delft University of Technology, The Netherlands, pp. 21-36.

Babović, V. (1995), *Emergence, Evolution, Intelligence; Hydroinformatics*, Ph.D Thesis, Balkema, Rotterdam.

Babović, V., Larsen, L.C. and Wu, Z.Y. (1994), Calibrating hydrodynamic models by means of simulated evolution, in Verwey *et al* (eds), *Hydroinformatics '94, Proc. 1st International Conf. on Hydroinformatics*, Balkema, Rotterdam, pp. 193-200.

Babović, V. and Minns, A.W. (1994), Use of computational adaptive methodologies in hydroinformatics, in Verwey *et al* (eds.), *Hydroinformatics '94, Proc. 1st International Conf. on Hydroinformatics*, Balkema, Rotterdam, pp. 201-210.

Beale, R. and Jackson, T. (1990), *Neural Computing: An Introduction*. Institute of Physics, Bristol.

Darwin, C. (1859), *The Origin of Species by Means of Natural Selection*, 6th Ed., John Murray, London.

Delleur, J.W. and Baffaut, C. (1990), An expert system for urban runoff quality modelling, *Proc. 5th Int. Conf. on Urban Storm Drainage*, Osaka, pp. 1323-1328.

Fogel, L.J., Owens, A.J. and Walsh, M.J. (1966), *Artificial Intelligence through Simulated Evolution*, Ginn, Needham Heights.

Hall, M.J. and Minns, A.W., 1993. Rainfall-runoff modelling as a problem in artificial intelligence: experience with a neural network, *Proc 4th Nat. Hydrol. Symp.*, Cardiff, British Hydrological Society, London, 5.51-5.57.

Harnad, S. (1990),The symbolic grounding problem, in *Emergent Computation*, Forrest, S. (ed), MIT Press, Cambridge, pp. 335-346.

Hertz, J., Krogh, A. and Palmer, R.G. (1991), *Introduction to the Theory of Neural Computation*. Addison-Wesley, Redwood City, Ca.

Holland, J.H. (1975), *Adaption in Natural and Artificial Systems*, Univ. Of Michigan Press.

Hopfield, J.J. (1994), Neurons, dynamics and computation. *Physics Today*, 47 (2): 40-46.

Hornik, K., Stinchcombe, M. and White, H. (1989), Multilayer feedforward networks are universal approximators, *Neural Networks*, 2: pp.359-366.

IAHR (1994), *Extra Issue Journal of Hydraulic Research*, Vol 32, IAHR, Delft, 214pp.

Khelil, A., Heinemann, A. and Muller, D. (1993), Learning algorithms in a rule-based system for control of urban drainage systems, *Proc. 6th Int. Conf. on Urban Storm Drainage*, Niagara, pp. 1401-1408.

Koza, J. (1992), *Genetic Programming: On the Programming of Computers by Means of Natural Selection*, MIT Press, Cambridge.

Minns, A.W. and Hall, M.J., 1995. Artificial Neural Networks as Rainfall-Runoff Models, submitted to *Hydrological Sciences Journal*.

Minns, A.W. (1996), Extended rainfall-runoff modelling using artificial neural networks, *to be published*.

Martin-Garcia, H. (1995), *Combined Logical-Numerical Enhancement of Real-time Control of Urban Drainage Networks*, Ph.D Thesis, Balkema, Rotterdam.

Masood-Ul-Hassan, K. and Wilson, G. (1995), Hydroinformatic applications in real-time control strategy selection, *Hydra 2000: Proc. XXVI Congress IAHR*, Vol. 5, Thomas Telford, London, pp. 85-90.

Rumelhart, D.E., McClelland, J.L. *et al* (1986), *Parallel distributed processing. Explorations in the microstructure of cognition, Vol 1, Foundations*. The MIT Press, Cambridge, Ma.

Rumelhart, D.E., Widrow, B. and Lehr, M.A. (1994), The basic ideas in neural networks. *Communications of the ACM*, 37 (3): pp.87-92.

Schwefel, H.-P. (1981), *Numerical Optimisation of Computer Models*, Wiley, Chichester.

Solomatine, D. (1995), The use of global random search methods for model calibration, *Hydra 2000: Proc. XXVI Congress IAHR*, Vol. 1, Thomas Telford, London, pp. 224-229.

Verwey, A., Minns, A.W., Babović, V. And Maksimović, C. (Eds.) (1994), *Hydroinformatics '94: Proc. 1st International Conference on Hydroinformatics*, 2 Vols, Balkema, Rotterdam.

Walker R., Gerrets, M. And Haasdijk, E. (1993), *A Genetic Algorithm for the Approximation of Formulae*, ESPRIT Project 6857, Deliverable D2.2.

Wilson, S.W. (1994), ZCS: A zeroth level classifier system, *Evolutionary Computation*, Vol. 2, No. 1, pp. 1-18.

Wittgenstein, L. (1953//1992), *Philosophical Investigations*, 3rd ed., translation from German by Anscombe, G.E.M., Blackwell, Oxford.

Subject Index

Accuracy criteria, 25, 43
Active systems, 172
Advection-dispersion, 60, 78
AGNPS, 132
Agricultural crop production, 123
Agrochemical modelling
    distributed physically-based catchment scale modelling, 132
    case study, 132
    fertilizers, 122
    spatial heterogeneity, 138
    leaching models, 123
    lumped conceptual field scale modelling, 131
    nitrogen modelling, 122
    pesticide modelling, 127
    phosphorous modelling, 126
    point scale, 122
    upscaling, 128
ANIMO, 124
ANIMO-P, 127
ANSWERS, 95, 126
API, 28
ARC/INFO, 220, 238
Area-time integral, 151
Artificial neural networks, 29, 303
Aquatic ecosystems, 5, 7, 9, 233
Aquifer heterogeneity, 196
ARIMA, 28

Backscattering coefficient, 175
Bacteria, 75
Biodegradation, 75
Black box model, 19
Borehole, 193
Bright band, 149

Calibration, 25, 41, 47, 110, 240, 310
Calibrated roughness parameter, 186
Calvin Rose, 95
Catchment, 18
Certification documentation, 20
Chemical, 71
Classification of model codes, 17, 27, 35, 73, 94, 121, 195
Climate change, 5, 7, 10
CLS, 28
Code
    definition, 19
    development, 25
    selection, 25
    verification, 25
Cohesive, 105
Complexation, 77
Component, 18
Conceptual model, 20, 24
Construction, 25, 41, 44, 110, 240
Contamination, 72

Contoured surface maps, 197, 202
CREAMS, 95, 131
Credibility of a model code, 53
Critique, 255, 279, 289
Cross sections, 193, 197
Crusting, 101

DAISY, 124, 236
Danube, 83, 233
Database, 194, 215, 238
Data quality, 204
Denitrification, 83
Deterministic model, 18, 27
Dialogue, 255, 279, 289
Differential split-sample test, 50
Disaggregation approach, 263, 283
Distributed model, 19
Distributed physically-based models, 30, 49, 55, 97, 132, 233, 255, 279, 289
Dublin Statement, 1
Drop size distribution, 147
Droughts, 5
Dual polarisation radar, 155

Effective parameter values, 65, 261
Electromagnetic radiation, 166
Empirical model, 19, 28, 96
EPIC, 126
Error sources, 41
EUROSEM, 95
Evolutionary algorithms, 29, 306
Expert system, 301

Fence diagrams, 193, 200
Fertilizer, 102
Field applications, 62, 81, 83, 111, 132, 180, 218, 233
Floodplain dynamics, 246
Floods, 5
Forecasting, 7, 9, 157
Fundamental problem, 255, 280

Gabcikovo hydropower plant, 233
Geochemistry, 72
Geological modelling
    aquifer heterogeneity, 196
    boolean (object) modelling, 207
    borehole, 193
    choice of modelling approach, 195
    connectivity, 211
    contoured models, 202
    contoured surface maps, 197, 202
    cross sections, 193, 197
    data extraction routines, 203
    data quality, 204
    fence diagrams, 193, 200
    geological data, 194
    geological data base, 194
    geological model, 193
    geological variability, 207

SUBJECT INDEX 315

    geophysical data, 197
    georadar, 197
    geostatistic models, 206
    ghost data, 204
    gravimetric, 197
    hard data, 194
    hydraulic parameters, 200
    lithological units, 200
    outcrop, 197
    probability maps, 209
    ranking, 211
    sequential Gaussian modelling, 207
    sequential indicator modelling, 207
    soft data, 194, 207
    sedimentary architecture, 196
    sedimentary facies models
    seismic, 197
    surface models, 202
    traditional deterministic models, 196
Geological models
    Geostat Toolbox, 207
    GSLIB, 207
    IMAGEX, 208
    ISIM3D, 207
    PREVAR2D, 208
    RC^2
    SPYGLASS, 208
    Storm, 207
    VARIO2DP & MODEL, 208
Geological variability, 207
Geophysical data, 197
Georadar, 197
Geostatistic models, 206
GIS
    ARC/INFO, 220, 238
    case study, 218
    databases, 216
    geographical information, 217
    management scenarios, 225
    modelling environment, 227
GLEAMS, 131
Global Circulation Models, 264
GLUE methodology, 266
Grey box model, 19
Groundwater pollution, 4, 7, 8, 60, 72, 121
Groundwater chemistry, 71

Hot start, 107
Hydrograph forecast, 157
Hydroinformatics, 13, 28, 297
Hydrological model, 18
Hydropower plant, 233

ICWE, 1
IHDM, 31, 56, 255
Inducing models, 299
Information technology, 297
Infiltration, 59, 101

Integral Equation Model (IEM), 174
Ion exchange, 77
Irrigation, 3, 7, 8

Kalman filtering, 32
KINEROS, 95
Kinetically-controlled, 75

LEACHM, 124, 129
Local Equilibrium Assumption (LEA), 73
Lumped model, 18
Lumped conceptual models, 29, 48, 96, 131

MACRO, 129
Management scenarios, 225
Mathematical model, 18
Mediating models, 256
Microbiological, 71
Micro-topography, 102
Microwave
    radars, 168
    radiation, 170
    scattering models, 173
MIKE 11, 36, 236
MIKE 21, 236
MIKE SHE, 11, 31, 56, 60, 76, 95, 129, 202, 236
Minerals, 77
MODANSW, 96
Model
    black box model, 19
    calibration, 25, 41, 47, 110, 240, 310
    conceptual model, 20, 24
    construction, 25, 41, 44, 110, 240
    certification, 22
    credibility, 53
    deterministic model, 18, 27
    distributed model, 19
    distributed physically-based, 30, 49, 55, 97, 132, 233, 255, 279, 289
    domain of intended application, 21
    domain of applicability, 21
    empirical model, 19, 28, 96
    grey box model, 19
    hydroinformatics based, 28, 297
    hydrological model, 18
    level of agreement, 20
    lumped model, 18
    lumped conceptual, 29, 48, 96, 131
    mathematical model, 18
    model, 19
    model code, 19
    modelling system, 19
    physically-based model, 19
    qualification, 21
    range of accuracy, 21
    statistical, 28
    stochastic model, 18
    validation, 21, 26, 41, 49, 110, 241, 259, 281, 291
    verification, 21, 25

SUBJECT INDEX

317

    white box model, 19
Model names
    AGNPS, 132
    ANIMO, 124
    ANIMO-P, 127
    ANSWERS, 95, 126
    API, 28
    ARIMA, 28
    Calvin Rose, 95
    CLS, 28
    CREAMS, 95, 131
    DAISY, 124, 236
    EPIC, 126
    EUROSEM, 94
    GLEAMS, 131
    IHDM, 31, 56, 255
    KINEROS, 95
    LEACHM, 124, 129
    MACRO, 129
    MIKE 11, 36, 236
    MIKE 21, 236
    MIKE SHE, 11, 31, 56, 60, 76, 95, 129, 202, 236
    MODANSW, 96
    MODFLOW, 60, 218
    MUSLE, 96
    NAM, 56, 62
    PELMO, 129
    PESTLA, 129
    RUSLE, 95
    RZWQM, 124, 129
    SEM, 95
    SHE, 11, 31, 255
    SHESED, 95
    SHETRAN, 11
    SLEMSA, 95
    SOIL-N, 124
    Stanford, 29, 56
    SWAGSIM, 56
    SWATRE, 58
    SWRRB, 131
    THALES, 31
    TOPMODEL, 34, 263
    USLE, 95
    WEPP, 95
    WAVE, 124, 129
Modelling environment, 13, 227
Modelling protocol, 17, 23
Modelling scales, 116, 128, 239, 263, 283
MODFLOW, 60, 218
Monte Carlo techniques, 33, 267
Multi-species reactive transport modelling
    advection-dispersion, 78
    bacteria, 75
    biodegradation, 75
    case study, 81, 83
    chemical, 71
    complexation, 77
    contamination, 72

    denitrification, 83
    geochemistry, 72
    groundwater chemistry, 71
    ion exchange, 77
    kinetically-controlled, 75
    Local Equilibrium Assumption (LEA), 73
    microbiological, 71
    minerals, 77
    Operator Splitting (OS), 80
    redox processes, 78, 84
    spatial heterogeneity, 84
MUSLE, 96

NAM, 56, 62
Natural system, 17
Neural networks, 29, 303
Nitrogen modelling, 102

Land degradation, 4, 93
Land use change, 7, 9, 31
Leaching models, 103
Level of agreement, 21

Operator Splitting (OS), 80
Overland flow, 59, 104

Parameter, 18
Parameter optimization, 47
Parametrisation, 46, 110
Passive systems, 172
PELMO, 129
Performance criteria, 25, 43
Pesticide modelling, 127
PESTLA, 129
Phosphorous modelling, 126
Physical process, 19
Physically-based model, 19
Polarisation diversity radar, 156
Postaudit, 26
Precipitation, 143
Process parameterisation, 261, 282
Proxy-basin test, 50
Proxy-basin differential split-sample test, 50

Radar, 143, 146
Raindrop impact, 98
Rainfall erosivity, 98
Rainfall-runoff models, 29
Range of accuracy, 21
Reactive, 71
Reality, 20
Redox processes, 78, 84
Remote sensing
    active systems, 172
    backscattering coefficient, 175
    calibrated roughness parameter, 185
    electromagnetic radiation, 166
    hydrological applications, 169

SUBJECT INDEX 319

      Integral Equation Model (IEM), 174
      microwave scattering models, 173
      passive systems, 172
      recent hydrological experiments, 180
      satellite systems, 168
      sensors, 167
      soil moisture, 170
      soil moisture retrieval, 173
      soil roughness parameters, 175
      Zwalm catchment, 182
Remote sensors, 167
Representative elementary volume (REV), 257
Rills, 103, 106
River branch system, 233
Rock fragments, 101
Routine, 18
Rule-based system, 301
RUSLE, 95
RZWQM, 124, 129

Salinization, 3
Satellite systems, 168
SCADA, 298
Scale problems, 65, 116, 128, 263, 283
Seismic, 197
SEM, 95
Sensors, 167
SHE, 11, 31, 255
SHESED, 95
SHETRAN, 11
Simulation, 18, 26, 241
SLEMSA, 95
Soil degradation, 93
Soil detachment, 98
Soil erodibility, 98
Soil erosion, 4, 7, 8, 93
Soil erosion modelling
      case study, 111
      coupling with hydrological models, 108
      cohesive forces, 105
      conceptual models, 96
      crusting, 101
      empirical models, 96
      EUROSEM, 94
      infiltration conditions, 101
      micro-topography, 102
      model calibration, 110
      model construction, 110
      model validation, 110
      overland flow, 104
      parametrisation, 110
      physically-based models, 97
      rills, 103, 106
      rock fragments, 101
      soil degradation, 93
      soil detachment, 98
      spatial variability, 115
      splash erosion, 98

tillage, 103, 105
USLE, 94
Soil moisture, 170
Soil moisture retrieval algorithms, 173
Soil roughness parameters, 175
SOIL-N, 124
Solute transport, 60, 71
Spatial heterogeneity, 65, 84, 138
Spatial variability, 32, 34, 115, 207
Splash erosion, 98
Split-sample test, 50
Stanford Watershed model, 29, 56
Statistically based methods, 28
Stochastic models, 18, 31
Stochastic-deterministic models, 32
Stochastic PDE's, 32
Storm smearing, 157
Submodel, 18
Surface soil moisture, 180
Surface water pollution, 4, 7, 8
SWAGSIM, 56
SWATRE, 58
SWRRB, 131
Symbolic paradigm, 299

Technological level, 35
Terminology, 17, 279
Terminology - definitions
    black box model, 19
    calibration, 25
    catchment, 18
    conceptual model, 20, 24
    construction, 25
    certification, 22
    deterministic model, 18,
    distributed model, 19
    domain of intended application, 21
    domain of applicability, 21
    empirical model, 19,
    grey box model, 19
    hydrological model, 18
    level of agreement, 20
    lumped model, 18
    mathematical model, 18
    model, 19, 279
    model code, 19, 279
    modelling protocol, 17,
    modelling system, 19, 279
    natural system, 17
    parameter, 18
    performance criteria, 25
    physical process, 19
    physically-based model, 19
    qualification, 21
    range of accuracy, 21
    reality, 20
    routine, 18
    statistical, 28

SUBJECT INDEX 321

    stochastic model, 18
    validation, 21, 279
    verification, 21, 279
    white box model, 19
THALES, 31
Tillage, 103, 105
TOPMODEL, 34, 263
Transcientific problem, 257
Transport capacity, 105
Trial-and-error calibration, 47

UNCED, 1
Uncertainty, 41, 265, 284
Universal Soil Loss Equation (USLE), 95
Unsaturated zone, 58, 121

Validation, 21, 26, 41, 49, 110, 241, 259, 281, 291
Validation tests
    differential split-sample, 50
    proxy-basin, 50
    proxy-basin differential split-sample, 50
    split-sample, 50
Value of data, 267, 285, 292
Verification, 21, 25
Vertical profile of reflectivity, 150

Water resources assessment, 6, 7
Water use, 2
Waterlogging, 3
Watershed smearing
WAVE, 124, 129
Weather radar precipitation data
    accuracy, 143
    area-time integral, 151
    bright band, 149
    distributed hydrological models, 144
    drop size distribution, 147
    dual polarisation radar, 155
    hydrograph forecasts, 157
    polarisation diversity radar, 156
    precision, 143
    radar, 146
    storm smearing, 157
    vertical profile of reflectivity, 150
    watershed smearing, 157
    window probability matching method, 154
    Z:R relationship
WEPP, 95
White box model, 19
Window probability matching method, 154
World Bank, 1

Z:R relationship, 150
Zwalm catchment, 182

Water Science and Technology Library

1. A.S. Eikum and R.W. Seabloom (eds.): *Alternative Wastewater Treatment.* Low-Cost Small Systems, Research and Development. Proceedings of the Conference held in Oslo, Norway (7–10 September 1981). 1982
   ISBN 90-277-1430-4
2. W. Brutsaert and G.H. Jirka (eds.): *Gas Transfer at Water Surfaces.* 1984
   ISBN 90-277-1697-8
3. D.A. Kraijenhoff and J.R. Moll (eds.): *River Flow Modelling and Forecasting.* 1986
   ISBN 90-277-2082-7
4. World Meteorological Organization (ed.): *Microprocessors in Operational Hydrology.* Proceedings of a Conference held in Geneva (4–5 September 1984). 1986
   ISBN 90-277-2156-4
5. J. Němec: *Hydrological Forecasting.* Design and Operation of Hydrological Forecasting Systems. 1986
   ISBN 90-277-2259-5
6. V.K. Gupta, I. Rodríguez-Iturbe and E.F. Wood (eds.): *Scale Problems in Hydrology.* Runoff Generation and Basin Response. 1986
   ISBN 90-277-2258-7
7. D.C. Major and H.E. Schwarz: *Large-Scale Regional Water Resources Planning.* The North Atlantic Regional Study. 1990   ISBN 0-7923-0711-9
8. W.H. Hager: *Energy Dissipators and Hydraulic Jump.* 1992
   ISBN 0-7923-1508-1
9. V.P. Singh and M. Fiorentino (eds.): *Entropy and Energy Dissipation in Water Resources.* 1992
   ISBN 0-7923-1696-7
10. K.W. Hipel (ed.): *Stochastic and Statistical Methods in Hydrology and Environmental Engineering.* A Four Volume Work Resulting from the International Conference in Honour of Professor T. E. Unny (21–23 June 1993). 1994
    10/1: Extreme values: floods and droughts   ISBN 0-7923-2756-X
    10/2: Stochastic and statistical modelling with groundwater and surface water applications   ISBN 0-7923-2757-8
    10/3: Time series analysis in hydrology and environmental engineering
    ISBN 0-7923-2758-6
    10/4: Effective environmental management for sustainable development
    ISBN 0-7923-2759-4
    Set 10/1–10/4: ISBN 0-7923-2760-8
11. S.N. Rodionov: *Global and Regional Climate Interaction: The Caspian Sea Experience.* 1994
    ISBN 0-7923-2784-5
12. A. Peters, G. Wittum, B. Herrling, U. Meissner, C.A. Brebbia, W.G. Gray and G.F. Pinder (eds.): *Computational Methods in Water Resources X.* 1994
    Set 12/1–12/2: ISBN 0-7923-2937-6
13. C.B. Vreugdenhil: *Numerical Methods for Shallow-Water Flow.* 1994
    ISBN 0-7923-3164-8
14. E. Cabrera and A.F. Vela (eds.): *Improving Efficiency and Reliability in Water Distribution Systems.* 1995
    ISBN 0-7923-3536-8

Water Science and Technology Library

15. V.P. Singh (ed.): *Environmental Hydrology*. 1995   ISBN 0-7923-3549-X
16. V.P. Singh and B. Kumar (eds.): *Proceedings of the International Conference on Hydrology and Water Resources* (New Delhi, 1993). 1996
    16/1: Surface-water hydrology   ISBN 0-7923-3650-X
    16/2: Subsurface-water hydrology   ISBN 0-7923-3651-8
    16/3: Water-quality hydrology   ISBN 0-7923-3652-6
    16/4: Water resources planning and management   ISBN 0-7923-3653-4
    Set 16/1–16/4 ISBN 0-7923-3654-2
17. V.P. Singh: *Dam Breach Modeling Technology*. 1996   ISBN 0-7923-3925-8
18. Z. Kaczmarek, K.M. Strzepek, L. Somlyódy and V. Priazhinskaya (eds.): *Water Resources Management in the Face of Climatic/Hydrologic Uncertainties*. 1996   ISBN 0-7923-3927-4
19. V.P. Singh and W.H. Hager (eds.): *Environmental Hydraulics*. 1996   ISBN 0-7923-3983-5
20. G.B. Engelen and F.H. Kloosterman: *Hydrological Systems Analysis*. Methods and Applications. 1996   ISBN 0-7923-3986-X
21. A.S. Issar and S.D. Resnick (eds.): *Runoff, Infiltration and Subsurface Flow of Water in Arid and Semi-Arid Regions*. 1996   ISBN 0-7923-4034-5
22. M.B. Abbott and J.C. Refsgaard (eds.): *Distributed Hydrological Modelling*. 1996   ISBN 0-7923-4042-6

Kluwer Academic Publishers – Dordrecht / Boston / London